T0180410

Lecture Notes of the Institute for Computer Sciences, Social Informatics and Telecommunications Engineering   **327**

More information about this series at http://www.springer.com/series/8197

Yu-Dong Zhang · Shui-Hua Wang ·
Shuai Liu (Eds.)

Multimedia Technology and Enhanced Learning

Second EAI International Conference, ICMTEL 2020
Leicester, UK, April 10–11, 2020
Proceedings, Part II

 Springer

Editors
Yu-Dong Zhang ⓘD
School of Informatics
University of Leicester
Leicestershire, UK

Shui-Hua Wang ⓘD
University of Leicester
Leicestershire, UK

Shuai Liu ⓘD
Human Normal University
Changsha, China

ISSN 1867-8211   ISSN 1867-822X   (electronic)
Lecture Notes of the Institute for Computer Sciences, Social Informatics
and Telecommunications Engineering
ISBN 978-3-030-51102-9   ISBN 978-3-030-51103-6   (eBook)
https://doi.org/10.1007/978-3-030-51103-6

This Springer imprint is published by the registered company Springer Nature Switzerland AG
The registered company address is: Gewerbestrasse 11, 6330 Cham, Switzerland

Preface

We are delighted to introduce the proceedings of the second European Alliance for Innovation (EAI) International Conference on Multimedia Technology and Enhanced Learning (ICMTEL). This conference has brought together researchers, developers, engineers, scientists, and practitioners from around the world. Participants shared their knowledge in leveraging and developing multimedia technology with machine learning, especially enhanced learning. The theme of ICMTEL 2020 was "Deep learning driven multimedia technology in academy, industry and society."

The proceeding of ICMTEL 2020 consists of 83 full papers, all presented in YouTube Live due to the impact of coronavirus. Aside from the high-quality academic and technical paper presentations, the technical program also featured three keynote speeches and five workshops. The three keynote speeches were presented by Prof. Lu Liu, Head of School of Informatics from University of Leicester, UK; Dr. Shui-hua Wang, Research Fellow at University of Loughborough, UK, and Prof. Shuai Liu, Head of School of Artificial Intelligence from Hunan Normal University, China. The five workshops organized were (i) International Workshop on Data fusion filter and Machine Learning for Statistical Signal Processing (DFMLSSP 2020); (ii) International Workshop on Intelligent Technology and Design for Special Education/Rehabilitation (ITD 2020); (iii) International Workshop on Intelligent Technology and Design for Special Education/Rehabilitation (ITD 2020); (iv) International Workshop on Weather Radar and Antenna Design (WRAD 2020); and (iv) International Workshop on Digital Image Processing, Analysis and Application Based on Machine Learning (DIPAA 2020).

Support from the steering chair Prof. Imrich Chlamtac was essential for the success of the conference. We sincerely appreciate his constant support and guidance. It was also a great pleasure to work with such an excellent Organizing Committee, and we are grateful for their hard work in organizing and supporting the conference. In particular, the Technical Program Committee (TPC), led by our TPC chair: Prof. Zhengchao Dong, and TPC co-chairs: Prof. Vishnu Varthanan Govindaraj, Prof. Raymond F. Muzic, Jr., Prof. Gautam Srivastava, Dr. Preetha Phillips, Prof. Amin Taheri-Garavand, Prof. Vikrant Bhateja, Prof. Seifedine Kadry, Prof. Yuriy S. Shmaliy, Prof. Mohammad R. Khosravi, Dr. Muhammad Attique Khan, Dr. Suresh Chandra Satapathy, Dr. Deepak Ranjan Nayak, Dr. Mohammad Momeny, Prof. Juan Manuel Górriz, Dr. Yeliz Karaca, Dr. Miguel Martínez García, and other co-chairs who contributed to the peer-review process of technical papers. We are also grateful to our conference manager, Lukas Skolek, for his support and all the authors who submitted their papers to the ICMTEL 2020 conference and workshops.

We strongly believe that the ICMTEL 2020 conference provided a beneficial forum for all researchers, developers, engineers, scientists, and practitioners to discuss scientific and technological aspects that are relevant to multimedia technology, machine learning, and artificial intelligence. We also expect that the future ICMTEL conferences will be more successful than the contributions presented in this volume.

May 2020   Yu-Dong Zhang
                                                                      Shuai Liu

Organization

Steering Committee

Imrich Chlamtac	University of Trento, Italy
Yu-Dong Zhang	University of Leicester, UK

Organizing Committee

General Chair

Yu-Dong Zhang	University of Leicester, UK

General Co-chairs

Shui-Hua Wang	Loughborough University, UK
Shuai Liu	Inner Mongolia University, China
Juan Manuel Górriz	University of Granada, Spain

TPC Chair

Zhengchao Dong	Columbia University, USA

TPC Co-chairs

Xinhua Mao	Nanjing University of Aeronautics and Astronautics, China
Mingwei Shen	Hohai University, China
Vishnu Varthanan Govindaraj	Kalasalingam Academy of Research and Education, India
Guodong Han	The 54th Research Institute of CETC, China
Pengjiang Qian	Jiangnan University, China
Raymond F. Muzic, Jr.	Case Western Reserve University, USA
Gautam Srivastava	Brandon University, Canada
Preetha Phillips	West Virginia School of Osteopathic Medicine, USA
Amin Taheri-Garavand	Lorestan University, Iran
Vikrant Bhateja	SRMGPC, India
Xianwei Jiang	Nanjing Normal University of Special Education, China
Nianyin Zeng	Xiamen University, China
Seifedine Kadry	Beirut Arab University, Lebanon
Yuriy S. Shmaliy	Universidad de Guanajuato, Mexico
Yuan Xu	University of Jinan, China
Kaijian Xia	The Affiliated Changshu Hospital of Soochow University, China

Mohammad R. Khosravi	Shiraz University of Technology, Iran
Zhihai Lu	Nanjing Normal University, China
Muhammad Attique Khan	HITEC University, Pakistan
Suresh Chandra Satapathy	KIIT Deemed to University, India
Deepak Ranjan Nayak	Sardar Vallabhbhai National Institute of Technology, India
Zhi-Hai Lu	Nanjing Normal University, China
Chenxi Huang	Xiamen University, China
Jin Hong Tan	National University of Singapore, Singapore
Honghao Gao	Shanghai University, China
Mohammad Momeny	Yazd University, Iran
Yeliz Karaca	University of Massachusetts Medical School, USA
Miguel Martínez García	Loughborough University, UK

Local Chair

Xiang Yu	University of Leicester, UK

Workshop Chair

Xinhua Mao	Nanjing University of Aeronautics and Astronautics, China

Publicity and Social Media Chairs

Cheng Kang	University of Leicester, UK
Qinghua Zhou	University of Leicester, UK
Rossi Kamal	Shanto-Marium University of Creative Technology, Bangladesh

Publications Chair

Yizhang Jiang	Jiangnan University, China

Web Chair

Lijia Deng	University of Leicester, UK

Technical Program Committee

Abdon Atangana	University of the Free State, South Africa
Aijun Liu	Arizona State University, USA
Amin Taheri-Garavand	Lorestan University, Iran
Arifur Nayeem	Saidpur Government Technical School and College, Bangladesh
Arun Kumar Sangaiah	Vellore Institute of Technology, India
Atiena Pereira	University of Campinas, Brazil
Carlo Cattani	University of Tuscia, Italy
Chenxi Huang	Tongji University, China

Chunlei Shan	Shanghai University of Traditional Chinese Medicine (SHUTCM), China
Chunxia Xiao	Wuhan University, China
Dan Sui	California State Polytechnic University, USA
Dang Thanh	Hue College of Industry, Vietnam
David Guttery	University of Leicester, UK
Debesh Jha	Chosun University, South Korea
Dimas Lima	Federal University of Santa Catarina, Brazil
Elijah Nguyen	Flinders University, Australia
Elizabeth Lee	Chattanooga State Community College, USA
Fidel Evans	University of Florida, USA
Foxen Cod	Middlesex University, UK
Frank Vanhoenshoven	University of Hasselt, Belgium
Gautam Srivastava	Brandon University, Canada
Ge Liu	Tennessee State University, USA
Gonzalo Napoles Ruiz	University of Hasselt, Belgium
Guangzhou Yu	Guangdong Ocean University, China
Guodong Han	The 54th Research Institute of CETC, China
Hari Mohan Pandey	Edge Hill University, UK
Heng Li	Henan Finance University, China
Hong Cheng	First Affiliated Hospital of Nanjing Medical University, China
Honghao Gao	Shanghai University, China
Jianfeng Cui	Xiamen University of Technology, China
Jitendra Pandey	Middle East College, Oman,
John Liu	Michigan State University, USA
Juan Manuel Górriz	University of Granada, Spain
Koji Nakamura	Kyushu Institute of Technology, Japan
Lei Ma	Beijing Polytechnic University, China
Leonid Snetkov	ITMO University, Russia
Liam O'Donnell	University of Limerick, Ireland
Liangxiu Han	Manchester Metropolitan University, UK
Linkai Niu	Taiyuan University of Technology, China
Logan Graham	Ryerson University, Canada
Mackenzie Brown	Edith Cowan University, Australia
Miguel Martínez García	Loughborough University, UK
Ming Pei	West Virginia University, USA
Mingwei Shen	Hohai University, China
Mohamed Elhoseny	Mansoura University, Egypt
Mohammad Momeny	Yazd University, Iran
Muhammad Bilal	Hankuk University of Foreign Studies, South Korea
Matben Suchkov	Kazan Federal University, Russia
Neeraj Kumar	Thapar University, India
Nianyin Zeng	Xiamen University, China
Peng Chen	Columbia University, USA
Pengjiqiang Qian	Jiangnan University, China

Praveen Agarwal   Anand International College of Engineering, India
Preetha Phillips   West Virginia School of Osteopathic Medicine, USA
Qingmei Lu   Bioengieering, University of Louisville, USA
Ravipudi Venkata Rao   Sardar Vallabhbhai National Institute of Technology,
                                   India
Rayan S Cloutier   Carleton University, Canada
Raymond F. Muzic, Jr.   Case Western Reserve University, USA
Rik Crutzen   Maastricht University, The Netherlands
Rodney Payne   Clemson University, USA
Seifedine Kadry   Beirut Arab University, Lebanon
Shipeng Xie   Nanjing University of Posts and Telecommunications,
                                   China
Shuai Yang   Changchun University of Technology, China
Shui-Hua Wang   Loughborough University, UK
Sunil Kumar   National Institute of Technology, India
Tanveer Hussain   Sejong University, South Korea
Tianming Zhan   Nanjing Audit University, China
Vikrant Bhateja   SRMGPC, India
Vishnu Varthanan   Kalasalingam Academy of Research and Education,
  Govindaraj   India
Wagner Quinn   University College Cork, Ireland
Weibo Liu   Brunel University London, UK
Weiguo Zhu   Huaiyin Institute of Technology, China
Weiling Bai   Inner Mongolia University, China
Wenbo Fu   Datong University, China
Xianwei Jiang   Nanjing Normal University of Special Education,
                                   China
Xinhua Mao   Nanjing University of Aeronautics and Astronautics,
                                   China
Xuanyue Tong   Nanyang Institute of Technology, Singapore
Yang Wang   School of Electronics and Information Engineering,
                                   China
Yeliz Karaca   University of Massachusetts Medical School, USA
Yi-Ding Lv   Nanjing Medical University, China
Yin Zhang   University of Economics and Law, China
Ying Shao   Harvard University, USA
Yongjun Qin   Guilin Normal College, China
Yuan Xu   University of Jinan, China
Yuankai Huo   Vanderbilt University, USA
Yuriy S. Shmaliy   Universidad de Guanajuato, Mexico
Zehong Cao   University of Tasmania, Australia
Zheng Zhang   University of Southern California, USA
Zhimin Chen   Shanghai Dianji University, China
Zhou Zhang   Stevens Institute of Technology, USA
Zhuo Tao   Nanyang Technological University, Singapore
Zhuqing Jiao   Changzhou University, China

Contents – Part II

Multimedia Technology and Enhanced Learning

Digital Image Processing, Analysis and Application Based on Machine Learning

Data Fusion Filter and Machine Learning for Statistical Signal Processing

Intelligent Technology and Design for Special Education/Rehabilitation

Transfer Learning Methods Used in Medical Imaging and Health Informatics

Weather Radar and Antenna Design

Contents – Part I

Multimedia Technology and Enhanced Learning

Heuristic Network Similarity Measurement Model Based on Cloud Computing

Yang Guo[1(✉)] and Jia Xu[1,2]

[1] Dalian Institute of Science and Technology, Dalian, China
yc306@le.ac.uk, jf125125@163.com
[2] Dalian Jiaotong University, Dalian, China

Abstract. In order to solve this problem, a heuristic network similarity measurement model based on cloud computing is proposed. First, the heuristic network data is collected, and then the spherical harmonic function method is used to match the network data similarity measurement. After the above work, the heuristic network similarity measurement model is built according to the structure balance theory. Thus, a heuristic network similarity measurement model based on cloud computing is constructed. In the experiment, the quality of service node genes obtained by the two models was tested. The experimental results show that the service node genes obtained by the model are better and meet the design requirements.

Keywords: Cloud computing · Heuristic network · Similarity measurement · Node · Evaluation criteria · Network data · Similarity propagation mode

1 Introduction

At present, the research on polar information network mainly focuses on link prediction. Among them, part of the work involves the correlation measurement of nodes, but only uses the correlation measurement method on the non-polar information network, there is no special research on the correlation measurement of nodes on the polar information network by using the negative edge information effectively. Based on the status theory, foreign scholars calculate the similarity between two nodes according to the positive degree, negative degree, positive degree and negative degree. Positive in and negative in can improve the status of nodes, while positive in and negative in can reduce the status of nodes. However, only the degree of two nodes is considered, and whether there is a relative preference relationship between two nodes is not considered [1].

In this paper, we focus on the measurement of the correlation between nodes in polar information networks, including the measurement of the similarity between nodes of the same type in homogeneous polar information networks and the measurement of the correlation between nodes of different types in heterogeneous polar information networks. The main work and contributions of this paper are summarized as follows: in order to test the similarity between nodes of the same type in homogeneous polar information networks, a heuristic network similarity measurement model based on

Y.-D. Zhang et al. (Eds.): ICMTEL 2020, LNICST 327, pp. 3–16, 2020.
https://doi.org/10.1007/978-3-030-51103-6_1

cloud computing is proposed. The model can make full use of the semantic information contained in the positive and negative sides of the isomorphic polar information network. The similarity between the two nodes is measured by comparing the direct neighbor sets of the source node and the target node. Because there are both positive and negative edges in polar information networks, we need to consider the positive and negative neighbor sets of nodes respectively. Furthermore, for the directed isomorphic polar information network, the neighbor set is subdivided into positive incoming neighbor set, negative incoming neighbor set, positive outgoing neighbor set, and negative outgoing neighbor set due to the more definite direction of the edge. For two nodes without common neighbors, similarity propagation is used to measure the similarity between the source node and the target node. Finally, experiments are carried out on real datasets, and the experimental results show the effectiveness of the proposed method.

2   Heuristic Network Data Collection

Heuristic network data collection refers to the process of constructing a heuristic network data with some algorithm for the original feature set of input data set. The heuristic network data extracted from the original feature variables is the most consistent with the set feature selection criteria [2]. The data with lighter weight in the original data can be filtered by feature selection, and the data that can best reflect the data features will be preserved. After feature selection, the data model will be more accurate and simple, and the processing efficiency of the model will be greatly improved. Feature selection model can be described as a simple mathematical model: Given the sample data set $S = \{F, C, D\}$, where F represents the feature sample set, C represents the category sample set, D is the data sample set. It is assumed that the feature algorithm $E(x) \in (0, 1)$ is a feature evaluation function, and its value size corresponds to the weight of data, which measures the importance of data. There are several types of functions in the process of selecting optimal heuristic network data:

A: Select a subset in feature set F to make E(x) maximum;
B: The minimum value E(x) of the given E_0 value. The corresponding function value is greater than the subset.
C: Find a subset in the feature set F so that the amount of E(x) is large and the number of features is as small as possible;

The above methods take different measures from the point of view of the number or weight of features, but ultimately can select the most favorable feature data [3].

Heuristic network data collection process (Fig. 1);

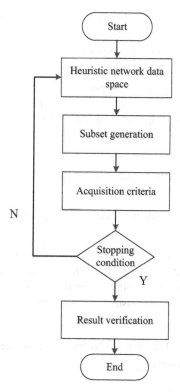

Fig. 1. Heuristic network data collection process

The subset generation part refers to the process of extracting a subset from the original feature set, which generally needs to search for a subset that can meet specific conditions in the feature space according to some search rules. This section contains two key concepts: search direction and search strategy. The search direction refers to the starting point of feature search and the direction of search [4]. Any subset can be used as the starting point of search, which determines the search direction: forward search (starting from the empty set), backward search (starting from the whole feature set), two-way search (starting from both directions above), and random search (starting from a randomly generated subset). Search strategies generally include complete search, randomized search and heuristic search [5]. Evaluation criteria refer to the criteria for evaluating the heuristic network data selected from the generated part, mainly focusing on the rationality of the results, the effectiveness of the algorithm in the problem and whether it is helpful to achieve the specified goals. Different evaluation criteria will lead to different heuristic network data, and the quality of evaluation criteria will directly affect the effect of the algorithm. Common evaluation criteria can be divided into distance measurement, information measurement, consistency measurement, classification error rate, etc. For different specific problems, different evaluation criteria can be selected [6]. Stop condition can also judge the rationality of current heuristic network data and send the unreasonable heuristic network data back to the subset generation process for screening again. Generally, the stop condition can be set as follows:

A: The number of subsets reaches the preset value.
B: According to the currently selected heuristic network data, the classification rate has reached the requirements or been improved.
C: The change of feature number no longer affects the function value of evaluation criteria.

The value of the D: evaluation criterion function reaches the inflection point or the threshold [7] already set.

E: The selected heuristic network data is the optimal solution of the evaluation criteria function.

The result verification part refers to comparing the optimal heuristic network data selected from the original features and the existing optimal subset, which has been used to verify the applicability and rationality of the algorithm, so that the optimal algorithm and the corresponding optimal heuristic network data can be determined.

3 Cloud Computing Heuristic Network Similarity Measurement Matching

Firstly, the heuristic network data is collected, and then to solve the problem of spherical harmonic function information loss, the cloud computing heuristic network similarity measurement matching is realized. At present, the proposed solutions can be roughly divided into three categories:

A: Some changes are added when using the spherical harmonic function method to retain some local information, such as the optical ray method and the improved optical ray method.
B: When Chebyshev points are extracted by spherical harmonic decomposition, the points containing global information are sampled.
C: Map the decomposed harmonic components to the Cartesian coordinate system, and add a coordinate [8].

These three methods have their own advantages and disadvantages: the first method has high execution efficiency, but it can not eliminate the problem of information loss in essence; the second method is novel and effective, but unstable; the third method is the best method among the current three-dimensional object shape matching methods, but the algorithm complexity is high, which is not suitable for large database query [9]. Because of the high information stored in 3D object model, the efficiency is very important in 3D object query system. Based on this consideration, this paper decided to adopt the first method and improve the matching effect as much as possible on the basis of high efficiency.

Cloud computing heuristic network similarity measurement matching process (Fig. 2):

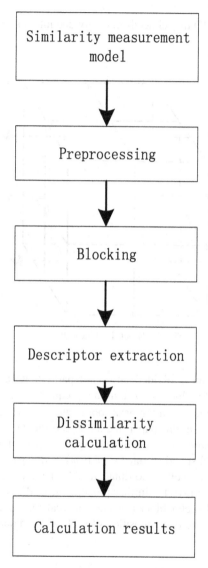

Fig. 2. Cloud computing heuristic network similarity measurement matching process

In the first step, the preprocessing process first needs to express the ball function in the form of $f(\theta, \varphi) = R$ where R is the distance from the sample point of Monte Carlo surface to the center of mass of the object [10].

Secondly, in order to find the corresponding sector blocks of two objects, the spherical function $f(\theta, \varphi) = R$ should be normalized by PCA coordinate axis transformation.

In the third step, the 3D model needs to be divided into n sector blocks, as shown in Fig. 3.

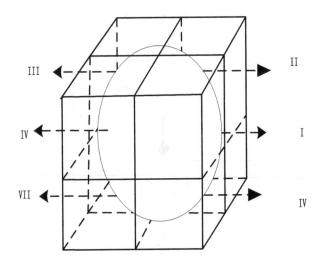

Fig. 3. Block by block diagram

First, the ball function is divided into eight parts, that is, eight quadrants, Then divide each quadrant into $\frac{n}{8}$ blocks by dividing φ equally.

As the block is carried out in the new coordinates, it is necessary to transform the coordinates of each point. In this paper, the affine coordinate transformation method is used, and the affine coordinate system is a special case of rectangular coordinate system. After the original spherical function $f(\theta, \varphi) = R$ is partitioned according to the coordinate value of the point after coordinate transformation, the points of the spherical function are stored in n arrays for further processing. In the fourth step, the spherical harmonic function of N sector blocks of the spherical function is decomposed. After spherical harmonic function decomposition of each sector block, the obtained rotation invariant descriptor is:

$$SH(f)_n = \{\|f_0(\theta, \varphi)\|\} \tag{1}$$

In practical operation, it is impossible for us to obtain the energy component with infinite frequency. As an approximation, we can decompose the spherical coordinate function into the sum of the finite terms of the spherical harmonic function with a certain bandwidth, B is used to specify the upper limit of the frequency, that is, the bandwidth, and the higher expansion coefficient exceeding the frequency will be ignored:

$$f_0(\theta, \varphi) = \sum_{I=0}^{B-1} a_{i,m} \tag{2}$$

Through a large number of tests, it is found that when $B = 64$ or greater, the accuracy of the system has generally reached about 5×10^3. For the requirements of 3D model retrieval, it is enough.

Finally, by summarizing the descriptors of these n sectors, we get a two-dimensional grid distribution, which is the final shape descriptor.

Fourth, there are many ways to match the final descriptor. A simple method is to directly subtract the descriptors of the corresponding sector block, and then overlay the results of the subtraction to get the final dissimilarity.

4 Heuristic Network Similarity Measurement Model

First, the heuristic network similarity data is collected, then the heuristic network similarity measurement is matched, and the heuristic network similarity measurement model is constructed by using the structure balance theory.

Specific implementation process:

In an undirected homogeneous polar information network, given two nodes v_i and v_j, by considering their direct neighbor set, we give the calculation method of measuring the similarity between these two nodes. For node VI on undirected homogeneous polar information network, its direct neighbor set, the direct neighbor set connected with VI positive edge and the direct neighbor set connected with VI negative edge are respectively:

$$\begin{aligned} N_i &= \{v_k | (v_i, v_k) \in \delta |\} \\ N_i^+ &= \{v_k | (v_i, v_k) \in \delta^+ | + \{v_i\}\} \\ N_i^- &= \{v_k | (v_i, v_k) \in \delta^- |\} \end{aligned} \tag{3}$$

Among them, $N_i = N_i^+ \cup N_i^-$.

To sum up, the more common neighbors with the same evaluation between two nodes, the more similar they are. Accordingly, the more common neighbors with opposite evaluation between two nodes, the more dissimilar they are, given two nodes v_i and v_j. We can get the set of neighbors with the same evaluation and the set of neighbors with different evaluation.

$$C_S(v_i, v_k) = \left\{ v_k \middle| \left(v_k \in N_i^+ \land v_k \in N_j^+ \right) \middle| \right\} \tag{4}$$

Where $C_S(v_i, v_k)$ represents the set of common neighbor nodes with the same evaluation held by v_k and Represents a set of common neighbor nodes with different evaluations held by v_k and v_i.

Then the similarity between the two nodes is defined as follows:

$$sim(v_i, v_k) = \frac{|C_S(v_i, v_k)| - C_d(v_i, v_k)}{|N_i \cup N_j|} \tag{5}$$

The value range of the similarity between the two nodes is in the range of $[-1, 1]$, $sim(v_i, v_k) > 0$ Indicates that nodes v_k and v_i are similar, $sim(v_i, v_k) < 0$ Indicates that the nodes v_k and v_i are not similar. The greater the similarity value is, the more similar the two nodes are, and the smaller the value is, the less similar the two nodes are. It should be noted that both the maximum value L and the minimum value -1 can be obtained. When two nodes have identical neighbors, and their evaluation of each neighbor is the same, the similarity between the two nodes is 11. When two nodes have identical neighbors, but their evaluation of each neighbor is not the same, then the similarity between the two nodes is -1, similarity 1 means that the positions of the two nodes are identical, and similarity -11 means that the positions of the two nodes are different.

The similarity measure is symmetric and satisfies $sim(v_i, v_k) = 1$, It is worth noting that if we ignore the information, we define the similarity on the undirected isomorphic polar information network, and the measurement method is the same as Jaccard coefficient.

Use formula (3) to calculate the similarity of nodes in Fig. 4, and consider users $u1$ and users $u3$. Then $N_1 = \{U1, U2, U3, U4, U6, U7\}$. The similarity between user $U1$ and user $U2$ is $sim(U1, U3) = -\frac{1}{3}$, According to the theory of structural equilibrium, it can be seen in Fig. 4.

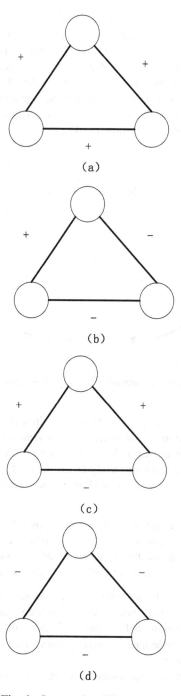

Fig. 4. Structural equilibrium theory

In the directed polar information network, the similarity between two nodes is measured. The directivity of the edge is more clear than that of the undirected information network. A user can express his attitude to other users, and can also accept the evaluation of other users. Therefore, we divide the user's neighbor set into two parts: (1) in neighbor set; (2) out neighbor set. In particular, the incoming neighbor set includes the positive incoming neighbor set and the negative incoming neighbor set, and the outgoing neighbor set includes the positive outgoing neighbor set and the negative outgoing neighbor set. As shown in Fig. 5;

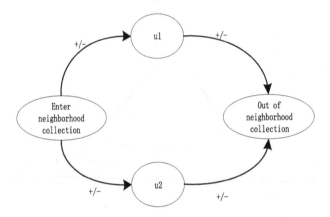

Fig. 5. Heuristic network similarity measurement model

If two users want to have a high degree of similarity, they should meet two conditions: the first condition is that as many users hold the same evaluation as possible; the second condition is that as few users hold different evaluation as possible. Then we can extend this basic idea to apply to the directed polar information network.

If two users want to have high similarity, they need to meet the following conditions:

A: We should try to evaluate as many users as possible.
B: We should try to evaluate as many users as possible.
C: The opposite users should be evaluated as little as possible.
D: Users with opposite evaluation should be as few as possible.

In other words, if two users have the same evaluation on many other people, and many people have the same evaluation on these two users, then the two users are likely to be highly similar. Correspondingly, if two users have different comments on many other people, and many people have different comments on these two users, then the two users may be different.

Thus, the heuristic network similarity measurement model based on cloud computing is completed.

5 Experimental Analysis

In order to verify the feasibility and effectiveness of the model, this chapter will carry out simulation experiments. The experimental environment is: the experimental host is configured with Intel Core2 2.93 GHz processor, 2.00 gb memory, Windows XP operating system, and the algorithm is implemented with MATLAB 9.

5.1 Experimental Parameters

The heuristic optimization algorithm in this paper consists of two stages. The following describes the setting of key parameters in the two stages:

The main parameters of the traditional model are population size n, crossover probability p_c, mutation probability p_m, where p_c and p_m represent the adjusted probability of white adaptation. The population size is n = 40. To facilitate calculation, the size of the element candidate service set C_i in the set C of the candidate service set is the same.

The heuristic network similarity measurement model based on cloud computing influences the behavior of artificial ants by changing the three parameters of information heuristic factor α, expectation heuristic factor β, and pheromone volatility coefficient ρ, we reduce the influence of pheromone on the movement of artificial ants by reducing the value of α and ρ and increase the exploration ability of artificial ants by increasing the value of β, $\beta = 5$.

5.2 Experimental Result

The business process contains four abstract services, and each candidate collection of abstract services is assumed to have the same scale, with five entity Web services respectively. In this paper, we consider two QoS parameters including overhead and response time, and define the global QoS constraint vector as $*R = (400, 90)$. First of all, in the preprocessing candidate service set of similarity measurement parameters, the list of entity service similarity measurement parameters is shown in Table 1. Because the cost and response time belong to the linear additive, negative heuristic network similarity measurement model, there is no need for type conversion, only a unified numerical order of magnitude, mapping to the same range of values. According to the preprocessing operation of application formula (5), the parameter list of entity service heuristic network similarity measurement model is shown in Table 1. In the calculation process of using formula (5), set the real number C1 = 0, C2 = 100 (Table 2).

Table 1. Entity service heuristic network similarity measurement parameter list

C1		C2		C3		C4	
Expenses	Response time	Expenses	Response time	Expenses	Response time	Expenses	Response time
204	5	10.5	10	33	42	94	20
203	5	10.3	5	43	43	95	30
205	4	19.2	7	51	51	84	13
206	2	14.5	8	46	34	75	21
212	3	6.2	5	42	41	123	16

Table 2. Application formula (5) list of heuristic network similarity measurement parameters after preprocessing

C1		C2		C3		C4	
Expenses	Response time	Expenses	Response time	Expenses	Response time	Expenses	Response time
52	50	15	45	40	65	54	50
56	30	80	0	65	16	100	60
100	0	100	15	46	0	22	45
80	50	60	100	30	162	84	0
5	60	0	90	60	54	0	152

According to the algorithm of initial distribution of pheromones, a group of feasible solutions are obtained, including $(cs_{11}, cs_{22}, cs_{32}, cs_{43})$ and $(cs_{11}, cs_{22}, cs_{32}, cs_{43})$, as shown in Table 3. The aggregation cost of heuristic network similarity measurement parameters is also listed in the table.

Table 3. List of feasible solutions of two models

Feasible solution	Traditional heuristic network similarity measurement model	Heuristic network similarity measurement model based on Cloud Computing
$(cs_{14}, cs_{22}, cs_{32}, cs_{43})$	70	302
$(cs_{11}, cs_{22}, cs_{32}, cs_{43})$	80	352

The comparison results show that feasible solution $(cs_{11}, cs_{22}, cs_{32}, cs_{43})$ is a better execution plan than feasible solution $(cs_{14}, cs_{22}, cs_{32}, cs_{43})$. We set pheromones according to this set of execution plans. At the beginning stage, the initial value of pheromones of all nodes is set to 1.00, and the two feasible solution parts are over-lapped, including entity services cs_{22}, cs_{32}, cs_{43}. It is clear that the genes of these nodes are better, and the pheromone value of these service nodes should be slightly higher, set to 1.20, cs_{11} as the starting point of the better solution, set cs_{11}, the pheromone value of the service node where is 1.15, and set the sub optimal solution starting service cs_{14}. The pheromone value of the service node is 1.10. The initial distribution of pheromones is obtained by calculation, as shown in Fig. 6.

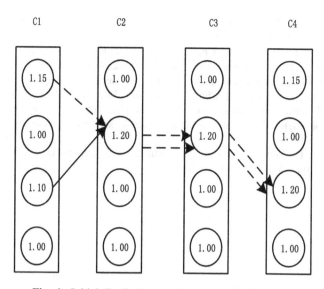

Fig. 6. Initial distribution of pheromones in the model

As shown in Fig. 6, using the heuristic network similarity measurement model of cloud computing, these node genes are better.

6   Conclusion

This paper studies the similarity measurement problem on heuristic networks and discusses it. This paper proposes a new method to measure the similarity between nodes of the same type on the same polar information network. This method can take into account the interaction of positive and negative sides on the polar information network at the same time, and make full use of the semantic information of the polar information network. The experimental results on real datasets also show the effectiveness of the proposed method.

References

1. Chen, X.: Research on vine type serial-parallel clustering method of network public opinion based on cloud computing. J. Intell. **36**(2), 122–126 (2017)
2. Li, H., Mao, Y.: Forecasting model of aquaculture production based on heuristic Johnson algorithm optimization and BP neural network. Fishery Modernization **44**(6), 19–23 (2017)
3. Yuan, J.M.: Research of collection model based on heuristic algorithm in MANET. Control Eng. China **24**(6), 1248–1253 (2017)
4. Song, J., Wen, L.: A similarity measure for process models based on task occurrence relations. J. Comput. Res. Dev. **54**(4), 832–843 (2017)
5. Han, X., Wang, Q.: Geographic ontology concept semantic similarity measure model based on BP neural network optimized by PSO. Comput. Eng. Appl. **53**(8), 32–37 (2017)

6. Huang, Y., Xie, F.: k-modes algorithm based on structural similarity. Comput. Eng. Appl. **53** (23), 102–107 (2017)
7. Yang, S., Liu, C.: Two stages of heuristic based algorithm for influence maximization in social network. J. Chin. Comput. Syst. **38**(10), 2268–2274 (2017)
8. 胡启志, 颜娜.: Influence maximization algorithm based on term frequency and node similarity in social networks. J. Chin. Comput. Syst. **38**(2), 259–263 (2017)
9. Ye, J.: Clustering algorithms using dimension root distances-based cosine similarity measures of single-valued and interval neutrosophic sets. Mach. Tool Hydraulics **46**(6), 199–208 (2018)
10. Yao, Z., Wu, L.: Network visual analysis method based on heuristic community detection. J. Comput. Appl. **37**(s1), 155–159 (2017)

Study on the Preparation of the Precursor of the Li-ion Screen Based on Big Data Analysis

Xiang Xiao[(⊠)], Zhuan Wei, and Pei Pei

Department of Information Science and Engineering,
Changsha Normal University, Changsha, China
xiaoxiang665366@163.com

Abstract. In order to improve the preparation ability of lithium ion screen precursor, an anti-interference high definition lithium ion screen precursor preparation data dynamic migration and information enhancement method based on dynamic migration equilibrium modulation is proposed. The dynamic migration transmission channel model of HD lithium ion screen precursor preparation is constructed, and the anti-interference design of HD lithium ion screen precursor preparation is carried out by using interference filtering algorithm. The dynamic migration state characteristic of the preparation data of HD lithium ion screen precursor is extracted, and the Porter interval equilibrium method is used to control the dynamic migration of HD lithium ion screen precursor preparation data. The dynamic migration optimization of preparation data of anti-interference HD lithium ion screen precursor is realized, and the anti-interference and safety of lithium ion screen precursor preparation are improved. The simulation results show that the preparation of HD lithium ion screen precursor by this method is safe and has strong anti-interference ability.

Index Terms: Big data analysis · Lithium ion · Preparation of screen precursor

1 Introduction

At present, many aspects of the country need to use lithium or its compounds. With the country paying more and more attention to high technology, the demand for lithium in the market is increasing rapidly, and the lithium resources are less and less, which requires lithium extraction from liquid lithium resources. The reserves of lithium resources in the ocean are about 260 billion tons, less than 1/10000 tons on land. Therefore, it is urgent to extract lithium from low lithium concentration seawater, geothermal may protect the environment and improve the efficiency. In this paper, while standing on the shoulders of giants, we continue to explore the preparation technology of lithium-ion screen membrane, and find out a better method. Generally speaking, water and salt lake brine [1], in which the application of ion screen adsorption we should find out better film-forming materials and focus on the amount of lithium ion screen precursor. With the development of preparation technology of lithium ion screen precursor, higher requirements are put forward for the security and anti-interference of the network, and it is necessary to control the dynamic migration of

© ICST Institute for Computer Sciences, Social Informatics and Telecommunications Engineering 2020
Published by Springer Nature Switzerland AG 2020. All Rights Reserved
Y.-D. Zhang et al. (Eds.): ICMTEL 2020, LNICST 327, pp. 17–26, 2020.
https://doi.org/10.1007/978-3-030-51103-6_2

the preparation data of high definition lithium ion screen precursor. The transmission link model of HD lithium ion screen precursor preparation is analyzed. The dynamic migration and enhancement of HD lithium ion screen precursor preparation data is carried out by anti-interference filtering method, the network design is optimized, and the adaptive control ability of the network is improved. The research on data dynamic migration and information enhancement of HD lithium ion screen precursors has attracted great attention [2].

Traditionally, the security enhancement methods for dynamic migration of high definition lithium ion screen precursors mainly include fuzzy control method, adaptive equilibrium control method and interval equilibrium control method. The embedded virtual network is used to realize the dynamic migration of the preparation data of the anti-interference high-definition lithium-ion screen precursor, the big data transmission control is carried out, and the dynamic forwarding and migration design of the anti-interference high-definition lithium-ion screen precursor preparation data is carried out [3]. The self-adaptability and fuzziness of the above methods for the dynamic migration safety control of anti-interference high-definition lithium-ion screen precursors are not good, in order to solve the above problems, A data dynamic migration and information enhancement method for anti-interference HD lithium ion screen precursors based on dynamic migration equilibrium modulation is proposed. The dynamic migration transmission channel model of HD lithium ion screen precursor preparation is constructed, and the anti-interference design of HD lithium ion screen precursor preparation is carried out by using interference filtering algorithm. The dynamic migration state characteristic of the preparation data of HD lithium ion screen precursor is extracted, and the Porter interval equilibrium method is used to control the dynamic migration of HD lithium ion screen precursor preparation data. The dynamic migration optimization of preparation data of anti-interference HD lithium ion screen precursor is realized, and the anti-interference and safety of lithium ion screen precursor preparation are improved. Finally, the simulation results show the superior performance of this method in improving the dynamic migration and information enhancement ability of HD lithium ion screen precursors [4].

2 Data Transmission Channel Model and Interference Suppression for Precursor of High-Definition Lithium Ion Screen

2.1 Data Transmission Channel Model for Precursor of Li-ion Sieve

In order to realize the dynamic migration and the information enhancement of the anti-interference type lithium ion sieve precursor preparation data, a data transmission channel model of an anti-interference type lithium ion sieve precursor system is first constructed, and the channel output structure of the anti-interference type lithium ion sieve precursor preparation data is combined and controlled, the self-adaptive random link forwarding control protocol is adopted to carry out the balance control of the anti-interference type lithium ion sieve precursor system standby data transmission, the baud interval equalization control technology is adopted [5], and the fuzzy self-adaptive Arabic-complete processing of the data output of the precursor of the lithium ion screen is carried out, the anti-interference filtering design is carried out on the weighting result, the balance control of the data transmission channel of the precursor of the lithium ion screen is realized, and the average mutual information characteristic quantity of the anti-interference type lithium ion sieve precursor preparation data is extracted, the time reversal processing is carried out on the data output bit sequence flow of the anti-interference type lithium ion sieve precursor preparation data output, and the balance of the data output of the anti-interference type lithium ion sieve precursor preparation data output is realized, the anti-interference type lithium ion sieve precursor preparation data distribution model is constructed by adopting a cooperative Kalman filtering algorithm to realize the anti-interference type lithium ion sieve precursor preparation data distribution, and the load of the preparation data of the precursor of the lithium ion sieve is $E_j = \sum_k |C_j(k)|^2$, According to the load of the data dynamic migration of the precursor of the lithium ion sieve, the dynamic migration control of the network is carried out by adopting a fuzzy self-adaptive equalization scheduling method[, and the learning function of the anti-interference type lithium ion sieve precursor preparation data transmission is as follows:

$$\begin{cases} \min \sum_{1 \leq i \leq K} \sum_{e \subseteq k(e)} \dfrac{f(e(i))}{C(e,i)} \\[2mm] 0 \leq f(e,i) \leq C(e,i) \\[2mm] F = const \\[2mm] \sum_{1 \leq i \leq K, e \subseteq k(e)} \dfrac{f(e(i))}{C(e,i)} + \sum_{e \subseteq k(e)} \dfrac{f(e'(i))}{C(e',i)} \leq k(v) \end{cases} \quad (1)$$

The filter detection of anti-interference lithium ion screen precursor preparation data is carried out by link random allocation method. The spatial sampling load of anti-interference lithium ion screen precursor preparation data is expressed as follows:

$$\begin{aligned} Computition(n_j) &= (E_{elec} + E_{DF})l\delta + E_{Tx(l,d_j)} \\ &= (E_{elec} + E_{DF})l\delta + lE_{elec} + l\varepsilon_{fs}d_j^2 \\ &= [(E_{elec} + E_{DF})\delta + E_{elec} + \varepsilon_{fs}d_j^2]l \end{aligned} \quad (2)$$

the load balance design of the anti-interference type lithium ion sieve precursor preparation data is carried out by adopting a link random distribution method [8], a self-correlation matched filter is designed, the multi-path suppression of the data transmission of the precursor system of the lithium ion sieve is realized through a direct sequence spread spectrum method [6], the output load and the balance scheduling formula of the obtained network are expressed as follows:

$$\eta_k^w(\omega) = E\big(T_k^w \big| T_k^w > \xi_k^w(\omega)\big), \ k \in R_w, w \in W \tag{3}$$

The channel model of data allocation prepared by lithium ion screen precursor is constructed. The subsequence of channel dynamic migration and information enhancement scheduling set is obtained by using time reversal mirror:

$$r_1(n) = r_2(n) \exp(-j\omega_0 T_p/2), \ n = 0, 1, \ldots, (N-3)/2 \tag{4}$$

The $(N-1)/2$ point Fourier transform of $r_1(n)$ and $r_2(n)$ is carried out respectively, and the optimal design of data transmission channel for lithium ion screen precursor preparation is realized [7].

2.2 Anti-interference Design of Lithium Ion Screen Precursor Preparation

The interference filtering algorithm is used to design the anti-interference of lithium ion screen precursor preparation, and the cooperative filtering method is used to sample the code element of lithium ion screen precursor system, and the output dynamic load is obtained as:

$$R_1(k) = R_2(k) \exp(-j\omega_0 T_p/2), \ k = 0, 1, \ldots, (N-3)/2 \tag{5}$$

$$R_2(k) = A_k \exp(j\varphi_k), \qquad\qquad k = 0, 1, \ldots, (N-3)/2 \tag{6}$$

Wherein, the dynamic transmission delay of the preparation data of the lithium ion screen precursor is the mean value of the linear equilibrium, the variance is the variance, and the received symbol sequence of the anti-interference filtering of the data prepared by the lithium ion screen precursor is expressed as follows:

$$t_a = E(T_a) = t_a^0 + \beta t_a^0 E((V_a)^n) E(1/(C_a)^n), \ a \in A \tag{7}$$

$$E\big((T_a)^2\big) = (t_a^0)^2 + 2\beta(t_a^0)^2 E((V_a)^n) E(1/(C_a)^n) + (\beta t_a^0)^2 E\big((V_a)^{2n}\big) E\big(1\big/(C_a)^{2n}\big), \ a \in A \tag{8}$$

$$\varepsilon_t^a = Var(T_a) = E\big((T_a)^2\big) - (E(T_a))^2, \ a \in A \tag{9}$$

According to the above analysis, the interference filter anti-interference model of dynamic migration of lithium ion screen precursor preparation data is constructed [8].

3 Dynamic Migration and Information Enhancement Processing

3.1 Equilibrium Control of Dynamic Migration of Lithium Ion Screen Precursor Preparation Data

On the basis of the above construction of the data dynamic migration transmission channel model of lithium ion screen precursor preparation and the anti-interference design of lithium ion screen precursor preparation by using interference filtering algorithm [9], the data dynamic migration and information enhancement optimization design of lithium ion screen precursor preparation are carried out. in this paper, a dynamic migration and information enhancement method of lithium ion screen precursor preparation data based on dynamic migration equilibrium modulation is proposed. The dynamic migration state characteristics of lithium ion screen precursor preparation data were extracted. The load in the dynamic migration process of lithium ion screen precursor preparation data was obtained as follows:

$$
\begin{aligned}
C_{T'}(f)Y_{T'}(f) &= C_{T'}(f)\sum_{n} x\left(f - \frac{n}{T'}\right) e^{j2\pi\left(f-\frac{n}{T'}\right)\tau_0} \\
&= C_{T'}(f)X(f)\, e^{j2\pi f \tau_0}
\end{aligned}
\tag{10}
$$

According to the channel characteristics, the equilibrium scheduling and the balanced allocation of big data are carried out, and the mean value of load distribution Q^w, V_a, F_k^w of lithium ion screen precursor preparation data is obtained by using coherence detection method]. From this, the average mutual information characteristic quantity of the preparation data of lithium ion screen precursor can be obtained as follows:

$$
q^w = E(Q^w) = \sum_{k\in R_w} f_k^w, \ w \in W
\tag{11}
$$

$$
v_a = E(V_a) = \sum_{w\in W} \sum_{k\in R_w} \delta_{ak}^w f_k^w, \ a \in A
\tag{12}
$$

$$
f_k^w \geq 0, \ k \in R_w, w \in W
\tag{13}
$$

The measurement distance for calculating the preparation data of lithium ion screen precursor is as follows:

$$
\|r\| = \sqrt{\sum_{i=0}^{m} r_i^2 \left\|v_{\sigma(i)}^*\right\|^2 + \|w\|^2} \geq |r_m| \left\|v_{\sigma(m)}^*\right\|
\tag{14}
$$

The dynamic migration and the information enhancement output iteration equation are obtained by adopting an adaptive feedback adjustment method to carry out channel equalization modulation of the data dynamic migration of the precursor of the lithium ion screen:

$$f_{ij}(n+1) = f_{ij}(n) + \mu_{MCMA} \frac{\partial J_{MCMA}(n)}{\partial f_{ij}(n)} \tag{15}$$

In which, μ_{MCMA} represents an initial load, and according to the algorithm and the model design, the balance control of the dynamic migration of the preparation data of the precursor of the lithium ion screen is realized, and the stability and the channel self-adaptive distribution performance of the data migration process are improved [10].

3.2   Optimization of Data Dynamic Migration for Precursor System of Li-ion Screen

The data distribution model of lithium ion screen precursor preparation was constructed, and the impulse response of lithium ion screen precursor preparation data was obtained by using decentralized control protocol under the condition of static mapping:

$$\begin{cases} y(t) = x(t - t_0) \Rightarrow W_y(t, v) = W_x(t - t_0, v) \\ y(t) = x(t)e^{j2\pi v_0 t} \Rightarrow W_y(t, v) = W_x(t, v - v_0) \end{cases} \tag{16}$$

The impulse response of the preparation data of lithium ion screen precursor is calculated. In the dynamic migration and information enhancement of big data prepared by lithium ion screen precursor, the link allocation is carried out according to the synthesis characteristics of lithium ion screen precursor preparation channel [11]. The dynamic migration normal distribution of the preparation data of the lithium ion screen precursor satisfies $X \sim S_\alpha(1, \beta, 0), 1 < \alpha < 2$, the inertia characteristic distribution of the dynamic migration of the preparation data of the lithium ion screen precursor is obtained by:

$$y(t) = \sqrt{k}x(kt), k > 0 \tag{17}$$

$$W_y(t, v) = W_x(kt, v/k) \tag{18}$$

Wherein, G represents the characteristic resolution of the preparation data distribution of the precursor of the lithium ion sieve, v represents the modulation frequency, and W_x is a combined state estimation of the data dynamic migration load of the precursor of the lithium ion sieve. The dynamic migration load response control model for the preparation data of the precursor of the lithium ion screen is calculated, and the formula is expressed as:

$$\Phi(\omega) = E[e^{j\omega X}]$$
$$= \begin{cases} \exp\{j\mu\omega - |\sigma\omega|^\alpha[1 - j\beta\mathrm{sgn}(\omega)\tan(\frac{\pi\alpha}{2})]\}, \alpha \neq 1 \\ \exp\{j\mu\omega - |\sigma\omega|^\alpha[1 + j\beta\mathrm{sgn}(\omega)\frac{2}{\pi}\ln|\omega|]\}, \alpha = 1 \end{cases} \tag{19}$$

In embedded environment, the output of the $j = 0, 1, \ldots, M$ sampling point to obtain the dynamic migration state characteristic quantity of lithium ion screen precursor preparation data is as follows:

$$x(t)e^{j\pi t^2 \cot \alpha} = \sum_{n=-N}^{N} x(\frac{n}{2\Delta x})e^{\frac{j\pi(\cot \alpha)n^2}{(2\Delta x)^2}} \sin c\left[2\Delta x\left(t - \frac{n}{2\Delta x}\right)\right] \qquad (20)$$

Porter interval balancing method is used to control the dynamic migration of lithium ion screen precursor preparation data, and the dynamic migration optimization of lithium ion screen precursor preparation data is realized [12]. the dynamic load migration scheduling model is obtained as follows:

$$G(U|\mu_k, \sum_k) = (2\pi)^{-d/2}\left|\sum_k\right|^{-1/2} \times \exp\left[-\frac{1}{2}(U - u_k)^T \sum_k^{-1}(U - u_k)\right] \qquad (21)$$

In the above formula, the $G(U|\mu_k, \sum_k)$ is expressed as the feature quantity of the dynamic migration of the data of the precursor of the lithium ion sieve and the information enhancement, U represents the bandwidth of the channel, and u_k represents the symbol sequence for dynamic migration of the data of the precursor of the lithium ion sieve. In this paper, the data dynamic migration and information enhancement of the precursor of the lithium ion screen are realized.

4 Simulation Experiment and Result Analysis

In order to test the application performance of this method in the dynamic migration and information enhancement of lithium ion screen precursor preparation, the experimental test was carried out, and the simulation analysis was carried out by MATLAB. VC casting liquid solution is to dissolve PVC (fixed quantity) in DMAc. Li1.6Mn1.6O4 is mixed with PVC casting liquid solution, then heated and stirred at 80 °C, Li1.6Mn1.6O4 is uniformly dissolved to form casting film solution. After cooling, a layer of liquid film (of a certain thickness) is scraped off the clean glass plate with a scraper and placed in ion water. The two are separated automatically. Finally, Li. Li1.6Mn1.6O4 was extracted from hydrochloric acid of 0.5 mol/L to change the mass concentration of PVC casting solution, and the effect of the concentration of casting solution on the adsorption capacity was studied. The concentration of casting solution remained unchanged and the effect of Li1.6Mn1.6O4 addition on the adsorption capacity was observed. PVC lithium ion screen precursor membrane and lithium ion screen membrane were marked as Li-MPVC-x-y and H-MPVC-x-y, respectively. Assuming that the distribution region of the preparation data transmission of the lithium ion screen precursor is the uniform square region of the 400 m * 400 m, and the sample sequence bandwidth of the sample output of the lithium ion screen precursor preparation data is 140 Gbit, The intensity of intersymbol interference of lithium ion screen precursor preparation data transmission is-10 × 10 dB, and the adaptive symbol transmission rate is 10 kBaud. according to the above parameters, a simulation model of data dynamic migration and information enhancement of lithium ion screen precursor preparation is constructed. The sampling time interval of the transmission symbol is 0.12 s, and the output of the dynamic migration of the preparation data of the lithium ion screen precursor is shown in Fig. 1.

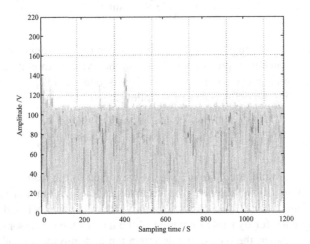

Fig. 1. Dynamic migration of preparation data of lithium ion screen precursors

Taking the data of Fig. 1 as the research object, the dynamic migration state characteristic of the preparation data of lithium ion screen precursor is extracted, and the Porter interval equilibrium method is used to control the dynamic migration of lithium ion screen precursor preparation data. The dynamic migration of data is realized, and the optimized preparation output is shown in Fig. 2.

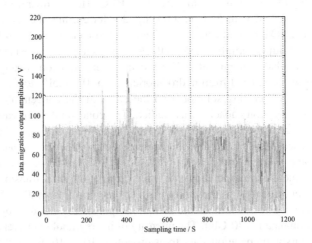

Fig. 2. Dynamic migration output of lithium ion screen precursor preparation data

Figure 2 shows that the output balance of dynamic migration of lithium ion screen precursor prepared by this method is better, the interference filtering performance is better, and the output bit error rate (BER) is tested. The comparative results are shown in Table 1, and the analysis shows that, in this paper, the output BER of dynamic migration of lithium ion screen precursors is low.

Table 1. Comparison of output BER

SNR/dB	Proposed method	Reference [3]	Reference [4]
−10	0.124	0.176	0.187
0	0.045	0.065	0.132
10	0	0.012	0.034

5 Conclusions

In this paper, an anti-interference high definition lithium ion screen precursor preparation data dynamic migration and information enhancement method based on dynamic migration equilibrium modulation is proposed. The dynamic migration transmission channel model of HD lithium ion screen precursor preparation is constructed, and the anti-interference design of HD lithium ion screen precursor preparation is carried out by using interference filtering algorithm. The dynamic migration state characteristic of the preparation data of HD lithium ion screen precursor is extracted, and the Porter interval equilibrium method is used to control the dynamic migration of HD lithium ion screen precursor preparation data. The dynamic migration optimization of preparation data of anti-interference HD lithium ion screen precursor is realized, and the anti-interference and safety of lithium ion screen precursor preparation are improved. The simulation results show that the preparation of HD lithium ion screen precursor by this method is safe and has strong anti-interference ability. The method has good application value in the preparation of lithium ion.

Acknowledgements. School-level Project of Changsha Normal University: XXZD20171103. Hunan Natural Science Foundation: 2018JJ3555.

References

1. Hao, S.G., Zhang, L., Muhammad, G.: A union authentication protocol of cross-domain based on bilinear pairing. J. Softw. **8**(5), 1094–1100 (2013)
2. Ma, Z., Chen, W.: Friction torque calculation method of ball bearings based on rolling creepage theory. J. Mech. Eng. **53**(22), 219–224 (2017)
3. Zhou, S.B., Xu, W.X.: A novel clustering algorithm based on relative density and decision graph. Control Decis. **33**(11), 1921–1930 (2018)
4. He, H., Tan, Y.: Automatic pattern recognition of ECG signals using entropy-based adaptive dimensionality reduction and clustering. Appl. Soft Comput. **55**, 238–252 (2017)

5. He, W., Tan, D., Kuang, H., et al.: Effect of yttrium barrier on the preparation of precursor powders of WC-Co cemented carbide and properties of sintered bulk. J. Alloy. Compd. **742**, S0925838818303979 (2018)

6. Wang, Z., Huang, M., et al.: Integrated algorithm based on density peaks and density-based clustering. J. Comput. Appl. **39**(2), 398–402 (2019)

7. Farnadi, G., Bach, S.H., Moens, M.F., et al.: Soft quantification in statistical relational learning. Mach. Learn. **106**(12), 1971–1991 (2017). https://doi.org/10.1007/s10994-017-5647-3

8. Tu, B., Chuai, R., Xu, H.: Outlier detection based on K-mean distance outlier factor for gait signal. Inf. Control **48**(1), 16–21 (2019)

9. Wei, X.S., Luo, J.H., Wu, J.: Selective convolutional descriptor aggregation for fine-grained image retrieval. IEEE Trans. Image Process. **26**(6), 2868–2881 (2017)

10. Tu, G.S., Yang, X.Y., Zohu, T.P.: Efficient identity-based multi-identity fully homomorphic encryption scheme. J. Comput. Appl. **39**(3), 750–755 (2019)

11. Ma, L., Zhang, T., Ma, D., Fu, Y.: Access network selection algorithm based on Markov model. Comput. Eng. **45**(5), 105–109 (2019)

12. Liu, S., Fu, W., Deng, H., et al.: Distributional fractal creating algorithm in parallel environment. Int. J. Distrib. Sens. Networks (2013). https://doi.org/10.1155/2013/281707

Big Data Fast Extraction Method of Lithium Ion Screen Exchange Feature in Cloud Computing

Xiang Xiao[✉], Zhuan Wei, and Pei Pei

Department of Information Science and Engineering,
Changsha Normal University, Changsha, China
xiaoxiang665366@163.com

Abstract. The characteristic distribution performance of big data, the exchange characteristic of lithium ion screen in cloud computing environment, quantitatively reflects the running state of lithium ion screen exchanger, in order to realize the effective monitoring of lithium ion screen exchange process. A fast extraction algorithm of Li-ion screen exchange feature big data based on big data is proposed. Big data acquisition of lithium ion screen exchange characteristics is realized in lithium ion screen exchange array, and the statistical analysis model of big data mining is constructed. In big data distribution subspace, the spectral feature extraction method is used to extract the spectral stripe feature of Li-ion screen exchange feature big data, and the extracted spectral stripe feature is fuzzy clustering and mining by adaptive neural network learning algorithm. Big data rapid extraction of exchange characteristics of lithium ion screen was realized. The simulation results show that the method has high accuracy in fast extraction of exchange features of lithium ion screen, strong resolution of exchange characteristics of lithium ion screen, and has good application value in high precision measurement of exchange characteristics of lithium ion screen.

Keywords: Cloud computing · Lithium ion screen · Exchange · Feature big data · Rapid extraction

1 Introduction

The exchange characteristic of lithium ion screen, big data, comes from the characteristic sensor of lithium ion screen exchange. The exchange characteristic of lithium ion screen is a dense, equal spacing parallel line engraving on a long strip of optical glass. The measurement data are collected and integrated by distributed lithium ion screen exchange characteristic sensor array, and big data rapid extraction of lithium ion screen exchange feature is carried out in cloud computing environment. To improve the ability of measuring information analysis of exchange characteristics of lithium ion screen, so as to improve the high precision measurement performance of exchange characteristics of lithium ion screen, the research of fast data extraction method has been paid great attention to [1].

The exchange characteristics of lithium ion screen are made of photosensitivity in optical fiber. In the rapid extraction of big data, it is easy to be interfered by the spatial

© ICST Institute for Computer Sciences, Social Informatics and Telecommunications Engineering 2020
Published by Springer Nature Switzerland AG 2020. All Rights Reserved
Y.-D. Zhang et al. (Eds.): ICMTEL 2020, LNICST 327, pp. 27–37, 2020.
https://doi.org/10.1007/978-3-030-51103-6_3

distribution of light intensity in doped optical fiber, which leads to the poor accuracy of mining. In the traditional method [1], the rapid extraction methods of big data for lithium ion screen exchange features include HPCC (High Performance Computing Cluster) mining method, irregular spectral stripe mining method, phase lithium ion screen exchange feature transmission mining method and fuzzy frequent itemset mining method. The spectral characteristic quantity of big data in lithium ion screen exchange feature sensor is extracted, combined with information recombination and fuzzy clustering method, the rapid extraction of lithium ion screen exchange feature big data is realized, and some mining efficiency is obtained. In reference [3], a feature extraction method of lithium ion screen exchange feature big data based on inter-class closed frequent itemsets mining is proposed. The multi-level distributed array grooming model of lithium ion screen exchange feature sensor is used to realize data mining. The feature directivity of data mining is improved, but this method has a large amount of computation in fast data extraction. In reference [4], the piecewise regression test method is used to extract the exchange feature of lithium ion screen, big data is used to extract the exchange feature of lithium ion screen, and the matched filter is used to filter redundant information, so as to improve the statistical analysis ability of big data fast extraction of exchange feature of lithium ion screen. This method is greatly interfered with each other in the feature information fusion of fuzzy sets, which can easily lead to the misclassification and leakage of big data characteristics of lithium ion screen exchange characteristics. In reference [5], a fast extraction method of Li-ion screen exchange feature big data based on improved chaotic partition algorithm is proposed, which uses chaotic partition algorithm for fast extraction and data clustering. The big data feature optimization mining of big data concentrated lithium ion screen exchange feature is realized. the reflection bandwidth range and additional loss of this method are large in the fast extraction of lithium ion screen exchange feature [5].

In order to solve the above problems, a fast extraction algorithm of Li-ion screen exchange feature big data based on big data is proposed in this paper. Firstly, the exchange characteristics of lithium ion screen in lithium ion screen exchange array, big data, were collected, and the characteristic distribution sequence of big data, the exchange feature of lithium ion screen, was reorganized, and then the spectral characteristic quantity of the exchange feature big data of lithium ion screen was extracted. The adaptive neural network learning algorithm is used for fuzzy clustering and mining, and the improved design of the fast extraction algorithm is realized. Finally, the simulation experiment is carried out. The superior performance of this method in improving the fast extraction ability of exchange features of lithium ion screen is shown.

2   Data Acquisition and Big Data Sequence Analysis of Lithium Ion Screen Exchange Characteristics

2.1   Lithium Ion Screen Exchange Characteristic Sensing Data Acquisition

In order to extract the exchange characteristics of lithium ion screen quickly by big data, firstly, the network model of lithium ion screen exchanger is constructed, and the distributed lithium ion screen exchanger array is used for big data acquisition. The exchange characteristics of lithium ion screen collected by big data are mainly scale lithium ion screen exchange characteristic big data, indicating lithium ion screen exchange characteristic big data [6]. The characteristic stripe data of lithium ion screen exchange and the electric pulse data of measurement system in optical path system. The big data acquisition model of lithium ion screen exchange characteristic sensor is constructed, as shown in Fig. 1.

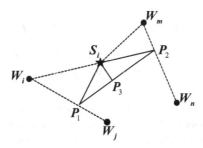

Fig. 1. Big data acquisition model of lithium ion screen exchange characteristic sensor

The big data acquisition model of lithium ion screen exchanger shown in Fig. 1 is divided into three layers: physical layer, transport layer and application layer. In the physical layer, the distributed lithium ion screen exchanger array is used to sample the lithium ion screen exchange characteristic measurement data, and the lithium ion screen exchange characteristic sensor network model is that N lithium ion screen exchanger nodes are randomly deployed in the monitoring area [7]. The running cluster head and computational cluster head of lithium ion screen exchange characteristic sensor network use chirped lithium ion screen exchange characteristics to control the periodic oscillations to improve the accuracy of data acquisition. In the lithium ion screen exchange characteristic sensor network model, each cluster has a cluster head node (SN) and several intra-cluster nodes (V0). The exchange characteristics of lithium ion screen can be divided into periodic structure and aperiodic structure. The distance between the characteristic nodes of lithium ion screen exchange is calculated by Euclidean distance formula, which is as follows:

$$d(i,j) = \sqrt{(x_i - x_j)^2 + (y_i - y_j)^2} \tag{1}$$

Wherein, $E_{Tx}(L,d)$ represents the Euclidean distance between the two lithium ion sieve exchanger nodes, and the distance $E_{Tx}(L,d)$ between the source and the Sink node is expressed as:

$$E_{Tx}(L,d) = \begin{cases} LE_{elect} + L\varepsilon_{fs}d^2, d < d_0 \\ LE_{elect} + L\varepsilon_{mp}d^4, d > d_0 \end{cases} \tag{2}$$

The energy characteristics of the sensing array are expressed as follows:

$$E_{Rx}(L) = LE_{elect} \tag{3}$$

Where E_{elect} represents the transfer energy of lithium ion screen exchange characteristic node (intermediate node) S in the range of reflection bandwidth, d is the transmission link set of big data characteristic of lithium ion screen exchange feature. The load information transmitted in the lithium ion screen exchange array to the Sink node is expressed as follows:

$$\begin{aligned} Computition(n_j) &= (E_{elec} + E_{DF})l\delta + E_{Tx(l,d_j)} \\ &= (E_{elec} + E_{DF})l\delta + lE_{elec} + l\varepsilon_{fs}d_j^2 \\ &= [(E_{elec} + E_{DF})\delta + E_{elec} + \varepsilon_{fs}d_j^2]l \end{aligned} \tag{4}$$

According to the results of load fusion, the attribute partition characteristics of big data, the exchange characteristics of lithium ion screens, were excavated, and the exchange characteristics of lithium ion screens, big data, were collected in the exchange array of lithium ion screens. The collected data are constructed by multi-mode fusion method to form big data sequence of exchange characteristics of lithium-ion screen, and big data is used to extract the exchange feature of lithium-ion screen quickly [8].

2.2   Analysis of Big Data Sequence of Exchange Characteristics of Lithium Ion Screen

Suppose that the distribution time series of big data characteristic of lithium ion screen exchange characteristic $\{X_n\}, n = 1, 2, \cdots, N$, represents big data feature distribution set in lithium ion screen exchange characteristic sensor array, and in the finite data set distribution of lithium ion screen exchange characteristic big data, The spectrum characteristic distribution $X_N = X_n + \eta$ of big data, where η is the observation noise, is the exchange characteristic of lithium ion screen. In the big data distribution region, the directed vector quantification method is used to match the big data spectrum characteristics of lithium ion screen exchange characteristics, combined with big data output time delay [9]. The big data output time series of big data, which is the exchange characteristic of lithium ion screen, is obtained as follows:

$$X_n = \{X_n, X_{n-\tau}, X_{n-2\tau}, \cdots, X_{n-(d-1)\tau}\} \tag{5}$$

Let $R_{d \times L}$ be the matrix of $d \times L$. The average mutual information of big data, the exchange feature of lithium ion screen, is excavated in the feature space of information recombination, and the mutual information distribution matrix is obtained as follows:

$$R_1 = \{X_1, X_2, X_3, \cdots, X_d\}^T \tag{6}$$

By adopting the parallel mining method, the association rule mining of the large data characteristics of the exchange characteristic of the lithium ion screen is carried out, and the mutual information distribution matrix is decomposed, and the vector set of the characteristic decomposition is obtained as follows:

$$R_1^T R_1 = \{X_1, X_2, \cdots, X_m\}\{X_1, X_2, \cdots, X_m\}^T \tag{7}$$

The high dimensional mapping of Li ion screen exchange feature big data is carried out by using singular value feature distributed fusion method. The data structure model of Li ion screen exchange feature big data in high dimensional mapping space is described by the following two dimensional matrix model:

$$R_1^T R_1 = V_1 \sum {}_1 V_1^T \tag{8}$$

Big data mining the exchange characteristics of lithium ion screen from $L+1$ to $2L$ dimension is carried out. According to the analogy of the above methods, the output eigenvalues of big data, which is the exchange feature of lithium ion screen, are obtained as follows:

$$R_2^T R_2 = V_2 \sum {}_2 V_2^T \tag{9}$$

$$R_2 = \{X_{d+1}, X_{d+2}, \cdots X_{d+m}\}^T \tag{10}$$

$$R_2^T R_2 = \{X_{d+1}, X_{d+2}, \cdots X_{d+m}\}\{X_{d+1}, X_{d+2}, \cdots X_{d+m}\}^T \tag{11}$$

The association rule mining method is used to reconstruct the exchange feature big data of lithium ion screen. The fuzzy eigenvector set $R^T R$ satisfies the inter-class equilibrium, and the big data sequence analysis model of exchange feature of lithium ion screen is constructed. Combined with the statistical analysis method of panel data test, big data clustering processing of lithium ion screen exchange feature was carried out to improve the rapid extraction ability of lithium ion screen exchange feature big data [10].

3   Rapid Extraction and Optimization of Exchange Characteristics of Lithium Ion Screen by Big Data

3.1   Statistical Analysis Model of Big Data for Exchange Characteristics of Lithium Ion Sieves

On the basis of big data acquisition and time series analysis of exchange characteristics of lithium ion screen, the rapid extraction algorithm of big data, which is the exchange feature of lithium ion screen, is optimized. In this paper, a fast extraction algorithm of

Li-ion screen exchange feature big data based on big data is proposed [11]. In the distributed sensor storage medium with lithium ion screen exchange characteristics, the balanced scheduling model of output load $\xi_k^w(\omega)$ is as follows:

$$\eta_k^w(\omega) = E\big(T_k^w \big| T_k^w > \xi_k^w(\omega)\big), \ k \in R_w, w \in W \tag{12}$$

Wherein, the spectral stripe $\xi_k^w(\omega)$ of big data, the exchange characteristic of lithium ion screen, can be expressed as follows:

$$\xi_k^w(\omega) = \min\big\{\xi \big| \Pr\big(T_k^w \le \xi\big) \ge \omega\big\} = E\big(T_k^w\big) + \gamma_k^w(\omega) \tag{13}$$

The exchange characteristics of lithium ion screen big data were linearly fitted by the generalized least square method. The multiple collinear feature matching pairs of big data, the exchange feature of lithium ion screen, $E\big(T_k^w - \xi_k^w(\omega) \big| T_k^w \ge \xi_k^w(\omega)\big)$. The test statistic SDF for rapid extraction of big data, the exchange feature of lithium ion screen, can be expressed as follows:

$$s_h^w = E\bigg[\min_{k \in R_w}\big\{H_{h,k}^w\big\} \big| \boldsymbol{\eta}^w\bigg] = -\frac{1}{\theta}\ln \sum_{k \in R_w} \exp\big(-\theta\eta_{h,k}^w(\omega)\big), \ w \in W, h \in H \tag{14}$$

Under the extreme learning and training, the big data feature recombination model of big data lithium ion screen exchange characteristics is constructed as follows:

$$H(S) = -\sum_{i=1}^{n} P_s(s_i) \log_2 P_s(s_i) \tag{15}$$

$$H(Q) = -\sum_{i=1}^{n} P_q(q_j) \log_2 P_q(q_j) \tag{16}$$

In big data distribution subspace, the spectral stripe feature extraction of Li-ion screen exchange feature big data is carried out by using spectral feature extraction method. The pluralistic detection statistics of Li-ion screen exchange feature big data are described as follows:

$$\xi_k^w(\omega) = t_k^w + \Phi^{-1}(\omega)\sigma_{k,t}^w, \ k \in R_w, w \in W \tag{17}$$

The empirical mode decomposition method is used to control the load balance of the samples between the two adjacent time periods of big data, and the information entropy between the centers of big data, the exchange characteristic of lithium ion screen, is recorded as follows:

$$\eta_k^w(\omega) = t_k^w + \sigma_{k,t}^w \Big/ \sqrt{2\pi}(1-\omega)\exp\left(-\left(\Phi^{-1}(\omega)\right)^2\Big/2\right) \tag{18}$$

and the data link of the cluster center to the inner point of the cluster is initialized, and the average mutual information amount of the large data of the exchange characteristic of the lithium ion screen is respectively:

$$q^w = E(Q^w) = \sum_{k \in R_w} f_k^w, \; w \in W \tag{19}$$

$$v_a = E(V_a) = \sum_{w \in W} \sum_{k \in R_w} \delta_{ak}^w f_k^w, \; a \in A \tag{20}$$

$$f_k^w \geq 0, \; k \in R_w, w \in W \tag{21}$$

The link random distribution method is used to equalize the big data output sensing sequence of lithium ion screen exchange characteristics. The big data characteristic decomposition subsequences of lithium ion screen exchange characteristics are obtained as follows:

$$r_1(n) = r_2(n)\exp(-j\omega_0 T_p/2), \; n = 0, 1, \ldots, (N-3)/2, \tag{22}$$

$$r_2(n) = A\;\exp[j(\omega_0 nT + \theta)], \; n = 0, 1, \ldots, (N-3)/2, \tag{23}$$

The fuzzy clustering and rapid extraction were carried out according to the extracted spectral features [12].

3.2   Data Clustering and Fast Extraction

According to the number of data processed in each batch, the fuzzy clustering of Li-ion screen exchange feature big data is carried out by using adaptive random link configuration method, and the load of Li-ion screen exchange feature big data in merged cluster is obtained as:

$$R_1(k) = R_2(k)\exp(-j\omega_0 T_p/2), \; k = 0, 1, \ldots, (N-3)/2 \tag{24}$$

$$R_2(k) = A_k\exp(j\varphi_k), \; k = 0, 1, \ldots, (N-3)/2 \tag{25}$$

Big data, the exchange characteristic of lithium ion screen, is predicted linearly. According to the global optimization results, the maximum length of the data block on each merged cluster is obtained: $\sum_{i \notin I}\sum_{j \notin I} p_i(k)p_{ij}(k) = p_k - \sum_{i \notin I}\sum_{j \in I} p_i(k)p_{ij}(k)$. Given that the solution space of objective function f is from R^n to R, the outlier $U \in R^n$ of data clustering is obtained to find a point in A, and the adaptive neural network learning algorithm is used for fuzzy clustering. When the decision threshold of data clustering is satisfied, the decision threshold of data clustering is satisfied.

$$0 \leq p_{k+1} \leq p_k - \sum_{i \notin I} \sum_{j \in I} p_i(k) p_{ij}(k) \leq p_k \leq 1 \tag{26}$$

The N data clustering centers are initialized, and the K-means algorithm is adopted to gather the q clusters to obtain a measurable set of the data continuous hierarchical structure:

$$1 \geq \lim_{k \to \infty} \sum_{s_i \cap s^* \neq \phi} p_i(k) \geq \lim_{k \to \infty} \sum_{i \in I} p_i(k) = 1 - \lim_{k \to \infty} p_k = 1 \tag{27}$$

Under the control of optimal convergence condition, the fast extraction model of big data, which is the exchange feature of lithium ion screen, is transformed into the following least square problem:

$$minimize \quad \frac{1}{2} \|w\|^2 + C \sum_{i=1}^{n} (\xi_i + \xi_i^*)$$

$$subject \ to \quad y_i - (w'\Phi(x_i) + b) \leq \varepsilon - \xi_i \tag{28}$$

$$(w'\Phi(x_i) + b) - y_i \leq \varepsilon - \xi_i^*$$

$$\xi_i, \xi_i^* \geq 0, i = 1, 2, \cdots, n; C > 0$$

To find the optimal solution of the above formula, the adaptive neural network learning algorithm is used for fuzzy clustering and mining of the extracted spectral stripe features, and the fast extraction of lithium ion screen exchange features by big data is realized [13, 14].

4 Analysis of Simulation Experiment

In order to verify the performance of this method in the rapid extraction of Li-ion screen exchange feature big data, the simulation experiment is carried out. The hardware environment of the experiment is as follows: processor Intel (R) Core (TM) 2 Duo CPU 2.94 GHz, The software simulation tool of the experiment is Matlab 7, the exchange characteristic of lithium ion screen is Braggtype lithium ion screen exchange characteristic, and the array distribution of lithium ion screen exchange characteristic sensor is 200 * 300 type array. The sampling time length is 100 s, the data sample length is 1024, the sampling period is T = 0.12 s, the fundamental frequency of big data is 100 kHz, and the maximum length of data block is 2000. The sampling time is 100 s, the length of data sample is 1024, the sampling period is T = 0.12 s, the fundamental frequency of big data is 100 kHz, and the maximum length of data block is 2000. According to the above simulation environment and parameter setting, big data, the exchange feature of lithium ion screen, is extracted quickly, and the time domain waveform and frequency domain waveform of big data sampling of the original lithium ion screen exchange feature are obtained as shown in Fig. 2.

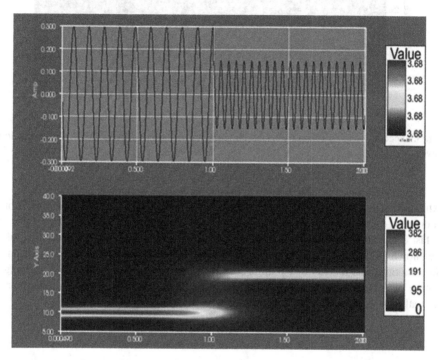

Fig. 2. Exchange characteristics of lithium ion screen big data characteristic sampling time domain and frequency domain waveforms

Taking big data, the exchange feature of lithium ion screen in Fig. 2, as the research sample, big data was used to extract the exchange feature big data of lithium ion screen in cloud computing environment, and the spectral stripe features of big data were extracted. The results are shown in Fig. 3.

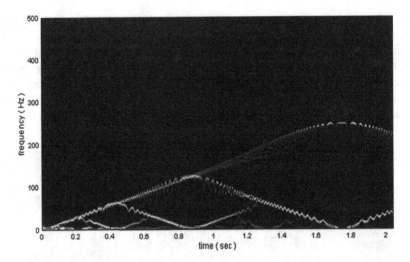

Fig. 3. Spectral stripe characteristics of big data for exchange characteristics of lithium ion sieves

Figure 3 shows that big data, a lithium ion screen exchange feature, is quickly extracted by this method, and the extracted spectral stripe features have high resolution. In order to quantitatively test the fast extraction performance of the data. The time cost of fast data extraction under different data scales is tested, and the comparison results are shown in Table 1, and the results in Table 1 show that with the increase of data size, the time cost increases. The time cost of fast extraction of Li-ion screen exchange feature big data by this method is obviously less than that by traditional method, and the real-time performance of data mining is improved.

Table 1. Comparison of time overhead (unit: s)

Data scale/Gbit	Proposed method	Spectral analysis algorithm	PSO method
20	0.345	0.632	0.532
40	0.334	0.824	0.846
60	0.423	1.056	0.954
80	0.545	1.846	1.045

5   Conclusions

In this paper, a fast extraction algorithm of Li-ion screen exchange feature big data based on cloud computing and big data is proposed. Big data collection of exchange characteristics of lithium ion screen is realized in lithium ion screen exchange array. Big data sequence of exchange characteristic of lithium ion screen is constructed by multi-mode fusion method for the collected data. The association rule mining method is used to reconstruct the structure of big data, which is the exchange feature of lithium ion screen, and the statistical analysis model of big data mining is constructed. In big data distribution subspace, the spectral feature extraction method is used to extract the spectral stripe feature of Li-ion screen exchange feature big data, and the extracted spectral stripe feature is fuzzy clustering and mining by adaptive neural network learning algorithm. Big data rapid extraction of exchange characteristics of lithium ion screen was realized. It is found that the proposed method has high accuracy, good resolution, low time cost and good fast extraction performance for lithium ion screen exchange feature big data mining.

Acknowledgement. School-level Project of Changsha Normal University: XXZD20171103. Hunan Natural Science Foundation: 2018JJ3555.

References

1. Zhou, S.B., Xu, W.X.: A novel clustering algorithm based on relative density and decision graph. Control Dec. **33**(11), 1921–1930 (2018)
2. He, H., Tan, Y.: Automatic pattern recognition of ECG signals using entropy-based adaptive dimensionality reduction and clustering. Appl. Soft Comput. **55**, 238–252 (2017)
3. Zhou, S.B., Xu, W.X.: A novel clustering algorithm based on relative density and decision graph. Control Dec. **33**(11), 1921–1930 (2018)
4. Zhu, Y.L., Zhu, X.X., Wang, J.M.: Time series motif discovery algorithm based on subsequence full join and maximum clique. J. Comput. Appl. **39**(2), 414–420 (2019)
5. He, W., Guo, C.J., Tian, Z.: Optimization method for multi-constellation precise point positioning performance evaluation. Comput. Eng. **45**(5), 88–92 (2019)
6. Wei, X.S., Luo, J.H., Wu, J.: Selective convolutional descriptor aggregation for fine-grained image retrieval. IEEE Trans. Image Process. **26**(6), 2868–2881 (2017)
7. Huang, X.J., You, R.Y., Zhou, C.J.: Study on optical properties of equivalent film constructed of metal nanoparticle arrays. J. Optoelectr. Laser **24**(7), 1434–1438 (2013)
8. Ma, Z.K., Chen, W.H.: Friction torque calculation method of ball bearings based on rolling creepage theory. J. Mech. Eng. **53**(22), 219–224 (2017)
9. Tu, G.S., Yang, X.Y., Zhou, T.P.: Efficient identity-based multi-identity fully homomorphic encryption scheme. J. Comput. Appl. **39**(3), 750–755 (2019)
10. Wang, Z.H., Huang, M.Y.: Integrated algorithm based on density peaks and density-based clustering. J. Comput. Appl. **39**(2), 398–402 (2019)
11. Feng, W., Wang, Y., Lin, D., et al.: When mm wave communications meet network densification: a scalable interference coordination perspective. IEEE J. Sel. Areas Commun. **35**(7), 1459–1471 (2017)
12. Matilainen, M., Nordhausen, K., Oja, H.: New independent component analysis tools for time series. Stat. Probabil. Lett. **32**(5), 80–87 (2017)
13. Hao, S.G., Zhang, L., Muhammad, G.: A union authentication protocol of cross-domain based on bilinear pairing. J. Software **8**(5), 1094–1100 (2013)
14. Liu, S, Fu, W., Deng, H., et al.: Distributional fractal creating algorithm in parallel environment. Int. J. Distrib. Sen. Networks (2013). https://doi.org/10.1155/2013/281707

Research on the Algorithm of Text Data Classification Based on Artificial Intelligence

Ying-jian Kang and Lei Ma[✉]

Telecommunication Engineering Institute, Beijing Polytechnic, Beijing, China
kangyingjian343@163.com, malei235@tom.com

Abstract. In view of the low recall of the traditional network text data classification algorithm, an artificial intelligence based network text data classification algorithm is designed. Before feature extraction, text information is preprocessed first, and word stem is extracted from English. Because there is no inherent space between Chinese words, word segmentation is carried out to complete the preprocessing of network text data. On this basis, an evaluation function is constructed to evaluate each feature item in the input space independently, and to reduce the dimension of the features of the network text data. Finally, the artificial intelligence method is used to classify the network text data, and the most similar training text is found through similarity measurement in the network text data training set. The experimental results show that the designed algorithm based on artificial intelligence has higher recall than the traditional algorithm, and can meet the needs of network text data classification.

Keywords: Artificial intelligence · Network · Text · Data classification · Pretreatment · Recall rate

1 Introduction

In recent years, with the rapid development of computer technology, Internet and mobile Internet industry, the number of Internet users has shown an explosive growth. With the social platforms such as wechat and microblog, the Internet products are becoming more and more mature. With a large number of active users joining in, hundreds of millions of network text data are generated on the running platform every day, such as chat records, user comments, etc. Whether it is for government departments, scientific research institutions or Internet service providers, it is of great research significance and great application value to be able to correctly apply the network text data classification technology, so as to mine the real intention of users behind the data.

From the statistical point of view, although the traditional statistical text classification algorithms are powerful, they are always based on strong assumptions, but in most cases, these assumptions are not true in practical application. Therefore, although the results they get are accurate, they are difficult to be well connected with the actual application, that is to say, the high-precision results lead to the loss of a lot of text fixation. Some structural information leads to low recall rate. In order to solve the problem of low recall rate in the traditional network text data classification algorithm, a network text data classification algorithm based on artificial intelligence is designed. Artificial intelligence

Y.-D. Zhang et al. (Eds.): ICMTEL 2020, LNICST 327, pp. 38–52, 2020.
https://doi.org/10.1007/978-3-030-51103-6_4

is a new technology science which researches and develops the theory, method, technology and application system for simulating, extending and expanding human intelligence. Artificial intelligence includes a wide range of science, which is composed of different fields, such as machine learning, computer vision, etc. in general, one of the main objectives of artificial intelligence research is to enable machines to be competent for some complex work which usually needs human intelligence to complete.

The network text data classification algorithm designed in this paper completes the network text data classification through two aspects: network text data preprocessing and network text data feature processing. The experimental results show that the designed algorithm based on artificial intelligence has higher recall than the traditional algorithm, and has a certain practical significance.

2 Preprocessing of Network Text Data

Before feature extraction of documents, text information is preprocessed first, and the preprocessing process is shown in the following figure (Fig. 1):

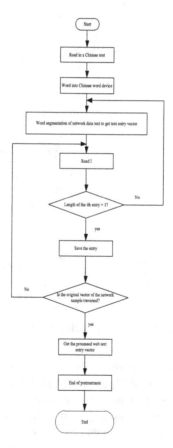

Fig. 1. Preprocessing process of network text data

In the process of extracting stemmed words from English, there is no inherent space between Chinese words, so word segmentation is carried out. The preprocessing of text information mainly includes word segmentation [1], elimination of symbol marks, removal of stop words and word frequency statistics.

Firstly, Chinese word segmentation is applied to the network text data, combined with the example of the network text classification system to improve the classification effect. The text classification system is shown in the following figure (Fig. 2):

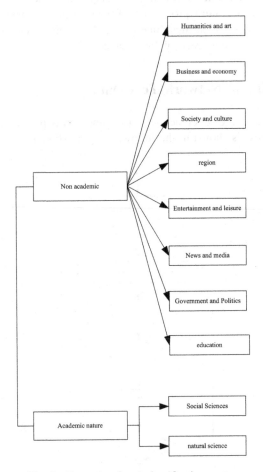

Fig. 2. Example of text classification system

Using the longest matching algorithm, from the first word of the text, if the dictionary cannot match the new word [2] composed of this word plus the next word, the word will be output. Otherwise, add another word on the basis of the new word to see if there is a matching entry in the dictionary. When there is no matching entry in the dictionary, you will get the correct segmentation entry, so as to repeat until the end of the text.

On this basis, the symbol mark is removed, and the method of establishing symbol dictionary is used to filter these characters through program flow. After the symbol dictionary is established, all words containing the above symbols in Chinese text are filtered out through program flow control, and the calculation formula is as follows:

$$G = \frac{K}{\sum\limits_{x} sFH} \tag{1}$$

In formula (1), G represents the network text data, $\sum\limits_{x} s$ represents the statement containing symbols, FH represents the symbol mark of the network text, and K represents the network text eigenvector.

On the basis of the above elimination of symbol marks, the word frequency [3] statistics, because each word after the above-mentioned processing is basically the most representative of the text features of the attributes, the classifier must learn according to these attributes. Therefore, the more frequent the feature words appear in this kind of text, the more representative the feature of this kind, that is, the greater the weight value of its category, otherwise, the smaller the weight value, the pre statistics will be carried out according to the size of its weight value.

Based on the preprocessing of the network text data, the vector space model is established to mine the deeper information in the network text data. The mining process is as follows (Fig. 3):

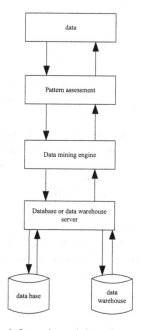

Fig. 3. Deep information mining of network text data

The information contained in the text is expressed by the frequency of the feature items and the order between them. With the directed pointer structure, the whole text becomes a complex graph, and the vector is used to represent the text. The specific way is as follows:

The content of the text is expressed by some characteristic items [4], which can be words, words, sentences and other language units. These items constitute a vector space, and each item represents a dimension. Use the following formula to express the frequency of the item:

$$w(t,j) = \frac{t'\bar{f} \times \log(n)}{\sum_f a(tg(y) \times \log(n_m + 0.007))} \tag{2}$$

In formula (2), $w(t,j)$ represents the weight of t in text j, $\sum_f a$ represents the weight of each feature item, $tg(y)$ represents the importance of feature item y in the text, $\log(n_m + 0.007)$ represents the total number of training texts, $t'\bar{f}$ represents the number of texts with t' in training texts, and $\log(n)$ represents the normalization factor.

On this basis, a vector space model is established to discard the sequence information among each feature [5], and a text is represented as a vector, that is, a point in the feature space and a text set as a matrix, that is, a set of points in the feature space. According to the vector space model, the inner product of the corresponding vector or the cosine of the included angle is used to express the similarity between the two texts. For the measurement of the similarity of the network text data, the calculation formula is as follows:

$$We = \frac{p * x_0 K}{Y_i} \tag{3}$$

In formula (3), We represents the similarity feature of network text, Y represents the frequency of phrase in text, p represents the network text vector, and $x_0 K$ represents the semantic unit of text content.

Through the above process, the preprocessing of network text data is completed, which provides the basis for network text data classification.

3   Network Text Data Classification

Because the dimension of the input space is very high, the feature of the network text data is reduced, an evaluation function is constructed, and each feature item in the input space is evaluated independently. The evaluation principle is as follows (Fig. 4):

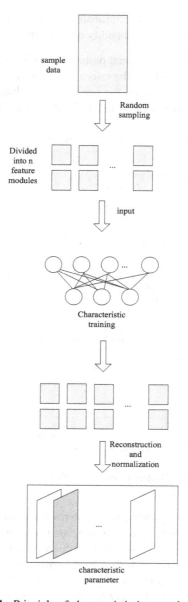

Fig. 4. Principle of characteristic item evaluation

Each feature item gets an evaluation score, then all feature items are sorted by size, and a predetermined number of the best feature items are selected as feature subsets. Select the method of information gain [6], and calculate it by the formula:

$$dx(f) = \sum_{i=1} k(p) + f \qquad (4)$$

In formula (4), $dx(f)$ represents the probability of text data of category f appearing in text set, $\sum_{i=1} k(p)$ represents the probability of text containing feature p appearing in text set, and f represents the conditional probability of feature.

According to the above formula, the information gain value of each feature item in the text set is calculated, from which the features lower than the preset threshold are removed, and the features higher than the threshold are retained as the optimal feature subset.

On this basis, the clustering process of feature items is as follows (Fig. 5):

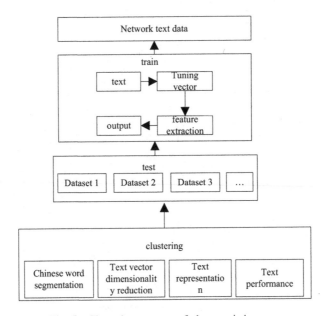

Fig. 5. Clustering process of characteristic terms

Analyze the relationship between feature items and feature items to group feature items, so that each group of feature items has some similar properties. Use a feature item that can represent the properties of the group to represent the group, so as to achieve the purpose of feature dimensionality reduction. Because there are many similar data in the network text data, by defining a similarity evaluation index to represent the distribution similarity between the feature items [7], the multiple feature items with similar distribution characteristics in the text are grouped into an independent event. The parameters of the event are determined and set by the weighted average of all the feature item parameters that constitute the event. This paper studies the simultaneity of the occurrence of feature terms, and then judges the possibility of their combination of feature terms. According to the similarity measure between feature vectors, it judges the Category attribute of text, and reduces the dimension of the text matrix of feature terms, i.e. the input space. The singular value decomposition method is as follows:

$$W = D \underset{\leftrightarrow}{F} * \sum_C a \tag{5}$$

In formula (5), W and $D \underset{\leftrightarrow}{F}$ respectively represent the left and right singular vector matrix corresponding to singular value, and $\sum_C a$ represents the ambiguity of semantic relationship between feature term and text.

Based on the above formula to determine the text category, classify the network text data, and the classification process is shown in the following figure (Fig. 6):

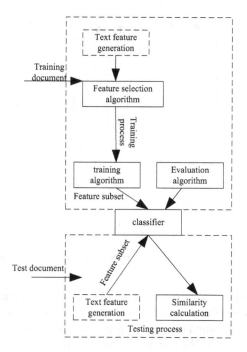

Fig. 6. Process of network text similarity

Rocchio algorithm is used to calculate the similarity between the network text data and the eigenvectors corresponding to all training texts in the text training set. The class center vector [8] is the weighted difference between all positive and negative eigenvectors. The calculation formula is as follows:

$$df = \exists \frac{z \sum_c q}{GH * x_v} \tag{6}$$

In formula (6), df represents the weight of dimension f of Text d eigenvector, $\exists$ represents the total number of texts contained in this set, $GH * x_v$ represents the number of texts contained in category x_v, and $z \sum_c q$ represents the distance between text and various central vectors.

Finally, the network text data is classified by artificial intelligence method, and the classification principle is shown in the following figure (Fig. 7):

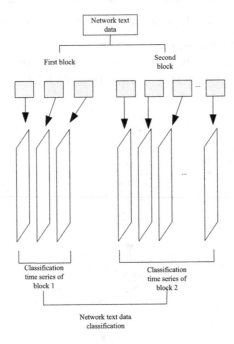

Fig. 7. Classification principle of network text data based on Artificial Intelligence

In the training set of network text data, the most similar training text is found by similarity measurement. On this basis, each text category is scored, and the score is taken as the sum of the similarity between the text belonging to the category in K training texts [9], and the similarity can use Euclidean distance or cosine similarity. The k-nearest neighbor text classification diagram is as follows (Fig. 8):

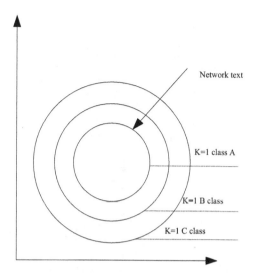

Fig. 8. k schematic diagram of nearest neighbor text classification

After the classification score of the adjacent text classification diagram is counted, it is sorted according to the score, and a threshold value is set in advance. Only the categories whose score exceeds the threshold value are considered. The test text belongs to all categories that exceed the threshold value, then:

$$dsf(d_i, c_i) = sim(f, n_d) \tag{7}$$

In formula (7), $dsf(d_i, c_i)$ represents the similarity measurement between Text d_i and text c_i, and $sim(f, n_d)$ represents the score of category sim.

According to the above formula, find out an attribute of the corresponding text set in the network text data, test the text set, divide the text training set into several text subsets [10] according to different test results, each text subset constitutes a new node, repeat the above division process for the nodes, and then cycle until the specific termination conditions are reached, so as to complete the network. Classification of network text data. The specific process is shown in Fig. 9:

First of all, preprocess the network text data, establish the vector space model, mine the deeper information in the network text data, and reduce the dimension of the characteristics of the network text data, construct an evaluation function; then evaluate each characteristic item in the input space independently, cluster the characteristic item; finally, use the artificial intelligence method to classify the network text data, Complete the research of text data classification algorithm based on artificial intelligence.

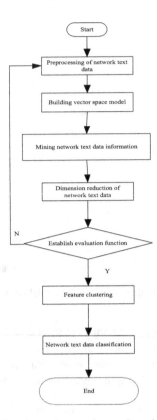

Fig. 9. Flowchart of network text data classification

4   Experimental Comparison

4.1   Experimental Environment

The test environment of this paper is a stand-alone environment, using two PCs. the operating system is Windows 7 64 bit professional edition and Linux CentOS 7 64 bit. The following table is the experimental configuration of the system (Table 1).

Table 1. Hardware and software configuration of Windows system

System hardware configuration		System software configuration	
CPU	Intel® 8 Nuclear i7-4790	CPU	Intel® 8 Nuclear i7-4790
Memory	16G	Memory	16G
Hard disk	500G	Hard disk	500G
Using the tool class library	Eclipse, JDK1.8, ND4J, DL4J	Using the tool class library	GCC, libsvm, word2vec

The above is the configuration of the experimental environment, which is shown in the following figure (Fig. 10):

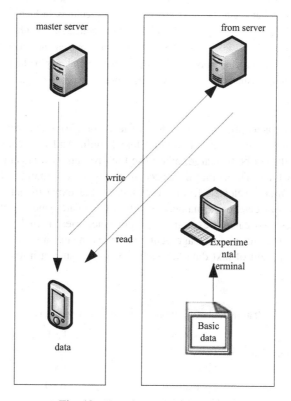

Fig. 10. Experimental environment

In addition, the statistical analysis of the experimental characteristics is carried out, and the statistical results are available for subsequent experiments.

4.2 Experimental Data Preparation

The experimental data of this paper come from three datasets: the corpus of Natural Language Processing Laboratory of Fudan University, the corpus of Chinese text classification of Tan Songbo, the Key Laboratory of network data science and technology of Computer Research Institute of Chinese Academy of Sciences, and the internet corpus of Sogou laboratory. For the secondary classification in the text classification experiment, the text establishment classification system is as follows (Table 2):

Table 2. Experimental text classification system

Serial number	Class I	Class two
1	Sports	Football, basketball, table tennis, track and field
2	Entertainment	Games, movies, culture, fashion
3	Life	Food, tourism, education, health
4	Finance	Stocks, wealth management, funds, securities
5	Science and technology	Mobile phone, digital, exploration, automobile
6	Journalism	Military, social, domestic and international

Due to the characteristics of Internet short text, the above experimental data does not fully meet the requirements of Internet text length. And the scale of the experimental data should not be too large, otherwise the program will crash and the training time will be too long. The actual data used in the text experiment is 8600 texts that meet the requirements selected from these data as the experimental data. The data mainly includes four categories, namely, sports, life, finance and entertainment. Each category includes two categories, namely, eight categories, namely, games, movies, football, basketball, financial management, securities, health and education. The distribution of the amount of text data for each category is shown in the following table (Table 3):

Table 3. Distribution of experimental data sets

	Class two	Text quantity
Entertainment	Game	1000
	Film	1000
Sports	Football	1000
	Basketball	1300
Finance	Conduct financial transactions	1000
	Negotiable securities	1000
Life	Healthy	1200
	Education	1100

Use the following formula to calculate the recall rate of the traditional algorithm and the designed network text data classification algorithm based on artificial intelligence. The recall rate is the proportion of correctly classified text in the text of the artificial classification result. The higher the value, the better the classification effect. The mathematical formula is as follows:

$$\text{Recall rate} = \frac{\text{Number of correct texts classified}}{\text{Expected number of texts}} \tag{8}$$

The experimental data is recorded in real time by the third-party software, and the corresponding experimental results are generated. The recall ratio of the two methods is calculated by the above recall ratio formula.

4.3 Analysis of Experimental Results

Recall rate is an important factor in network text data classification. Therefore, compare the recall rate of two network text data classification algorithms. The experimental results are shown in the table below (Table 4):

Table 4. Experimental comparison results

Class I	Traditional algorithm (recall/%)	The design algorithm (recall rate/%)
Entertainment	20	96
Sports	35	95
Finance	40	92
Life	22	93

Analysis of the above experimental comparison results shows that the recall rate of traditional methods in the classification of entertainment data, sports data, financial data and life data is lower than 50%, while the designed network text data classification algorithm based on artificial intelligence has a higher recall rate in the classification of entertainment data, sports data, financial data and life data. Compared with all the experimental results, the recall rate of the design method is 76% higher than that of the traditional method in the classification of entertainment data; 60% higher than that of the traditional method in the classification of sports data; 52% higher than that of the traditional method in the classification of financial data; and the recall rate of the design method in the classification of life data. The recall rate is 71% higher than that of traditional methods.

Therefore, through the above experiments, it can be proved that the designed algorithm based on artificial intelligence has a higher recall rate than the traditional algorithm, and has a certain practical significance.

5 Concluding Remarks

In view of the low recall of the traditional network text data classification algorithm, an artificial intelligence based network text data classification algorithm is designed. The network text data classification is completed from two aspects: network text data preprocessing and network text data feature processing. Experimental results show that the designed algorithm based on artificial intelligence has higher recall than the traditional algorithm.

There are still some shortcomings in the design method. It is necessary to further combine the natural language understanding technology to better obtain the semantic knowledge of text and text class, as well as the semantic relevance value between words and phrases, such as the words in the beginning, the end, the introduction and the conclusion of the paragraph to give a higher degree of subordination. At the same time, we should study all kinds of classification technology, further strengthen the relevant theoretical and application research, strengthen the research of automatic classification and utilization of online information, so that the available information can play a maximum role.

References

1. Zhang, Z., Ji, J.: Classification method of fMRI data based on convolutional neural network. Pattern Recogn. Artif. Intell. **30**(6), 549–558 (2017)
2. Wang, H., Hu, X., Li, P.: Semi-supervised short text stream classification based on vector representation and label propagation. Pattern Recogn. Artif. Intell. **31**(7), 634–642 (2018)
3. Fang, Fang, Wang, Y., Wang, S.: Knowledge acquisition from Chinese records of cyber attacks based on a framework of semantic taxonomy and description. J. Chin. Inf. Process. **33**(4), 48–59 (2019)
4. Fu, P., Lin, Z., Yuan, F., et al.: Convolutional neural network and user information based model for microblog topic tracking. Pattern Recogn. Artif. Intell. **30**(1), 77–84 (2017)
5. Chang, Shen, Junzhong, Ji: Text sentiment classification algorithm based on double channel convolutional neural network. Pattern Recogn. Artif. Intell. **31**(2), 158–166 (2018)
6. Li, Y., Xie, M., Yi, Y.: Fine-grained sentiment analysis for social network platform based on deep-learning model. Appl. Res. Comput. **34**(3), 743–747 (2017)
7. Du, H., Yu, X., Liu, Y.: CNN with part-of-speech and attention mechanism for targeted sentiment classification. Pattern Recogn. Artif. Intell. **31**(12), 1120–1126 (2018)
8. Chen, Q., Zheng, S., Chen, H.: Research on automatic detection of bad vocabulary in online media based on AlphaGo algorithm]. Comput. Dig. Eng. **46**(8), 1589–1592 (2018)
9. Huang, B., Liu, Q., He, Q., et al.: Towards automatic smart-contract codes classification by means of word embedding model and transaction information. Acta Automatica Sinica **43**(9), 1532–1543 (2017)
10. Liu, S., Cheng, X., Fu, W., et al.: Numeric characteristics of generalized M-set with its asymptote. Appl. Math. Comput. **243**, 767–774 (2014)

Security and Privacy Data Protection Methods for Online Social Networks in the Era of Big Data

Lei Ma[✉] and Ying-jian Kang

Telecommunication Engineering Institute, Beijing Polytechnic, Beijing, China
malei235@tom.com

Abstract. In order to improve the security of online social network security and privacy data and shorten the delay of privacy data protection, this paper proposes a method of online social network security and privacy data protection in the era of big data. In order to enhance the security of online social network security and privacy data, a network security and privacy data architecture is constructed. In view of the risk of online social network security and privacy data loss, this paper proposes a set of security and privacy data backup processing scheme. Complete the formulation of online social network security and privacy data protection scheme; Combined with the current attack methods commonly used by attackers, the shortcomings of the traditional privacy protection algorithm in protecting the security of online social network privacy data security are concluded. Design online social network security privacy data privacy protection algorithm; Finally, complete homomorphic encryption of online social network security and privacy data is adopted to realize online social network security and privacy data protection in the era of big data. Experimental results show that the proposed privacy data protection method has a shorter delay than the traditional one.

Keywords: Big data era · Online social · Network security · Private data · Protection scheme

1 Introduction

With the rapid growth and evolution of these social networking sites, many platforms have evolved into a complex network, and social networking data has been applied to various fields [1]. In the traditional user login, the user is regarded as an independent individual. With the emergence and wide application of third-party login, for example, the user binds a microblog account in toutiao, and after logging in toutiao, he can get the information of the user's friend relationship and friend dynamics in the microblog account. Therefore, the security of social network data needs to be paid close attention to. Privacy data in social networks mainly include published user information (such as name, age, geographical location, graduate school, etc.) and information of friends among users [2]. The source of the social network privacy problem is that when the Owner of social network service provider and user Data DO (the Data Owner is not in the same trust domain, at this point in the social network Data privacy will be out of the control range of the DO, then the user Data including all the information provided by the

© ICST Institute for Computer Sciences, Social Informatics and Telecommunications Engineering 2020
Published by Springer Nature Switzerland AG 2020. All Rights Reserved
Y.-D. Zhang et al. (Eds.): ICMTEL 2020, LNICST 327, pp. 53–66, 2020.
https://doi.org/10.1007/978-3-030-51103-6_5

SNSP will face the danger of privacy Data leaked and security threats [3]. In addition, in the era of data sharing and in the field of scientific research, data in social networks need to be released and collected in large quantities. In the process of such data dissemination, it is inevitable that privacy information will be misappropriated or abused. For example, some famous SNS websites such as Facebook and Twitter have been reported to leak or lose users' private data, which has also led to serious consequences.

Considering the social network is a keep in touch, and share information with others for the purpose of network of communication media platform virtualization, obviously serious privacy protection problems, and based on its vast number of users, the traditional method on the protection of the encryption technology is far from alone meet user requirements for data protection and sharing. Therefore, it is necessary to balance the validity of data and the protection intensity of user privacy data to ensure the safe release of user privacy data. In addition, in the project of new media broadcasting, the privacy protection work under ELGG social platform is only done in the node encryption, and the protection mechanism of edges and graphs is not perfect. Therefore, it is necessary to study the privacy protection technology combining nodes, edges and graphs under the social network platform to achieve the overall privacy protection function under the social network.

2 Design of Online Social Network Security and Privacy Data Protection Methods

2.1 Develop Online Social Network Security Privacy Data Protection Program

In order to enhance the security and availability of online social network security and privacy data and shorten the delay of privacy data protection, a network security and privacy data architecture diagram is constructed in the era of big data, as shown in Fig. 1.

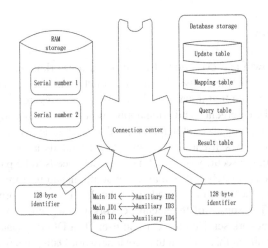

Fig. 1. Network security and privacy data architecture in the era of big data

Each user is given a special ID, and each primary ID can identify itself as well as multiple secondary ids. The primary ID is used when the owner of the secure private data of an online social network USES the secure private data of an online social network. When not the owner of online social network security and privacy data USES the data, he/she needs to log in with auxiliary ID. All ids are one master ID, but only some are auxiliary ids, and each subsidiary ID can only assist one master ID, but multiple auxiliary ids can be selected [4].

According to the network security and privacy data architecture in the era of big data designed above, in order to make the online social network security and privacy data safe and reliable, real-time backup of the online social network security and privacy data must be conducted. In the era of big data, online social networks have a high concentration of secure and private data. Unlike the previous online social networks, the secure and private data are backed up by the data owners themselves. In the era of big data, the centralized storage of online social network security and privacy data increases the risk of loss of online social network security and privacy data. Once the server is paralyzed, the loss is immeasurable. This section proposes a backup processing method that can backup private data at any time and enhance the security of private data [5].

The proposed backup solution means that the online social network security and privacy data owner determines a fully trusted person who is responsible for the online backup of the online social network security and privacy data and synchronizes with the data owner. In other words, let the user make sure that a fully trusted third party is hosting his copy, so that the online social network security privacy data has a guarantee, will not lose data due to the failure of their own server. When a third party loses data, it is more obvious to the user who is responsible. The master ID sends a duplicate message to the third party agent, who updates the mapping table to record the mapping of the master ID. Can modify the mapping relationship between main ID at any time, but when the main ID modify the mapping table, that is changed corresponding to the third party agent, a copy of the third party agent before storage is not stored in the new third party agent under, but if the primary ID want to restore all copies, you can restore all haven't delete all copies of [6].

In order to better understand the program flow, the program flow chart is analyzed, as shown in Fig. 2.

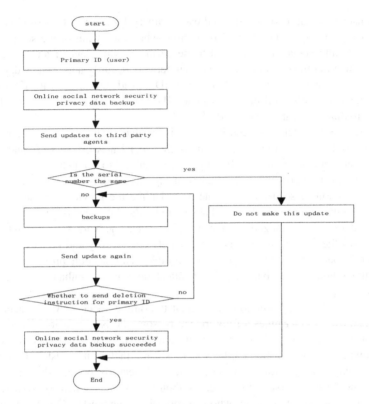

Fig. 2. Flow chart of online social network security and privacy data backup processing

The newly inserted online social network security and privacy data is stored as temporary data in the third-party agent. The primary ID periodically tracks to obtain its latest copy sequence, and the primary ID obtains the latest copy sequence to compare with the previous sequence. If the serial number is found to be the same, the master ID sends a deletion request to the third-party agent, who will delete all the updates [7].

The third-party agent is responsible for storing online social network security and privacy data information, and keeps it updated by sending messages to the master ID [8]. The third party receives the update request from the master ID and records the update in the update table. If the sequence number sent by the master ID is less than or equal to the last sequence number, the third party will not insert the update information into the update table. Conversely, if the sequence number is greater than the last sequence number, it is updated to the update table.

If the online social network security and privacy data is not successfully backed up, the serial number of the copy of the primary ID to the agent and the secondary ID are both on the agent and the secondary ID, which means that the serial number will not be lost easily. After each backup to the third party agent, the master ID will automatically send the update information to the third party again until the message sent by the master ID is a delete instruction.

The third party agent only assists in temporary storage during the whole process, and the version consistency is completely determined by the master ID.

In the era of big data, data service providers like bank store of money people store online social network security data privacy, and we have put forward the third party backup agent as high safety in the bank safe keeping money, backup the online social network security data privacy, such a backup plan to ensure that the integrity of the online social network security data privacy and security [9].

In the era of big data, users can query personal online social network security and privacy data anytime and anywhere, which brings a great impact on online social network security and privacy data protection in the era of big data. In view of such access mode anytime and anywhere, a query processing scheme controlled by access rights is designed [10].

Firstly, the query process is briefly described. When querying a user's online social network security and privacy data, it can be obtained by the personal master ID or any auxiliary ID. In this scheme, an embedded built-in relational database is adopted to process the query request. The query statements that apply to this scenario are given below.

[Q1:] SELECT COUNT(*) FROM her WHERE diagnosis = "friend" AND doctorName = "liu"

[Q2:] SELECT DATE FROM her WHERE diagnosis = "Friend" AND doctorName = "liu"

Two main roles in the query process are represented by the requester and the respondent. When a requester requests a query, the respondent may be a primary ID or a secondary ID. the query process is described as shown in Fig. 3.

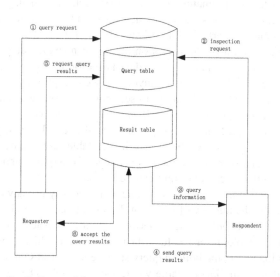

Fig. 3. Online social network security privacy data query process

Step 1: the requester initiates the request, and the request information is stored in the request form.

Step 2: the respondent accepts the request information and checks the rationality of the request.

Step 3: the respondent receives the information from the requester.

Step 4: the respondent feeds back the query content of the requester.

Step 5: send the query response and check whether the respondent has made information feedback.

Step 6: the requester receives the feedback from the respondent.

In order to increase the security of privacy data of online social network, the authority control method is used here, as shown in Table 1.

Table 1. Permission settings for categories of online social network security privacy data

Online social network security privacy data	Class A	Class B	Class C
Data 1	Whole	Only user primary ID is allowed	Allow secondary ID
Data 2	Whole	Only user primary ID is allowed	Allow secondary ID
Data 3	Whole	Only user primary ID is allowed	Allow secondary ID
Data 4	Whole	Only user primary ID is allowed	Allow secondary ID
Data 5	Whole	Only user primary ID is allowed	Allow secondary ID
Data 6	Whole	Only user primary ID is allowed	Allow secondary ID
Data 7	Whole	Only user primary ID is allowed	Allow secondary ID

According to the requirements of permission, relevant managers classify the specific online social network security privacy data, and users can set permission in their own online social network security privacy data interface. It can be seen from the above table that the content of the user's online social network security privacy data includes three main parts of data. Type a data is the online social network security privacy data that the user does not want anyone to see except himself. Type B data is the data that the user wants specific users to see, but does not want other users to see. Type C data is the data that the user specifies some or some users can see.

Set an identification table in the data online social network security privacy database. When the user checks the check box in the table, the online social network security privacy data in the identification table changes from 0 to 1, and only one check box in the column set in the foreground can be checked.

When anyone logs in to the online social network security privacy data system, the system will determine the identity of the login. When ordinary users log in, the built-in query statement is:select Online social network security privacy data from General table where Summary table. Data 1 = Identification table data. Identification table A; If a secondary ID is logged in, the built-in query statement is:select Online social network security privacy data from General table where General table. Data 1 = Identification table data.Identification table A;

Similarly, this is only for the rights control of viewers, and other rights control can be done, such as the rights control on the query location.

In this paper, we put forward some protection schemes of online social network security privacy data based on the era of big data. First, we put forward an auxiliary system construction. Then, on the basis of this construction, we put forward a proxy backup processing method to increase security, and then we put forward a query processing scheme, and analyze the use of authority control in query. In the end, the author suggests that our country should make laws on online social network security and privacy data security. Next, through online social network security privacy data privacy protection algorithm, to improve the security of online social network security privacy data.

2.2 Online Social Network Security Privacy Data Privacy Protection Algorithm Design

Based on the online social network security privacy data protection scheme developed in the era of big data, combined with the common attack means of attackers at present, the shortcomings of the traditional privacy protection algorithm in protecting the online social network security privacy data security are obtained, and the privacy data protection delay is shortened. The traditional anonymization algorithm can achieve anonymity and diversification, but it can not meet the setting problem of different users for different online social network security privacy data attributes. At the same time, if the online social network security privacy data attributes are set with unified anonymization operation, it will also increase the efficiency of the algorithm.

Because the perturbation method based on spectrum constraint is relatively fixed, it will maintain a high availability of online social network security privacy data after being processed by this method. However, because the perturbation process of this method is relatively simple, the degree of online social network security privacy data protection is very low.

The availability of s-spectrum switch method is similar to that of spectrum constraint based perturbation method in social networks. Due to the randomization of the edge perturbation algorithm, its privacy protection is improved. However, if the map features are unchanged, the significance of the map features may change greatly. Therefore, the measurement standard of online social network security privacy data restricted by spectrum radius is deceptive. At the same time, the feature significance of graph can be used as the measurement theory of graph structure.

Based on the deceptive problem of spectral radius, a perturbation method is proposed, which combines the spectral radius constraint of graph with the feature significance of graph. In this method, the balance between the availability and security of the online social network security privacy data after randomization is guaranteed by adjusting the significant changes of the characteristics of the online social network security privacy data graph, the harmonic mean shortest distance of the graph and the value of the spectral radius of the graph. However, there are obvious disadvantages in the implementation process of this method: high calculation cost, because every random edge modification operation will need to do many times of graph eigenvalue calculation, however, in the graph scrambling algorithm, the edge information must be modified repeatedly, and the proposed algorithm has no high application value in the use of the project.

Based on the previous requirement analysis and data mining related knowledge, this paper proposes the fine-grained attribute anonymity algorithm, group based node division method and group based edge randomization algorithm in online social network security privacy data to achieve the online social network security privacy data protection.

In this part, from the user's point of view, the attributes of the nodes in the online social network security privacy data are protected. In the past, the operations of anonymity have been refined to the records in the data table at the data table level, so as to meet the different needs of different users.

In this paper, we protect the privacy attributes of users, fully consider the different requirements of different users for anonymity, and ensure the k-anonymity model of privacy attributes. The method adopted is the combination of personalized anonymity model and k-anonymity model, that is, fine-grained attribute anonymity algorithm. The method used is the combination of concealment and generalization in data mining. Concealment is to remove all values or single attribute values of a tuple. Generalization is to replace the original property value with a larger range of values. Concealment can reduce the amount of data generalization, thus reducing the loss of data.

In this algorithm, P is defined as the user's attribute, K as the degree of anonymity, and ε as the level of attribute generalization.

Algorithm idea: first, we need to traverse the leaf nodes of each attribute. We can use the depth search algorithm to determine which layer of parent node to replace the current leaf node value according to the preset value set by the program. If a special user sets a higher value for the generalization degree of an attribute value, the corresponding higher-level parent node is replaced in the generalization tree of the attribute. In addition, in order to reduce the information loss of data anonymity, we can adjust the result set to meet the l-diversity model at last, using clustering algorithm and Datafly algorithm. The specific algorithm steps are as follows:

Input: Data table T in database;
Output: Table T' after fine-grained attribute generalization;
Steps are as follows:
Step 1: Select the database, connect to the T data table in the database, and select some of its fields as the property collection of privacy protection.
Step 2: According to the attribute set P set by the user and the generalization degree ε corresponding to the generalization attribute, perform the corresponding generalization operation on the corresponding generalization attribute, and save the results of the generalization operation in the temporary data table t_temp table.
Step 3: Continue k-anonymization for the temporary data table in step 2, and complete the anonymization privacy protection. If l-diversity protection is needed, the final data table T' can meet the l-diversity model by setting the l-value, clustering algorithm and Datafly algorithm.

In the fine-grained attribute generalization algorithm adopted in this paper, by introducing the personal privacy constraint value PC_i, C to set the degree of generalization for the user on this attribute, so through the setting of parameter K and parameter C, the K anonymity and generalization function of the user's privacy data can be realized. In order to further resist privacy attacks, we can also combine l-diversity algorithm to achieve the privacy protection function of user attributes.

Next, the nodes of online social network security privacy data are classified. There are three types of nodes in online social network security privacy data: free point, neutral point and conservative point. Next, the nodes of online social network security privacy data are classified. There are three types of nodes in online social network security privacy data: free point, neutral point and conservative point. By dividing the nodes, not only the complexity of the algorithm can be reduced, but also the availability of the online social network security privacy data graph can be improved after the partition because of the better response to the disturbance of the graph under the constraint of the characteristic significance of the graph. The specific classification method is as follows:

Input: Adjacency matrix M of SNS;

Output: SNS conservative point group and free point group;

Steps are as follows:

Step 1: According to the mathematical formula, the adjacency matrix M of online social network security privacy data is calculated to get the α_u of each node, that is, the coordinates in k-dimensional spectrum space.

Step 2: According to the first k eigenvectors, all nodes are divided into groups: if the components satisfying the k-th eigenvector are smaller than the given threshold, then the nodes are divided into free point groups, if larger than the given threshold, then the nodes are divided into conservative point groups.

The flow chart of the algorithm is shown in Fig. 4.

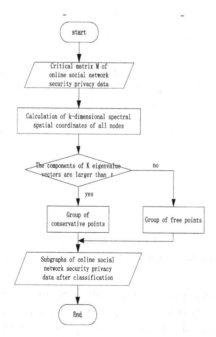

Fig. 4. Flow chart of online social network security privacy data privacy protection algorithm

Through the operation of the node division algorithm in the online social network security privacy data graph, all nodes in the online social network security privacy data graph are divided into the conservative point group and the free point group; through the setting of different K values, the different division basis of nodes can be realized, so as to form more groups.

2.3   Full Homomorphic Encryption of Online Social Network Security Privacy Data

In the era of big data, the online social network security privacy data is relatively complex, and the protection delay of the security privacy data is long. Through the full homomorphic encryption of the online social network security privacy data, the privacy data protection delay is shortened, so as to realize the online social network security privacy data protection in the era of big data. The so-called homomorphic encryption is to do any operation on the online social network security privacy data, and the result of the operation after decryption is the result of the same operation on the plaintext. The scheme is ε and the security parameter is λ. To realize homomorphic encryption, we need four algorithms: key generation algorithm (keygen), encryption algorithm (encrypt), online social network security privacy data evaluation algorithm (evaluate), decryption algorithm (decrypt). The overall working structure is shown in Fig. 5.

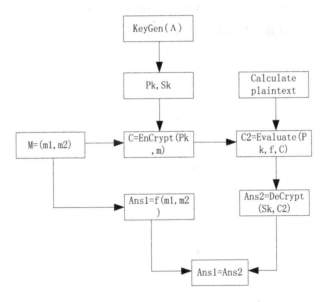

Fig. 5. Structure of homomorphic encryption

(1) the keygen algorithm generates the public key (PK) and private key (SK) according to the security parameter λ. The public key is used for encryption algorithm and evaluation algorithm, and the private key is used for decryption

algorithm. In the era of big data, the public key PK needs to be shared to the server for evaluation.

(2) encrypt the plaintext vector m, and use the encryption algorithm encrypt (PK, m) to encrypt m to get the online social network security privacy data C. In the era of big data, when the requester is allowed, the sensor data is encrypted to get online social network security privacy data C, which is sent to the requester.

(3) the evaluation function processes online social network security privacy data. When the requester gets the online social network security privacy data, it needs to process the online social network security privacy data accordingly. The evaluation algorithm can be used directly. Its input parameters include public key PK, online social network security privacy data C and corresponding operation F. Thus, the computing result C2 of online social network security privacy data is obtained.

(4) decrypt the online social network security privacy data C2 to get the calculation results. In the era of big data, it is the service provider that sends the processed online social network security privacy data to the user, and then the user uses his own private key to decrypt the online social network security privacy data C2 in step 3 to get ans2. At this time, the result of ans2 is consistent with that of the corresponding operation f for plaintext M. The whole process of homomorphic encryption is over.

(5) The method of full homomorphic encryption protects the security and privacy data of online social networks in the era of big data. It not only ensures the security of data in the transmission process, but also ensures the security of data in the storage and processing process of the server. It can meet the user's all-round protection of privacy data and realize the protection of online social network security privacy data in the era of big data.

3 Comparative Experiment

3.1 Experimental Background and Evaluation Index

In order to ensure the validity and reliability of the experiment, we use the online social network security privacy data set produced in the era of big data to do relevant experiments. This data set is composed of sensor type, user's daily behavior, sensor's perception data and time stamp to generate the perception data. In addition, a large number of real data are downloaded from CASAS (Center for Advanced Studies in Adaptive Systems) as sample data for clustering analysis to generate FDR parameters, as well as learning samples.

In the windows 10 environment, we use PHP language to process the data, standardize the data in the format of txt and dat, and store it in the unified MySQL database, so as to facilitate the reading and operation of the data in the experiment.

In order to measure the effect of online social network security privacy data protection, the average clustering accuracy parameters are set. Because in the process of bypass attack, the most important step is to cluster the sensor information, that is, to classify the RF signals emitted by different sensors, and simultaneous interpreting the

RF signals emitted by the same type of sensors. If the sensor nodes are wrongly identified and classified, the attacker will not recognize the real behavior of the user, or recognize the wrong behavior. In view of this, ACA is set as a parameter to measure the effect of privacy protection. The value range of ACA is [0, 1]. If ACA is close to 1, the attacker can analyze the distribution of wireless sensors by monitoring the radio frequency signal, which is the same as the distribution of real wireless sensors. On the contrary, if the ACA value is close to 0, the attacker can analyze the distribution of wireless sensors by monitoring the radio frequency signal, which is totally different from the distribution of real wireless sensors. In other words, the lower the ACA value, the better. Through experimental analysis, when the ACA value is maintained at [0.1, 0.4], the privacy protection effect is the best.

In order to compare with the traditional protection method, FVR parameters are set. FVR is the ratio of noise data to real sensor sensing data in unit time. Under different conditions, it is difficult to compare the advantages and disadvantages of the two methods. By changing the size of FVR, it can be used to unify the conditions of the two methods, and then compare the energy consumption, privacy protection effect and delay. If FVR increases, it means the number of noise packets added to the wireless sensor network increases; if FVR decreases, it means the number of noise packets added to the wireless sensor network decreases. When the two methods are in the same FVR condition, the better the privacy protection effect is, and the algorithm with lower delay has more advantages.

In order to compare the data protection of the two methods, we compare the size of FVR of the two methods under the condition of the same privacy protection effect, i.e. the same ACA.

3.2   Experiment Implementation

During the experiment, limited by the existing experimental environment, the two protection methods can not be deployed to the real reproduction social network environment. Only with the open real data set, combined with the algorithm idea, the algorithm can be simulated to achieve the overall privacy protection plus noise method. The specific experimental steps are as follows:

STEP1: The FDR parameters are obtained by analyzing the sample data set. In the process of analyzing and obtaining sample data sets, because the format of each open data set is different, we need to normalize different data sets, and then call the analysis to get FDR parameters.

STEP2: FDR parameters are used in the process of supervised learning to get the learning parameters.

STEP3: According to the real transmission data of the wireless sensor and the learning parameters in step 2, a random noise packet is generated and added to the transmission sequence of the wireless sensor.

STEP4: Take the execution result of step 3 as the fingerprint information of the wireless sensor monitored by the bypass attack, execute the bypass attack algorithm, and calculate the privacy data protection delay.

For the traditional privacy data protection method and the online social network security privacy data protection method in the era of big data, two algorithms are implemented through the same experimental data set, and then the noise data is added to the experimental data set according to the privacy data protection delay and FVR, and finally the data is statistically compared.

3.3   Analysis of Experimental Results

Using the above experimental background and implementation scheme, the following experimental results are obtained, as shown in Fig. 6.

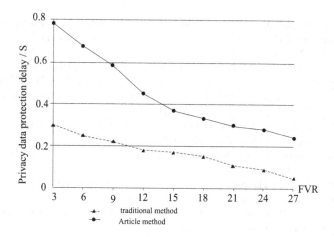

Fig. 6. Experimental results

As shown in Fig. 6, according to the privacy data protection delay comparison of the two methods, the method is to compare the amount of noise data added when the privacy protection effect is the same 当. When the delay value of privacy data protection is lower than 0.4, the effect of privacy protection is the best. Compared with the traditional privacy data protection method, the delay value of privacy data protection method in the era of big data is lower. Without considering the delay, it is obvious that the energy consumption of online social network security privacy data protection method is the smallest in the era of big data. For each privacy data protection delay is the same, the online social network security privacy data protection method is shorter than the traditional privacy data protection method in the era of big data.

4   Conclusion

In this paper, we propose a secure privacy data protection method for online social networks in the era of big data. First of all, combined with the characteristics of wireless sensor networks, this paper analyzes the problems that need to be considered in the design of noise model, that is, the design principles of privacy data protection. Then, from the construction of data protection and specific implementation steps, the specific implementation steps of privacy data protection methods are introduced in detail. Through experimental comparison and analysis, this paper compares the protection method of this paper with the traditional method from the perspective of privacy data protection delay value. The results show that the privacy data protection delay value of online social network security privacy data protection method is low in the era of big data, which has the advantage of good privacy protection effect.

References

1. Huang, W., Huang, J., Li, Y.: An analysis of the relationship between anti-terrorist intelligence work and the protection of personal privacy information in the era of big data. Library Inf. **182**(4), 43–50 (2018)
2. Lina, Z.: Personal information privacy protection in the informatization construction of university personnel files in the era of big data. Arch. Shanxi **238**(2), 70–72 (2018)
3. Zhang, C., Liu, C., Guo, Q.: Optimization of user information security protection in large data. Comput. Simul. **34**(7), 154–157 (2017)
4. Wang, T., Liu, Y., Jin, X.: Research on k-anonymity-based privacy protection in crowd sensing. J. Commun. **39**(S1), 176–184 (2018)
5. Huang, R.: Simulation research of network user privacy information protection. Comput. Simul. **11**, 319–322 (2017)
6. Huang, R., Long, L.: The problems and countermeasures of individual privacy protection under open government data in China. Library **10**, 1–5 (2017)
7. Lin, Y., Duan, X.: Digital information encryption technology for network privacy protection. Modern Electron. Tech. **41**(9), 45–48 (2018)
8. Li, X., Luo, X.: Simulation of user information security protection for communication transmission under big data. Comput. Simul. **35**(05), 178–182 (2018)
9. Wei, D., Ma, H.: Research on the influencing factors of personal data storage security and privacy protection under network environment. Library Theory Pract. **1**, 89–95 (2018)
10. Wang, X.: Computer network data security encryption technology in internet environment. Mobile Commun. **3**, 49–53 (2019)

Research on Security Enhancement Algorithm of Data Dynamic Migration in Optical Fiber Network

Yan-song Hu[✉]

Liaoning Petrochemical Vocational and Technology, Jinzhou, China
hys961@126.com

Abstract. Aiming at the problem that the transmission error of the traditional algorithm is too large, a dynamic data migration security enhancement algorithm for optical fiber network communication is proposed. Firstly, the migration data in the optical fiber network is collected. According to the data transmission process, the electric domain compensation, optical domain compensation and optical domain electric domain hybrid compensation in the PMD compensation unit design scheme are used to calculate the path of data safe migration. The dynamic migration data is collected by the variable electric delay algorithm, and then the amplitude of migration data is preprocessed according to the transmission characteristics of optical pulse in the optical fiber, to the preprocessing formula, use the PSO search mechanism to calculate the migration security feature data, and calculate the dynamic migration amount of n input layers, and finally get the dynamic migration security enhancement algorithm. The experimental results show that compared with the traditional methods, the error of dynamic data migration security enhancement algorithm is smaller when transmitting enhanced data.

Keywords: Optical network · Communication data · Dynamic migration · Security enhancement algorithm

1 Introduction

With the development of new network applications such as cloud computing and large-scale distributed scientific computing, the optical data center network has attracted the attention of academia [1]. The data center network has a high peak throughput and a great deal of burst, and the data center network has a large demand for bandwidth. The data center network urgently needs a physical layer technology to enhance the transmission security of this kind of data. Recent research in elastic optical network shows that it has very good performance of spectrum resource management and control, and it has dynamic migration on the optical path layer. At the same time, elastic optical network can provide huge bandwidth and strong stability. Therefore, elastic optical network is the inevitable choice of data center network interconnection. In order to better study the anycast routing and spectrum allocation of optical data center, the security enhancement algorithm of data dynamic migration in optical network is studied [2].

© ICST Institute for Computer Sciences, Social Informatics and Telecommunications Engineering 2020
Published by Springer Nature Switzerland AG 2020. All Rights Reserved
Y.-D. Zhang et al. (Eds.): ICMTEL 2020, LNICST 327, pp. 67–78, 2020.
https://doi.org/10.1007/978-3-030-51103-6_6

Single path and multi-path are used to solve the problem of unicast routing and spectrum allocation. For single path routing, some dynamic routing and spectrum allocation algorithms are proposed to achieve high throughput when the network is in dynamic service. In the calculation of dynamic data migration security enhancement algorithm, two types of algorithms in network dynamics are mainly calculated, namely, dynamic path migration data and spectrum allocation algorithm. In the pre-processing stage, the single path routing algorithm is extended to multi-path routing algorithm, including two types of multi-path algorithm, hybrid single path/ multi-path real-time computing path algorithm and hybrid single path/ multi-path algorithm using fixed path set. The simulation results show that, compared with the two traditional enhancement algorithms, the dynamic data migration security enhancement algorithm in optical fiber network has smaller transmission error in the transmission of enhanced communication data. It has a very low bandwidth blocking rate and is more suitable for enhancing the security of data dynamic migration in optical fiber network.

1.1   Security Enhancement Algorithm of Dynamic Data Migration in Optical Fiber Network

The security enhancement algorithm of communication data migration firstly collects the migration data in the optical fiber network. Aiming at the PMD problem in the high-speed optical fiber communication system in the process of data transmission, analyzes the law of dynamic migration, compares various PMD online monitoring and compensation methods, and collects the dynamic migration data in the optical fiber network.

By using the basic equation of optical transmission in single-mode fiber and the nonlinear schrondinger equation of pre-processing PMD effect, the vector models of first-order, second-order and high-order PMD are established in Stokes space to deal with the enhanced security relationship between DGD and DOP in optical fiber network. The PC of optical domain compensation scheme is combined with the variable (or fixed) delay line to make the network communication data in the optical fiber be placed in two dynamic migration processes of polarization disturbance and orthogonal polarization to complete the dynamic data preprocessing.

Then using the search algorithm of PMD compensation system, on the basis of the standard PSO algorithm, the security feature data is selected, the migration security inertia weight is adjusted, the immune clone principle in the immune mechanism is introduced, and the migration security feature data feature is calculated [3].

Using the improved SM as the tracking algorithm, using the fast and fine search ability of the tracking algorithm, through the reflection and expansion operation of the feature migration data, the tracking algorithm can quickly approach the extreme point in the iterative process, and estimate the dynamic migration security. Finally, the search algorithm and tracking algorithm are used to calculate the speed and position of ions in the optical fiber network, calculate the optimal fitness, get the compensation response time of the enhancement algorithm and the amount of particle data, and form the security enhancement algorithm.

1.2   Collect Dynamic Migration Data

When collecting the dynamic migration data in the optical fiber network, three kinds of PMD compensation unit are used: electric compensation, optical compensation and optical electric hybrid compensation. The optical domain compensation adopts the method of adjusting the PC in the optical path, changing the polarization direction of the light by adjusting the angle of three wave plates in the PC, making the fast and slow axis of the optical fiber at the end correspond to the fast and slow axis of the back polarization maintaining optical fiber, then dividing the two polarization states of the optical signal into two channels through the polarization beam splitter, using the polarization maintaining optical fiber as the delay line to introduce the appropriate delay for compensation, and finally using the polarization combiner combine the two optical signals. Introduce the fast axis into the delay of $\Delta t/2$ and the slow axis into the delay of $-\Delta t/2$, so that the delay difference between the two polarization modes is zero, so as to achieve the function of eliminating PMD and obtain a safe migration data path, as shown in Fig. 1 below:

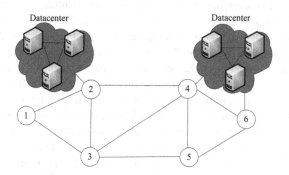

Fig. 1. Secure migration data path

The resulting security path is transmitted according to the sequence numbers 1–6 in Fig. 1 above. The dynamic compensation of data in the transmission process is to compensate the electric signal in the receiver by using the electric equalization compensator. However, in the process of data security migration, the hybrid compensation of optical domain and electric domain separates the two main polarization states of the optical signal, using the PC and polarization beam splitter contained in itself, and separates the optical data and electric data [4]. Let each optical signal change into electrical signal through photodiode, and collect the dynamic migration number of this part, as shown in the figure below, the rectangle of d changes into the ellipse of D, which is the dynamic migration data of the optical signal and electrical signal collected. In order to prevent the collected data from not dynamically changing data, the variable electric delay line is used to screen out the electric signal data mixed in the transmission of optical signal, and finally the acquisition process as shown in Fig. 2 is obtained:

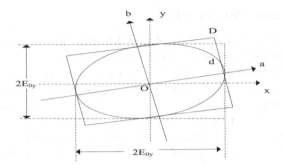

Fig. 2. Dynamic data collection process

After the acquisition, the dynamic data error needs to be calculated, and the electric field compensation calculation method is used to balance and reshape the electric signal. For the dynamic data collected by optical compensation, due to the limitation of electronic bottleneck, real-time extraction of line PMD change information is very easy to cause dynamic error value change. However, the sensitivity of PMD change information extraction is high, that is, it has the ability to reflect small PMD changes [5]. The sensitivity of the enhanced control signal is calculated, and the correlation between the enhanced precision of the acquisition PMD and the dynamic BER is obtained. According to the acquisition principle that the stronger the correlation is, the better the acquisition precision is, the response time of the acquisition is calculated. The response time of PMD and the time required by the control algorithm to process information are far lower than the time required by the PMD change rate, otherwise the optical fiber network will not work effectively.

In the optical fiber network communication, the electric Q value or eye pattern opening and migration process affect the transmission error, but a high-speed decision gate needs to be added, which has low sensitivity to PMD. Using SHB effect, the spectrum monitoring method of baseband radio frequency (RF) signal can accurately measure the DGD value which is larger than one bit period. It has high sensitivity to the smaller DGD value and short response time, but it needs the expensive narrow-band filter. DOP monitoring method has nothing to do with transmission rate, dispersion and modulator. Therefore, DOP value is used as PMD on-line monitoring signal for optimal control. By changing the control voltage to adjust the angle of wave plate in PC, an optimal angle is found to minimize the error of collected data.

1.3   Preprocessing Dynamic Migration Data

Preprocessing the dynamic migration data, using the electromagnetic theory of light, according to the transmission characteristics of optical pulse in the optical fiber, using the max equation group to calculate the vertical component, we get:

$$B_m = \frac{|\beta_x - \beta_y|}{k_0} = n_x - n_y \tag{1}$$

In the formula, B_m is the vertical component, k_0 is the fiber coefficient β_x, β_x is the component of X and Y axis n_x, n_y is the mapping component of X and Y axis. The component is not a constant, but fluctuates. When transmitting, the two vertical components will produce delay, which will widen the signal pulse. When the spread caused by PMD is too large, the performance of the system will be seriously degraded, and the error rate of the receiving system will rise. Mathematically, the propagation constant is slightly different for the two polarization modes in the X and Y directions, which is called mode birefringence. The mode birefringence level P_k is defined as:

$$P_k = \frac{H_k + |H_k|}{2} \tag{2}$$

Where H_k is the refractivity. The pretreatment PMD is random, and the PMD of any section of optical fiber is a random variable subject to the Maxwell distribution function, which is different from the chromatic dispersion with certainty, and its value is closely related to factors such as fiber manufacturing process, material, transmission line length and application environment [6].

In the optical fiber link, the delay difference generated by two orthogonal polarization modes follows a certain probability density distribution. PMD and chromatic dispersion have the same effect on the performance of the system, that is to say, they cause pulse broadening and limit the transmission rate. However, for the low-speed optical fiber transmission system, the PMD effect is much smaller than the chromatic dispersion, which can be ignored, even not included in the earlier fiber performance index. However, with the increase of system transmission rate, the influence of PMD gradually appears, which becomes the key factor to limit transmission speed and distance after attenuation and chromatic dispersion. PMD cannot be avoided, it can only be minimized. Control PMD value to achieve the effect of preprocessing transmission distance, and calculate the maximum transmission distance to get:

$$L_{\max} = \frac{10000}{(D_{PMD} \times BR)^2} \tag{3}$$

Where BR is the bit rate, and the unit is GB/s. D_{PMD} is the transmission polarization coefficient in PMD, and the influence of PMD value and transmission rate on transmission distance is calculated. As shown in Table 1 below:

Table 1. PMD value and transmission rate

PMD (ps/$\sqrt{km}$)	Maximum transmission capacity		
	2.5 Gbit/s	10 Gbit/s	40 Gbit/s
3.0	180	11	<1
1.0	1600	100	6
0.5	6400	400	25
0.1	160000	10000	625

Since the amplitude of the pre-processing migration data is between zero and the maximum residual amplitude $\max|D_k|$, the main calculation methods of the migration data are as follows: in the first case, if the amplitude of meaning is greater than the maximum residual amplitude $\max|D_k|$, then we will keep the transmission speed of the transmission optical path unchanged, because this is generally caused by the communication signal of data dynamic migration.

In the second case, if the amplitude of $\hat{S}_k$ is equal to the maximum migration data amplitude $\max|D_k|$, then we will take the smaller value, that is, we will keep the migration data unchanged. Although it is caused by the communication signal of dynamic migration, such processing can retain the transmission signal of small transmission rate for the amplitude of migration data.

The third case: if the amplitude of the migration data is less than the maximum migration residual amplitude $\max|D_k|$, then we generally take the minimum value of the adjacent frames as the enhanced communication signal value, because this case is generally caused by network communication.

To sum up, according to the idea of migration data attenuation method, we can get the specific preprocessing formula as follows:

$$|\hat{S}_k(k)| = \begin{cases} |\hat{S}_k(k)| & |\hat{S}_k(k)| \geq \max|D_k| \\ \min\left[|\hat{S}_k(k)|\right], j = i-1, i, i+1 & |\hat{S}_k(k)| < \max|D_k| \end{cases} \tag{4}$$

Among them, $\hat{S}_k$ is the electric field compensation coefficient, K is the control coefficient and D_k is the equilibrium coefficient. The final preprocessing results are used to calculate the data migration characteristic data.

1.4   Calculate Migration Security Feature Data

Use the search mechanism of PSO to get the transmission characteristics of characteristic data first. The search mechanism of PSO is shown in the following Fig. 3:

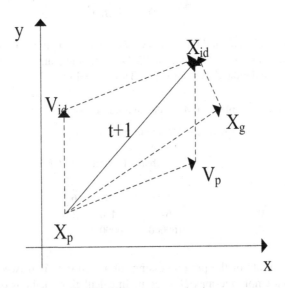

Fig. 3. PSO search mechanism

For each communication data, each feature data will have a fixed transmission feature. The transmission path in the figure above corresponds to a window sampling mechanism. Using dynamic consistency, the binary function of feature data is defined.

$$\partial(i, \hat{S}_k, D_k) = \begin{cases} 1 : \omega(i, \hat{S}_k, D_k) \leq T \\ 0 : \omega(i, \hat{S}_k, D_k) > T \end{cases} \tag{5}$$

In the formula, T is the possibility of parameter t in the consistency stage, and ω is the mechanism coefficient. The calculated point-to-point delay of data package is shown in Table 2 below:

Table 2. Point to point delay feature

Network type	Point to point delay	
	Normal value	Maximum value
10 M-CE	0.427	0.534
10 M-TE	0.238	0.350
100 M-CE	0.0728	0.0736
100 M-TE	0.0534	0.0543

The algorithm of real-time path calculation needs to establish a virtual topology every time, which has high computational complexity. The dynamic RMSA problem can also be realized by using the fixed path set method. In the topology $G(V, E, B, D)$, K shortest paths are calculated for each s-d node pair in advance, so the complexity of dynamic service is greatly reduced.

Formula (5) describes the detailed steps of this fixed path set HSMR algorithm [7].

Then in the service of an optical path request LR (s, d, c), the path set of s-d is sorted based on the path selection strategy mentioned later. When allocating spectrum c, we use a single path best effort service model, that is, we use the largest available spectrum block to allocate spectrum c. only when the largest spectrum block resources can not meet c, we can enable multi-path routing.

When using multi-path routing, calculate the migration security characteristic data according to the following algorithm. The detailed algorithm is shown in Table 3 below:

Table 3. The algorithm of using migration data in multipath computing

Reference algorithm	EPS-RMSA	Exhaustive path search RMSA
	Split-spectrum	Split-spectrum RMSA
Online algorithms	HSMR-OPC	HSMR with online path computation
Fixed path sets algorithms	HSMR-FPS-SPF	HSMR with the shortest path first
	HSMR-FPS-MSF	HSMR with the most slots path first
	HSMR-FPS-MLSF	HSMR with the most left slots path first
	HSMR-FPS-LSoHF	HSMR with the path has the 1 of slots-over- hops weight
	HSMR-FPS-LSoSHF	HSMR with the path has the largest weight of slots-over-square-of-hops

1.5   Estimating Dynamic Migration Security

The following steps are required to estimate and implement the whole dynamic migration security algorithm:

First, the initial data dynamic migration population size is n populations, including their location and speed data.
Second, calculate the fitness value of each particle according to updating the speed and position of the transmission particles in the optical fiber network.

Finally, for each particle, the fitness value is calculated and compared with the experienced safe position P. When the calculated result is 1, it is taken as the current best position P, and the distance between the two particles is calculated [8].
For each particle, its fitness value is compared with the best position P experienced by the whole, and the data amount of projection position of these two particles is calculated. The calculation process is as follows (Fig. 4):

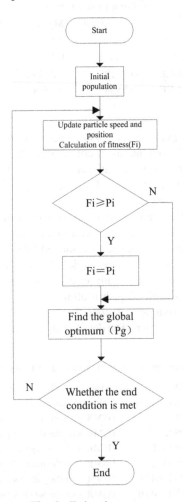

Fig. 4. Estimation process

The deviation method is used to estimate the deviation. The deviation method is to add a small deviation to the input of the input layer of the deep neural network structure, and then transfer the deviation in the network to calculate the change of the output of each hidden layer, that is, compared with the change of the output value of each hidden layer of the optical network when there is no deviation in the input layer [9]. The output value of the optical network layer changes as shown in the following Fig. 5:

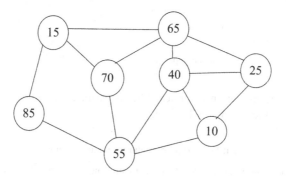

Fig. 5. Deviation

It is assumed that the input deviation value of input layer produces a small deviation value on the basis of the original input layer. Therefore, in the n-th layer, the dynamic migration amount of N input layers can be calculated cumulatively. The process of cumulative calculation is the process of data dynamic migration security enhancement algorithm [10].

2 Simulation Experiment

2.1 Experimental Preparation

Prepare two computers with the same parameters of high computing power, as shown in the following Fig. 6:

Fig. 6. Computer laboratory environment

The computer parameters in the laboratory are shown in Table 4 below:

Table 4. Computer parameters

Serial number	Name	Parameter
1	Memory capacity	8 GB
2	Mechanical hard disk capacity	DDR4-2666 MHz
3	Video card capacity	1 TB
4	Solid state disk	6G
5	CPU	i7-9750H
6	Resolving power	1920*1080

Prepare the data migration of optical fiber network communication in the experiment, as shown in the following Table 5:

Table 5. Experimental preparation data

Transmission efficiency(%)	Communication node	Data volume (MB)
15	1	10
20	2	20
25	3	30
30	4	40
35	5	50
40	6	60

Using data prepared in the above table, each data node represents the amount of data of 10 MB, and simultaneous interpreting the transmission error of data transmission under the same transmission efficiency.

2.2 Experimental Results

Simultaneous interpreting data transmission error under the same transmission rate is achieved by using the traditional security enhancement algorithm and the dynamic migration security enhancement algorithm of optical network communication data. The results of the following Fig. 7 are as follows:

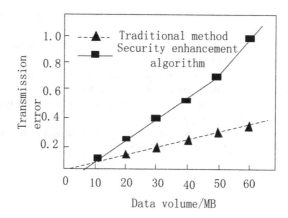

Fig. 7. Error comparison results

As can be seen from the above figure, when the two algorithms transmit 0–10 mb of data, there is no error in transmission, while when the amount of data is more than 10 MB, up to 60 MB, the transmission error of the traditional algorithm is far greater than the dynamic migration security enhancement algorithm of optical network communication data, which is beneficial to the transmission of security data in dynamic migration.

3 Concluding Remarks

In order to effectively reduce the space occupied by communication data resources and enhance the security of communication data dynamic migration, this paper studies the integrity and security protection of data communication security issues, optimizes the method of data encryption of communication data privacy information, so as to improve the security of data dynamic migration. The simulation results show that the dynamic data migration security enhancement method can effectively improve the network communication effect and ensure the quality of optical fiber communication.

References

1. He, Y., Xie, L.: Selection of the optimal laser sensor node after the optical fiber communication network is invaded. Laser J. **39**(4), 129–133 (2018)
2. Shi, Y., Zhang, H., Zhao, B., et al.: Selection of the optimal laser sensor node after the optical fiber communication network is invaded. Laser J. **39**(4), 65–75 (2018)
3. Ma, S., Wang, B., Wang, B., et al.: Research on real-time monitoring algorithm for tunnel safety status based on fiber Bragg grating strain sensors. Chin. J. Sci. Instrum. **38**(2), 304–311 (2017)
4. Mao, L., Cheng, G.: Dynamic OpenFlow switch migration algorithm based on the attractor selection. Inf. Control **46**(5), 525–530 (2017)
5. Wang, F.: Dynamic controllable flexible resource scheduling algorithm under cloud computing. Sci. Technol. Eng. **18**(3), 291–296 (2018)

6. Ma, S., Nie, G., Ye, T., et al.: Dynamic distribution selection algorithm for post mobile relay nodes after optical fiber network instrusion. **42**(5), 44–47 (2018)
7. Zhao, S., Ji, X., Mao, Y., et al.: Research on dynamic migration of virtual machine based on security level. **38**(7), 165–174 (2017)
8. Liu, H., Wang, T.: Research on breakpoint fault detection method of optical fiber communication LAN. Modern Electron. Tech. **40**(16), 81–85 (2017)
9. Ren, H., Shi, X.: Research on data management for smart grid based on dynamic migration. Chin. J. Power Sour. **42**(2), 291–293 (2018)
10. Zhao, C., Yan, L., Cui, Y., et al.: Dynamic adjusting threshold algorithm for virtual machine migration. J. Comput. Appl. **37**(9), 2547–2550 (2017)

Anti-interference Algorithm of Broadband Wireless Communication Based on Embedded Single Chip Microcomputer

Yan-song Hu[(✉)]

Liaoning Petrochemical Vocational and Technology, Jinzhou, China
hys961@126.com

Abstract. In order to solve the problems of the traditional anti-jamming algorithm of broadband wireless communication, such as poor anti-jamming performance and high bit error rate, an anti-jamming algorithm of broadband wireless communication based on Embedded MCU is proposed. In the broadband wireless communication based on embedded single-chip microcomputer, the m-sequence of communication signal data is constructed, encoded and decoded. Finally, CRC redundancy test and error correction are carried out for the decoded communication data, so far the design of broadband wireless communication anti-interference algorithm based on embedded single-chip microcomputer is completed. Through the contrast experiment, compared with the traditional anti-jamming algorithm of broadband wireless communication, the experimental results show that compared with the traditional anti-jamming algorithm of broadband wireless communication, the proposed anti-jamming algorithm of broadband wireless communication based on Embedded MCU has lower bit error rate, which shows that it has better anti-jamming ability.

Keywords: Singlechip · Wireless communication · Anti-interference

1 Introduction

The research of interference and anti-interference in broadband wireless communication system is more important than that in general wireless communication system. In fact, the modern digital mobile communication system with high communication rate and throughput is broadband to some extent. Broadband wireless communication system occupies a wide range of spectrum, but the spectrum resource of wireless communication system is limited, and the use of wireless spectrum has certain openness. If the spectrum is not used in a standard or unreasonable way, the wider the spectrum range, the easier it is to introduce various broadband or narrow band interference signals, reduce the transmission reliability of the communication link, and form a bad impact on the communication system. Ring. In order to ensure the quality of communication, the research on the interference immunity of broadband wireless communication can not be ignored. It is of great significance to improve the signal quality of broadband wireless communication by using some anti-interference algorithm to implement anti-interference operation for broadband wireless communication.

Y.-D. Zhang et al. (Eds.): ICMTEL 2020, LNICST 327, pp. 79–91, 2020.
https://doi.org/10.1007/978-3-030-51103-6_7

Based on embedded single chip microcomputer, it can complete wireless communication of data, realize wireless control, data acquisition, alarm and other functions, and has a wide application prospect in many fields of oil, power, light industry, environmental protection and other industries. Generally, the serial I/O port of single chip microcomputer is used to complete data communication, but there are many shortcomings in this method. For example, when using the serial port of single chip microcomputer to send and receive data, the transmitted data can only be parity checked. However, a large number of electromagnetic interference signals inevitably exist around wireless communication equipment, which makes the transmitted data prone to distortion, and parity generally can not detect the burst errors caused by such interference. Therefore, this method can not meet the high requirements of error detection. Moreover, the communication anti-interference ability of SCM serial port is poor, so it is difficult to overcome the errors caused by interference. In order to solve the above problems, an anti-interference algorithm based on embedded single chip microcomputer is proposed. The software coding method is used to realize the anti-interference of broadband wireless communication, and the effectiveness of the method is verified by experiments.

2 Anti-interference Algorithm of Broadband Wireless Communication Based on Embedded Single Chip Microcomputer

Among many coding techniques, Hamming code is intuitionistic and simple, with the highest frequency (information rate) of transmitting information number. It has the ability of error detection and error correction, and is also relatively simple to realize. When the receiving end finds that there is a bit error in the received information, it decodes and completes the error correction at the same time; when there are two bit errors, it can detect and discard in time.

Aiming at the broadband wireless communication based on embedded single chip microcomputer, by constructing the data sequence of broadband wireless communication signal, coding and decoding m sequence, after decoding, CRC redundancy test and error correction are carried out for the communication data, and the anti-interference algorithm of broadband wireless communication based on embedded single chip microcomputer is designed.

2.1 Construct M-Series

In the channel of broadband wireless communication, due to the complexity of wireless channel, especially the large Doppler frequency expansion caused by relative motion and multipath, inter code interference is caused. The time-domain coding technology based on forged sound sequence is adopted to overcome the problem of inter code interference. In the DS-SS system of broadband wireless communication, the transmission signal is expanded by pseudo-random sequence, which is used to compress the signal when it is received, and to spread the interference signal power, so as to improve the anti-interference ability of the system; in the frequency hopping system, the frequency generated by the frequency synthesizer is controlled by pseudo-random sequence to avoid interference; in the time hopping system, the pulse is controlled by pseudo-random sequence the time and duration of the transmission. The performance of pseudo-random sequence is directly related to the performance of the whole broadband wireless communication system. To construct the m-sequence of broadband wireless communication signal, we must first construct the linear shift register that generates the m-sequence, first determine its primitive polynomial, and its register logical structure is shown in Fig. 1.

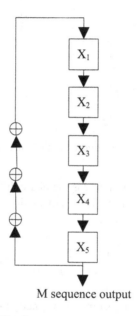

M sequence output

Fig. 1. m-sequence generator

The m-sequence of broadband wireless communication signal data is generated by the m-sequence generator in Fig. 1.

The maximum possible period of the sequence generated by the sequence generator composed of n-level shift registers is 21 nl. Under a certain n, the sequence with the maximum period is called the maximum length sequence, also called the m sequence. The autocorrelation characteristic curve of m series is a periodic triangle curve with sharp correlation peak, as shown in Fig. 2.

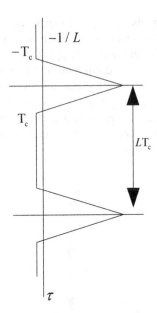

Fig. 2. Normalized autocorrelation curve of *m*-series

In Fig. 2, T_c is the symbol duration; τ is the phase difference of two sequences, τ is continuous, and L is the code length. The mathematical expression of the normalized autocorrelation function is as follows:

$$R(\tau) = \begin{cases} 1 - \frac{|\tau|}{T_c}\left(1 + \frac{1}{L}\right) & |\tau| \leq T_c \\ -\frac{1}{L} & T_c < |\tau| \leq \frac{LT_c}{2} \end{cases} \tag{1}$$

Formula (1) shows that the autocorrelation function of *m*-sequence only has peak value when the phase difference of two sequences is in the range of positive and negative symbols, showing sharp autocorrelation characteristics. Therefore, it is possible to determine whether the relative positions of the two sequences are in a positive or negative symbol according to whether there is a peak value in the correlation output.

m-sequence is a kind of pseudo-random sequence. According to the theory of stationary random process, its average power spectral density is its autocorrelation Fourier transform, that is, it is easy to get its power spectral density from the auto-correlation function of *m*-sequence through Fourier transform: $S_R(\omega)$

$$S_R(\omega) = \int_{-\infty}^{+\infty} R(\tau)e^{-j\omega\tau}d\tau \tag{2}$$

In formula (2), $R(\tau)$ is the autocorrelation function, in a period, that is, $0 \leq \tau \leq pT_c$ j represents the parameter, ω represents the spectral density parameter.

The discrete spectrum can be rewritten as follows:

$$S_R(f) = \sum_{m=-\infty}^{\infty} p_m \delta(f - mf_0) \tag{3}$$

In formula (3), band f represents frequency, band p represents period and band δ represents spectrum parameter. Among them,

$$\begin{cases} p_m = \begin{cases} \frac{1}{p^2} & m = 0 \\ \frac{1+p}{p^2} \sin c^2\left(\frac{m}{p}\pi\right) & m \neq 0 \, (m = \pm 1, \pm 2, \cdots) \\ f_0 = \frac{1}{pT_c} \end{cases} \end{cases} \tag{4}$$

Through the above formulas, we can get the characteristics of m-sequence of broadband wireless communication data, and determine the best m-sequence according to its characteristics.

2.2 m-sequence Coding

In order to change the information of all wideband wireless communication signals into deformed codes conveniently, firstly, the m-sequence is encoded. The generation matrix of the code is required. According to the relationship between the generation matrix and the consistent supervision matrix, and through the consistent supervision matrix, the generation matrix is obtained. The specific method is as follows:

The consistent supervision matrix is recorded as $H=(Q, 1)$, where 1 is the unit matrix and Q is the remaining part matrix. Then the generated matrix is $G = (1, P)$, where, $P = Q^T$. It is known that the 4-bit information in the broadband wireless communication signal information is C_1, C_2, C_3, C_4, then the corresponding 8-bit codeword can be obtained by the following formula:

$$(C_1 C_2 \cdots C_8) = (C_1 C_2 C_3 C_4) \tag{5}$$

H^* is transformed into a standard form by matrix equivalence.

$$H^* = \begin{pmatrix} 0 & 1 & 1 & 1 & 1 & 0 & 0 & 0 \\ 1 & 0 & 1 & 1 & 0 & 1 & 0 & 0 \\ 1 & 1 & 0 & 1 & 0 & 0 & 1 & 0 \\ 1 & 1 & 0 & 0 & 0 & 0 & 0 & 1 \end{pmatrix} = (Q, 1) \tag{6}$$

Among,

$$Q = \begin{pmatrix} 0 & 1 & 1 & 1 \\ 1 & 0 & 1 & 1 \\ 1 & 1 & 0 & 1 \\ 1 & 1 & 1 & 0 \end{pmatrix} \tag{7}$$

$$P = Q^T = \begin{pmatrix} 0 & 1 & 1 & 1 \\ 1 & 0 & 1 & 1 \\ 1 & 1 & 0 & 1 \\ 1 & 1 & 1 & 0 \end{pmatrix} \tag{8}$$

Corresponding, generate matrix

$$G = (1, P) = \begin{pmatrix} 1 & 0 & 0 & 0 & 0 & 1 & 1 & 1 \\ 0 & 1 & 0 & 0 & 1 & 0 & 1 & 1 \\ 0 & 0 & 1 & 0 & 1 & 1 & 0 & 1 \\ 0 & 0 & 0 & 1 & 1 & 1 & 1 & 0 \end{pmatrix} \tag{9}$$

Substitute formula (5) to get:

$$(C_1 C_2 \cdots C_8) = (C_1 C_2 C_3 C_4)$$
$$= \begin{pmatrix} 1 & 0 & 0 & 0 & 0 & 1 & 1 & 1 \\ 0 & 1 & 0 & 0 & 1 & 0 & 1 & 1 \\ 0 & 0 & 1 & 0 & 1 & 1 & 0 & 1 \\ 0 & 0 & 0 & 1 & 1 & 1 & 1 & 0 \end{pmatrix} \tag{10}$$

It can be seen from formula (10) that the first four bits of the code word are the information of the broadband wireless communication signal, and the last four bits are the supervision bits, and there are $C_5 = C_2 \oplus C_3 \oplus C_4$, $C_6 = C_1 \oplus C_3 \oplus C_4$, $C_7 = C_1 \oplus C_2 \oplus C_4$, $C_8 = C_1 \oplus C_2 \oplus C_3$. The coding program block diagram is shown in Fig. 3:

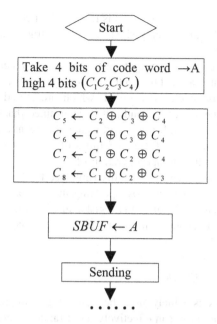

Fig. 3. Coding flow chart

As shown in Fig. 3, the coding flow of m sequence of broadband wireless communication data is completed.

2.3   Implementation of Decoding

After encoding the data of broadband wireless communication, when the data is transmitted to the receiver, the encoded data should be decoded. At the data receiving end, the program is designed so that the MCU at the receiving end can detect the received data bit by bit at a fixed time interval (3.6), and record the number of high and low level detected in each data bit cycle (906) (this number actually reflects the different durations of high and low level of the waveform). Finally, it is determined that the received data bits are "0", "1" or start bits by judging the occurrence times of high and low levels.

This method overcomes the disadvantage that when using serial port for data communication, each data bit is sampled only three times, which is easy to be misjudged. Because it does not determine that the received data bits are '0', '1' or start bits by simply sampling three times. It needs to sample hundreds of times (the period of data bits divided by the interval of sampling time), which greatly avoids the possibility of data misjudgment due to the interference signal in the process of data transmission.

When sending data, the sender uses the circular right shift with carry to realize, so as long as the value of carry flag bit is read, the data of 8 bits can be read successively from low position to high position, which facilitates the reading of data bits and the subsequent operation. Each frame of data starts with a start signal and then transmits 8-bit data bits. The number of data bits to be sent is stored in register R6 in advance, and the sending operation of 8-bit data bits is realized by performing 8 times of cyclic right shift operation with carry. Finally, the last bit of the transmitted data is determined to be '0' or '1' and sent.

The receiver also uses the cyclic right shift with carry to receive data. The received data is detected bit by bit at a fixed time interval, and the number of high and low levels detected in each data bit cycle is recorded, which are stored in variables I and I respectively. Through the values in i and j, it is determined that the received data bits are '0', '1' or the start bit signal. If it is a start signal, the count unit num (received data bits) is assigned an initial value of 8, and 8 times of cyclic right shift operation with carry is started to realize the storage operation of data bits. Before each cycle to the right, the data bits need to be detected and discriminated by the same method as above. The discriminating result is achieved by operating the flag bit STATUS. The flag bit STATUS is the 0-bit status <0> of status register in PIC single chip microcomputer: carry and borrow flag. It controls the value of the carry flag bit of the rotate right operation to be '0' or '1'.

2.4   CRC Redundancy Check

CRC redundancy check is widely used because of its simplicity and strong anti-interference ability. It not only can effectively resist random interference, but also can effectively resist burst interference. Therefore, it is not only applied to ordinary wireless channels, but also has been successfully applied in broadband wireless communication. Especially by nesting with convolutional coding, it can greatly improve the performance of broadband wireless communication under the condition of limited power.

The basic steps of CRC verification are as follows: using linear coding theory, according to the k-bit binary code sequence to be transmitted (converted from m-bit sequence), a supervision code (CRC code) R-bit for verification is generated with certain rules, and attached to the information, forming a new binary code sequence with a total of $(k + r)$ bits, and finally sent out. At the receiving end, it checks according to the rules between the information code and the CRC code to determine whether there is an error in the transmission. In order to reduce the storage space and calculation, a 16 bit CRC redundancy check based on the look-up table method is adopted. For single byte CRC-16 coding, in order to shorten the coding time, the check code words corresponding to 256 numbers of 0 $\sim$ FFH are often calculated in advance and stored in a table. In the case of CRC check coding, double byte check codeword can be obtained by looking up the table according to the single byte content, and then the single byte source information and double byte check codeword are combined to form CRC-16 coded codeword.

For the CRC-16 verification codeword table with single byte source information, the space needed to store the table can be reduced by using the regularity. Take the CRC check code word (32b) of L6 numbers corresponding to 0 $\sim$ F in the first row and the CRC check code word (32b) of 16 numbers corresponding to 0 $\sim$ F in the first column. For any single byte number, it can be taken as 14. Make the binary value corresponding to the high 4 bits as the row number of the table and the CRC corresponding to the low 4 bits as 1231, and the binary value corresponding to the column number of the table as 4084, then the CRC corresponding to 14 is I XOR J = 52b5, and

52b5 is the check code word of the single byte number. The calibration block diagram is shown in Fig. 4.

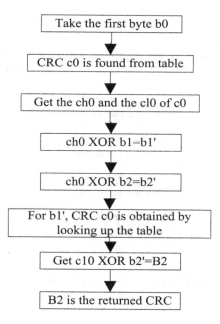

Fig. 4. CRC block diagram

For multi byte CRC-16 encoding, set the information of any byte asb0, b1, b2, b3, b4, b5, …. First of all, the check word c0 is obtained by looking up the table of b0. Take the high byte ch0 and b1 of c0 for XOR operation to get b1′, and take the low byte cl0 and b2 of C0 for XOR operation to get b2′. Then the information to be coded after the CRC Coding of B0 is b0′, b1′, b2′, b3′, b4′, b5′,…. After that, an operation similar to b0 is performed on b1′. In this way, until all bytes are executed, the final check code word is CRC check word of all information bytes. According to the above rules, the coding verification of broadband wireless communication signal data is obtained.

2.5   Error Correction

In broadband wireless communication, to correct the communication signal of the receiver is to discard the interference signal. The data packet of broadband wireless communication signal is copied twice (3 times in total). The first coded communication information is detected for error. If there is any error, the remaining two backups are used to correct the error. Error correction is by comparing the bits in the three backup data. If two or more bits are '0', the correct one should be '0'.

For example:

00001110CP1(Error byte)
11111010CP2
01011010CP3
01011010Corrected bytes

Its logical expression is: correction byte = = (CP1&CP2CP3)I(CP1&CP2).The program block diagram for correcting the error is shown in Fig. 5.

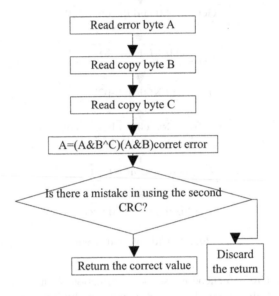

Fig. 5. Error correction program chart

Through Fig. 4, the wireless communication signal is corrected and the interference signal is removed.

So far, the design of anti-interference algorithm for broadband wireless communication based on embedded single chip microcomputer has been completed.

3   Experiment

By comparing the anti-interference algorithm based on Embedded MCU with traditional anti-interference algorithm, it is verified that the anti-interference algorithm based on Embedded MCU can effectively improve the anti-interference ability of broadband wireless communication.

3.1   Experimental Process

First, set up the experimental environment, and the parameters of the experimental environment are shown in Table 1.

Table 1. Experimental environment parameters

Experimental parameters	Tester parameters	Interference source
Signal bandwidth	1 MHz	20, 40, 80 MHz
Sensitivity	−81 dBm	
Agreement	FHSS	Noise
Connection rate	1 Mbit/s	
Working condition	Single channel frequency locking: 1 MHz	20 MHz
	Limited bandwidth: 34 MHz	
	Full frequency band: 79 MHz	75 MHz
	Limited frequency band: 79 MHz	
Loading rate	125 byte/500 ms	

In the experimental environment of Table 1, the effectiveness of the anti-interference algorithm of broadband wireless communication based on Embedded MCU is tested.

By using the coding anti-interference algorithm, the anti-interference algorithm is implemented for the broadband wireless communication based on the embedded MCU, and compared with the other two uncoded anti-interference algorithms, the anti-interference ability of broadband wireless communication is obtained.

3.2 Experimental Result

The comparison results of BER of the three wideband wireless communication anti-interference algorithms with SNR of −9 dB and Sir of −35 dB are shown in Fig. 6 and Fig. 7 respectively.

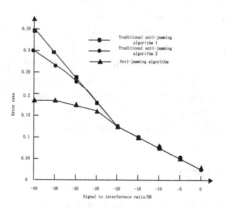

Fig. 6. Comparison results of bit error rate under different signal to noise ratio of −9 dB and signal to interference ratio

It can be seen from Fig. 6 that when the SNR is set to −9 dB and the SNR is lower than −25 dB, the proposed anti-interference algorithm for broadband wireless communication based on Embedded MCU has obvious advantages, and its bit error rate is significantly reduced compared with the other two traditional anti-interference algorithms. When the signal to interference ratio is −40 dB, the bit error rate is reduced by about one time by using the anti-interference algorithm of broadband wireless communication based on Embedded MCU. The experimental results show that the anti-jamming algorithm of broadband wireless communication based on Embedded MCU has better anti-jamming ability.

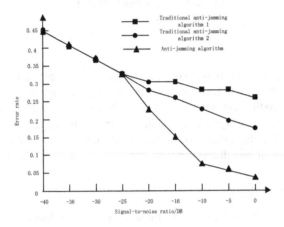

Fig. 7. Comparison results of bit error rate under different SNR and snr −35 db

As shown in Fig. 7, when the signal to interference ratio is set to −5 dB and the signal-to-noise ratio is greater than −25 dB, the proposed anti-interference algorithm for broadband wireless communication based on Embedded MCU has obvious advantages, and its bit error rate is significantly reduced compared with the other two traditional anti-interference algorithms. With the increase of signal-to-noise ratio, the bit error rate of Wideband Wireless Communication Anti-jamming algorithm based on embedded single-chip microcomputer is reduced more and more.

In conclusion, the experimental results show that the anti-interference algorithm based on Embedded MCU has better anti-interference ability.

4 Concluding Remarks

In view of the high bit error rate of the traditional anti-jamming algorithm for broadband wireless communication based on embedded single chip microcomputer, this paper proposes an anti-jamming algorithm for broadband wireless communication based on embedded single chip microcomputer. Compared with the traditional anti-interference algorithm of broadband wireless communication based on Embedded MCU, the experimental results show that the proposed anti-interference algorithm of broadband wireless communication based on Embedded MCU has better anti-interference ability.

References

1. Zhang, X., Cao, G., Liang, F., et al.: Design of embedded smart wireless sensor based on SCM control. Modern Electron. Tech. **40**(13), 80–82, 86 (2017)
2. Yang, N.: Design of embedded multi-node network communication system based on single chip microcomputer. Modern Electron. Tech. **41**(11), 13–16 (2018)
3. Li, Y., Wang, S.: Improved SRV Wideband anti-jamming algorithm under TDL structure. J. Signal Process. **35**(1), 125–131 (2019)
4. Zhang, X., Zhang, B., Guo, D., et al.: Communication anti-jamming technique based on enhanced oblique projection operator in polarization domain. Comput. Eng. **44**(4), 140–144 (2018)
5. Tan, L., Zhuang, Y.: Anti-interference algorithm of wireless channel for IoT communication. Telecommun. Sci. **33**(10), 58–64 (2017)
6. Fang, W., Zhang, W., Hu, M., et al.: Improved LDPC-based short-range frequency-hopping Wirel. Commun. Syst. **38**(12), 34–47 (2017)
7. Li, H.: Wireless communication channel equalization algorithm based on adaptive bit modulation. J. Hennan Inst. Eng. (Nat. Sci. Edn.) **30**(3), 64–69 (2018)
8. Gaiyan, B.A.I.: Laser network anti-interference routing communication algorithm design. Laser J. **39**(4), 115–118 (2018)
9. Yuan, Y., Ma, X., Zhou, C., et al.: An eEmbedded MCU-based algorithm and technology for improving analogy-to-digital precision. J. Hunan City Univ. (Nat. Sci.) **28**(1), 62–65 (2019)
10. Wei, L., He, X., Zhang, B.: The jamming efficiency analyze of multiply GPS broadband sources interference for GPS receiver with multiply module. J. Signal Process. **33**(12), 1631–1636 (2017)

Research on Task Driven Basketball Teaching Mode Based on ITbegin Cloud Platform

Ning-ning Zhang[⊠]

Liaoning Petrochemical Vocational and Technical College, Jinzhou, China
znn961@163.com

Abstract. In the traditional basketball teaching mode, teachers could't solve the problems in the training of students in time, which affected the training effect of students. Therefore, the basketball teaching mode based on ITbegin cloud platform is designed. According to the difficulty of basketball teaching and training content, it is divided into basic content and advanced content. On this basis, made basketball teaching and training ITbegin course. Published it in the ITbegin cloud platform through the Internet. Students watched the course by downloading or online, complete the training and timely feedback the training progress and problems encountered in the training, adjusted the teaching mode in time, ensured training quality. The test results showed that: In the case of similar physical quality of the test personnel, the overall level of the basketball teaching mode based on the ITbegin cloud platform was higher than that of the traditional teaching mode. It is proved that the design of basketball teaching mode based on ITbegin cloud platform was more suitable for practical training projects.

Keywords: ITbegin cloud platform · Basketball teaching · Teaching model

1 Introduction

ITbegin is based on the Internet, which is opened and promoted by leading figures of various disciplines, and relevant system courses are released on corresponding platforms. It is a new teaching method and method that users can choose freely, study independently and discuss with each other to complete certain teaching tasks with a certain course schedule [1]. In the field of education, the research on the nature, characteristics and development trend of ITbegin platform is gradually mature. However, in the field of sports, the research results of ITbegin are relatively few, and most of them focus on the theoretical research of the current situation, future prospects and inspiration for physical education of "sports ITbegin", lacking the research of practical application of ITbegin platform.

In the course of learning physical education, the basketball course is more popular with students. After years of reform and development, the teaching of basketball in Colleges and universities has basically formed two teaching modes, namely, basketball option teaching and basketball club teaching. The integration of these two teaching modes ensures the realization of the teaching objectives of theoretical knowledge learning and sports skills mastering of college students, as well as the lifelong physical

Y.-D. Zhang et al. (Eds.): ICMTEL 2020, LNICST 327, pp. 92–104, 2020.
https://doi.org/10.1007/978-3-030-51103-6_8

education of college students. The cultivation of sports consciousness. However, throughout the process of basketball teaching activities in Colleges and universities, the introduction of modern education technology and teaching methods is very limited, and the tumor disease of traditional teaching mode still exists [2].

This paper introduces the teaching mode of ITbegin into the basketball teaching mode, makes full use of the advantages of the teaching mode of ITbegin, and combines with the actual situation of basketball physical education, carries out the reasonable teaching reform of basketball physical education, so as to provide some reference for the current reform of basketball teaching mode. The introduction of ITbegin cloud platform will become an important opportunity to promote the reform of basketball teaching and improve the quality of basketball teaching.

2 Design of Basketball Teaching Mode Based on ITbegin Cloud Platform

2.1 Determine the Training Content of Basketball ITbegin

Basketball is an antagonistic sport, which can last for a long time or a short time, but it requires participants to run fast, take off suddenly and continuously, react quickly and compete with strength [3]. In the actual training, we need to do different special skills training. The details are shown in the table below (Figs. 1 and 2).

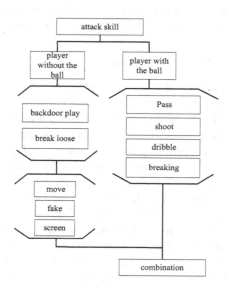

Fig. 1. Classification of basketball attack techniques

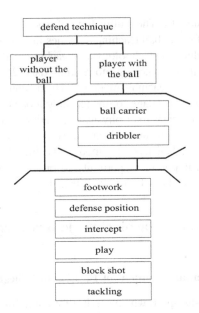

Fig. 2. Basketball defense technology classification

According to the actual situation of the students, adjust the training schedule appropriately. The students with poor basic skills choose basic training. The students with good basic training can choose relatively difficult advanced content training.

2.2   Making Basketball Teaching ITbegin

According to the characteristics of ITbegin and physical education courses, the design of college basketball ITbegin mainly refers to the following principles: first, the video design time of basketball ITbegin should be controlled within 12 min, so as to ensure the optimization of teaching effect. Because the time of students' excitement point and attention entering the highest peak of physiology and psychology is about 10–12 min, students can learn ITbegin content independently offline, lack of teachers' on-site management and guidance, and it is difficult to concentrate for a long time. Second, design the teaching plan reasonably. Because there are individual differences in students' basketball knowledge and skill base, students' differences should be considered in the selection of content, and the teaching plan should be designed reasonably. The starting point should not be too high to ensure that students actively participate in the basketball curriculum learning [4]. Third, design a diversified teaching evaluation system. When designing the basketball ITbegin course, in addition to the regular indicators such as the final skill test, quality test and the usual class assessment, the assessment indicators such as the online student assignment, technical action analysis, theoretical topic, group discussion, online time length and unit test should be added to make the teaching assessment diversified and ensure the rationality and scientificity of the assessment. After establishing the framework of basketball ITbegin teaching platform, it is necessary to make and design ITbegin. According to the requirements of the basketball course syllabus in Colleges and universities, combined with the cognitive characteristics of students, the basketball ITbegin teaching unit is designed in different levels.

Determine the teaching objectives and contents of the basic teaching unit. According to the analysis of learning situation, master the overall characteristics and individual differences of teaching objects, and then determine the teaching objectives. After determining the teaching objectives, analyze and determine the teaching contents. For example, the teaching objectives are to let students learn and master the basketball dribbling technology. The teaching contents must be explained around the dribbling technology and skills. If it is about the dribbling rules, it deviates from the theme [5]. The third is to make videos of teaching contents and collect teaching auxiliary materials. After determining the teaching content, the next step is to design the specific teaching content. The teaching design should pay special attention to the details of the teaching content. Each ITbegin course can not exceed 12 min at most, and can only explain a specific knowledge point or skill. If there are more contents, they should be designed separately. For example, the common technical actions of basketball dribbling are as follows: high dribbling, low dribbling, quick stop of dribbling, forward to hand changing dribbling, back dribbling, turn of dribbling, crotch transportation. The ball is [6]. Every action is a knowledge point. Make teaching video for knowledge points, strive to make every skill and action have video explanation and demonstration, collect and sort out teaching auxiliary database, and store rich and diverse information. The fourth is online interactive communication design. The purpose of online communication is to strengthen the interaction between teachers, students and students, including online learning, interactive discussion and feedback evaluation. The specific way is to design a video window for situational dialogue, which is conducive to the demonstration of technical action points. There are also difficult questions, teachers can preset several difficulties and give answers [7]. Another is to organize the basketball match, release the arrangement of the match and the judgment of the result in the interaction, increase the atmosphere of communication, and stimulate the interest of students. Finally, the evaluation system is designed. Students' evaluation of teaching can be carried out in the interaction, or leave a message in the evaluation part. Teachers should establish a reasonable evaluation system, such as setting up online time and online class hour requirements. Each basic unit is designed with test questions, and the scores of students' learning evaluation are calculated [8]. The effect is shown in Fig. 3.

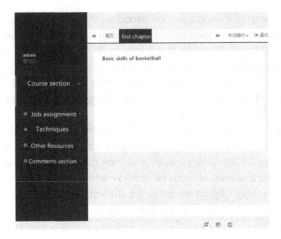

Fig. 3. Effect of basketball teaching mode ITbegin

As shown in the figure, each chapter has both video learning resources and courseware in the form of PPT. At the same time, the key and difficult points of teaching, teaching design, evaluation and assessment, and the content of teaching materials are available to the students who need to download in the form of files [9]. Different groups can unify the learning progress according to their basketball skill level. The discussion area of the basketball ITbegin course includes three sub sections: the teacher's question answering area, the note taking area and the discussion area. The students in the experimental group can consult the teachers or classmates at any time through the online communication platform of the discussion area for problems encountered in the process of independent learning. Students watch the audio resources of the intuitive and outstanding decomposition demonstration actions. Through the independent practice in the group and the learning of theoretical knowledge after class, they can answer and upload the ITbegin video as a sign in means to ensure the frequency of the students' after class practice. Before class, the instructor gives the learning task, and the team leader organizes the team members to watch the short video on the ITbegin platform within the specified time, make notes, and make records at any time for those who do not know the knowledge, learn the basic technical actions through the video resources, and preliminarily form the understanding and under-standing of basketball [10]. After watching the video, the students were organized to do outdoor exercises. The students in the experimental group were required to practice twice a week for 90 min each time. Each group can arrange training time and teaching time for other students. The instructor shall organize a unified guidance course once a week, conduct field tour guidance and evaluation, assist students in teaching, let stu-dents practice and teach by themselves, and irregularly share the successful experience of the group or individual with better performance as a learning example under the condition of ensuring normal teaching progress, so as to cultivate students' organiza-tional ability and self-confidence. The students help each other and communicate with each other. The girls in the group with poor basketball level can consult the boys with relatively good basketball level to help them learn.

2.3   Training with ITbegin Cloud Platform

ITbegin resource is an open resource, which is made by excellent professionals for basketball teaching and training. It is uploaded to ITbegin cloud platform. Through the network learning platform, videos and learning materials are published for students to download or watch on line, and discussion modules are organized to facilitate students to discuss and learn from each other, solve problems, and have doubts. Online consulting services are provided at the place where you ask. The display page in the ITbegin cloud platform is as follows (Fig. 4).

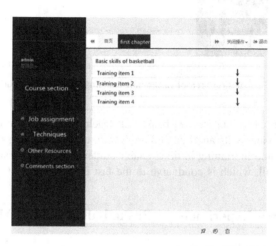

Fig. 4. Download interface of ITbegin courses in ITbegin cloud platform

After the course is released, the students are required to watch the ITbegin video and related auxiliary materials before the actual training. According to the training tasks set in the course, they are required to complete one by one in order. After the training, according to the students' own training situation, summarize the experience, sort out the problems found in the training process, record the problems that are still unsolved after the discussion between the students, and bring back the problems. To the classroom, through communication with teachers to solve the problems found.

In the actual classroom, teachers test the effect of basketball teaching and training in a certain way, discuss and solve the problems existing in the students, and then study the basketball teaching and training content in the ITbegin video in depth, put forward new problems to discuss and solve.

According to the test of students' training results, the training files of each student in each stage are established for quantitative evaluation. The content of the evaluation includes self-evaluation and others' evaluation. By synthesizing the two evaluation results, we can master the real-time training state of students and the mastery degree of basketball skills. The student's training evaluation page is shown below (Fig. 5).

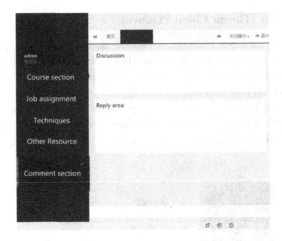

Fig. 5. Discussion area of basketball teaching and training students

Using ITbegin cloud platform in basketball teaching mode is convenient for students to learn relevant skills anytime and anywhere, arrange training time reasonably, solve problems found in training in time, enable students to better master the skills and actions of basketball, which is conducive to the fast improvement of students' sports.

3   Comparative Experiment of Basketball Teaching Mode

3.1   Pre Experiment Preparation

The comparative experiment of basketball teaching mode is carried out in an outdoor basketball court of a university, and the subjects of the experiment are the students who choose basketball skills in the University, Limited by the curriculum, The basic contents of basketball course in college physical education are selected as the experimental items. Its basic contents include: 1 min in-situ two handed chest pass and catch; 1 min in-situ one handed shoulder shot and back and forth dribble and lay up. Taking shooting skills as an example, the two groups of students can have the same test standards and evaluation standards at the end of the semester. This can ensure the objective test of the teaching effect of the two teaching methods.

The experimental training period is 32 h, 42 people are divided into two groups, one group is the experimental group, 21 people in each group. Before the basketball class begins, the initial physical fitness test is carried out first, then the experimental group uses the basketball teaching mode based on the ITbegin cloud platform to carry out the teaching and training of basketball technology, and the final test and evaluation are carried out at the end of the semester according to the examination standard. One group is the control group, 21 people, also in the basketball class before the beginning of the initial test of physical fitness, and then using traditional teaching methods for basketball technology teaching, in the end of the semester according to the test criteria for the final test and evaluation.

The content of basketball teaching and training used in the experiment is exactly the same. The teaching time, training time, assessment standard and assessment method of the two groups are completely the same. The only variable is the teaching method. Due to the weather, the original 32 class hours are actually 28 class hours, of which 2 class hours are occupied by testing the basic physical fitness of students, and 2 class hours are occupied by the final comparative test.

3.2 Basic Situation of Physical Quality of Students in Control Group and Experimental Group

Basketball project requires athletes to have good speed quality, endurance quality and strength quality. Based on the above three quality requirements, 100 m test, 3000 m long run and solid ball throw are selected as the test indexes of basic physical quality of students in the control group and the experimental group. After 2 h of testing, the test data is shown in the table below.

Table 1. Physical fitness test results of control group

	100 m	1000 m	A solid ball is thrown
Average	14.62 s	4.56 min	10.25 m
Maximum	19.23 s	5.74 min	13.67 m
Minimum	12.45 s	3.57 min	6.41 m
Standard deviation	1.05	0.81	0.96

Table 2. Test results of physical fitness of experimental group

	100 m	1000 m	A solid ball is thrown
Average	14.59 s	4.71 min	10.36
Maximum	18.24 s	5.96 min	14.21
Minimum	13.65 s	3.42 min	6.08
Standard deviation	1.03	0.82	0.98

Compared with the data in Table 1 and Table 2, the standard deviation of 100 m test in the control group is 0.81 for 1.051000 m test and 0.96 for solid ball throw test; the standard deviation of 100 m test in the experimental group is 0.82 for 1.031000 m test and 0.98 for solid ball throw test. The standard deviation between the two groups is not more than 0.02, which shows that the two groups of students have similar physical quality and meet the standard of normal physical quality, which has no significant impact on the subsequent basketball skill test, and can be tested normally.

3.3   The Test and Analysis of Basketball Shooting on the Shoulder with One Hand in Place

The two groups completed basketball teaching and training through different teaching methods to prepare for shooting test.

According to the progress of teaching, the control group adopted the traditional teaching and teaching mode, and the teacher demonstrated the standard action, and explained the action structure and method. The specific action method is: (take the right hand as an example) the specific action method is: the feet are opened in place, the right foot is slightly forward, the body center of gravity falls in the middle of the two feet, the elbows are bent, the wrists are tilted back, the palms are upward, the fingers are naturally opened, the ball is held in the upper right eye, the left hand is on the ball side, the knees are slightly bent, the upper body is relaxed and tilted later, and the eye is on the basket point. When shooting, the lower limbs should be stretched out, the waist should be stretched out according to the potential, the forearm should be raised and extended, the wrist should be bent forward and the fingers should be used to move the ball. Finally, the ball should be thrown out with gentle force through the index finger and middle finger. The right arm should follow up naturally after the ball leaves the hand. After the teacher's explanation, demonstrate the complete action again.

The experimental group, through the ITbegin cloud platform, through watching the teaching video and the teacher's demonstration actions, establishes the overall concept of the in-situ one hand shoulder shooting action, obtains the preliminary impression, divides the learning content into different steps according to the technical action essentials in the video, studies step by step, and gradually improves their basketball skills.

When the two groups completed all the teaching tasks, the final test was carried out.

Measurement and evaluation content: quantitative evaluation of basketball in situ one hand shoulder shooting technique

Index selection: 60 s continuous shooting ability. The unit is number.

Index meaning: Students' skill level and shooting effect of one hand shoulder shooting in situ in 60 s. Data collection method: one teacher clocked with a stopwatch, and the other one counted the number of hits made by the students and recorded them. Test method: when the tester stands under the basket and hears the "start" command (open the watch), he uses the method of one hand shoulder layup in situ to shoot at different points continuously, and calculates the score by the number of hits in 60 s.

Test requirements: the tester shall take the projection point of the ground as the center of the circle at the midpoint of the basket, jump up and shoot in the area beyond the semicircle within 4 m, and shall not be fixed on the same side or at the same point on the same side for repeated shooting, shall not grab the rebound, shall not run with the ball, or the shot will be invalid. Test twice for each person, with the best score.

The test results are shown in Table 3.

Table 3. Comparison test results of in-situ single hand shoulder shooting

Group	Control group	Experimental group
Number of people	21	21
Average	9	15
Maximum	15	21
Minimum	6	9
Standard deviation	1.13	0.46

It can be seen from Table 4 that the average number of shots in the control group is 9, the standard deviation is 1.13, and the average number of shots in the experimental group is 15. The standard deviation is 0.46. From the above two groups of data, the average number of shots in the experimental group is higher than that in the control group, and the standard deviation is lower than that in the control group, indicating that the overall shooting level of the experimental group is higher than that in the control group. It proves that the basketball teaching mode based on ITbegin cloud platform is better than the traditional teaching mode.

3.4 The Test and Analysis of Shooting from One Hand to Shoulder in Basketball Jump

The relevant requirements for the basketball jump up one hand shoulder shooting test are as follows:

Measurement and evaluation content: quantitative evaluation of one hand shoulder shooting technique in basketball jumping.

Index selection: 60 s continuous shooting ability. The unit is number.

Index meaning: Students' skill level and shooting effect of one hand shoulder shooting in situ in 60 s. Data collection method: one teacher clocked with a stopwatch, and the other one counted the number of hits made by the students and recorded them. Test method: the tester stands under the basket, hears the "start" command, uses the way of jumping up the one handed shoulder to lay up the basket, shoots continuously at different points (grabs the rebound by oneself), and calculates the score by the number of hits in 60 s.

Test requirements: the tester shall take the projection point of the ground as the center of the basket circle, jump up and shoot in the area beyond the semicircle within 4 m, and shall not repeat shooting at the same side or at the same point on the same side. After grabbing the rebound, he shall not run with the ball, or the shot will be invalid. Test twice for each person, with the best score.

The test results are as follows (Table 5):

Table 4. Basketball on the spot jump shoulder shooting test results

Group	Control group	Experimental group
Number of people	21	21
Average	8	13
Maximum	13	18
Minimum	5	8
Standard deviation	0.93	0.41

It can be seen from Table 4 that the average number of shots in the control group is 8, the standard deviation is 0.93, the average number of shots in the experimental group is 13, and the standard deviation is 0.41. From the above two groups of data, the average number of shots in the experimental group is higher than that in the control group, and the standard deviation is lower than that in the control group, indicating that the overall shooting level of the experimental group is higher than that in the control group. It proves that the basketball teaching mode based on ITbegin cloud platform is better than the traditional teaching mode.

3.5   Test and Analysis of One Hand Shoulder Shooting in Basketball

The related requirements of the test of one hand shoulder shooting between the two groups are as follows:

Measurement and evaluation content: quantitative evaluation of one hand shoulder shot during basketball March.

Index selection: the speed of half court round-trip dribbling, running and shooting. The unit is time.

Index meaning: the proficiency and ability to finish two round-trip fast running shooting techniques in a row.

Data collection method: the teacher uses the stopwatch to time and take charge of recording, and makes score assignment according to the score table.

Test method: the tester starts from the middle point of the right sideline of the court, and starts to lay up the basketball for the right dribbler at the same time. After the ball hit the basket, he grabbed the rebound and dribbled the ball with his right hand to the midpoint of the left sideline. Then the switch player continues to dribble and lay up the basket. After the throw, he dribbles back to the original starting point with his right hand. Repeat the dribble and lay up again, and then return to the original starting point.

Test requirements: the tester should dribble continuously and not push the dribble far to lay up the basket; the dribble can only be continued after the shot is hit, if the shot is not hit, make up the basket until the hit. Test twice for each person, with the best score.

The test results are as follows:

Table 5. Test results of one hand shoulder shooting during basketball movement

Group	Control group	Experimental group
Number of people	21	21
Average	40 s	35 s
Maximum	45 s	42 s
Minimum	33 s	26 s
Standard deviation	2.23	1.54

It can be seen from Table 4 that the shooting time in the control group is 40 s, the standard deviation is 2.23, the shooting time in the experimental group is 35 s, and the standard deviation is 1.54. From the above two groups of data, the average time of the experimental group is less than that of the control group, and the standard deviation is lower than that of the control group, indicating that the overall shooting level of the experimental group is higher than that of the control group. It proves that the basketball teaching mode based on ITbegin cloud platform is better than the traditional teaching mode.

4   Conclusions

ITbegin cloud platform is widely used in the education industry, but it is rarely used in physical education. Design the basketball teaching mode based on ITbegin cloud platform, solve the problems existing in the traditional basketball teaching mode, solve the problems encountered by students in basketball teaching and training in time, adjust training skills, and improve basketball teaching and training skills. The basketball teaching mode based on ITbegin cloud platform is superior to the traditional basketball teaching mode through the test of one hand on shoulder shooting in place, jumping up one hand on shoulder shooting and one hand on shoulder shooting in the process. The implementation of this method is of great significance to the development of sports in the future.

References

1. Su, P.: Dynamic analysis on specific physical condition of Chinese U15 men's basketball players in the year 2012–2016. J. Shandong Inst. Phys. Educ. Sports (05), 97–102 (2017)
2. Cheng, Y., Tan, A.: Discussion on teaching design of blended learning based on MOOC. Adult Educ. **38**(07), 38–42 (2018)
3. Wang, L., Shen, Z., Huang, J., et al.: Monitoring training and competition load based on player LoadTM and IMA in basketball. J. Guangzhou Sport Univ. (01), 73–76, 128 (2018)
4. Chen, C., Zhou, W.: The construction and application of "MOOC+ flipped classroom" teaching mode for university badminton. J. Phys. Educ. **24**(05), 98–101 (2017)
5. Wang, R., Xiao, L.: The building of Athlete Learning Gateway by the International Olympic Committee and inspirations therefrom. J. Phys. Educ. **25**(06), 36–41 (2018)

6. Lei, M.: An analysis of the current situation of construction of university physical education course informatization in China. J. Phys. Educ. **24**(3), 101–105 (2017)
7. Sun, Y.: Effects of distinct recovery methods on exercise performance and energy metabolism in high intensity interval training of young male basketball athletes. J. Shenyang Sport Univ. **36**(04), 106–112 (2017)
8. Li, N., Ma, X., Zhou, Y.: Research status and comments on technical characteristics of single-handed shoulder shooting. J. Guangzhou Sport Univ. **39**(03), 94–100 (2019)
9. Liu, J.: Research on the application and construction of MOOC platform in colleges and universities basketball courses. J. Nanyang Inst. Technol. **17**(03), 78–81 (2018)
10. Lu, G., Cai, J.: MOOC and teaching reform of basketball curriculum in sports colleges of China. Hubei Sports Sci. **37**(12), 96–98 (2018)

Method of Sports Assistant Teaching Based on Multimedia Network

Ning-ning Zhang[✉]

Liaoning Petrochemical Vocational and Technical College, Jinzhou, China
znn961@163.com

Abstract. In view of the poor teaching effect of the traditional sports teaching method, this paper designs the sports auxiliary teaching method based on the multimedia network. By introducing extended knowledge, using visual method and language method, improving teaching means, making multimedia courseware with flash, combining the teaching content and teaching needs of sports, changing teaching mode, applying advanced teaching facilities, and completing the design of auxiliary teaching method of sports based on multimedia network. The results show that, compared with the traditional teaching methods, the application of multimedia network in physical education teaching, its teaching effect has been significantly improved.

Keywords: Multimedia network · Sports · Auxiliary teaching

1 Introduction

As an important tool of modern information processing, computer has played an important role in many fields, of course, it is also applied to the field of education. With the application of Internet and the design and application of various course software, multimedia technology has entered the daily teaching and learning. The making and application of multimedia courseware has also become one of the important skills that teachers and educators must master [1]. Physical education includes not only the basic knowledge of physical education in the classroom, but also the technical skills of physical exercises in the gym and the combination of classroom and playground. This requires physical education teachers to explain the technical essentials with concise language, demonstrate and demonstrate the process with standard actions, so that students can understand and see clearly, and perceive the technical process from it, otherwise it will directly affect the learning effect of students. In the use of traditional teaching methods, influenced by many factors, to effectively solve these problems in the actual teaching has become the key to the reform of sports teaching. As a modern teaching method, multimedia teaching is a vivid intuitive teaching, which enriches the teaching methods, and its superiority is much greater than the role of intuitive teaching aids. Wu ying [2] proposes the application of autonomous learning based on multimedia network environment in universities Taking students as the center, cultivating students' autonomous learning ability, making students really improve their English application ability and improving the quality of college English teaching, has an important influence on our future English teaching work. Duan Zhenya et al. [3] took

© ICST Institute for Computer Sciences, Social Informatics and Telecommunications Engineering 2020
Published by Springer Nature Switzerland AG 2020. All Rights Reserved
Y.-D. Zhang et al. (Eds.): ICMTEL 2020, LNICST 327, pp. 105–115, 2020.
https://doi.org/10.1007/978-3-030-51103-6_9

the process equipment and control engineering course "process equipment design" as an example to introduce the online classroom construction and practice process based on superstar web teaching platform, and summarize the results of process equipment design in 2014. In 2015, as an evaluation indicator, students analyzed and summarized the effect of using online classroom assisted teaching in order to provide useful help and reference for the online classroom teaching of professional courses and the current classroom teaching reform. With the enrichment of material level, modern teaching methods become simple and easy. Many students have mobile phones, many mobile phones have clear video recording function. Students' actions are recorded in class, and then analyzed by teachers, so that students can clearly see the good and bad of their actions, and correct them. After class, the students can also copy the exercises into the computer, carefully analyze and recall the demonstration actions of the physical education teachers at that time. When the details of the actions and the key areas of the actions cannot be displayed, they can use such means as playing movies, videos, watching TV, computer demonstration, etc. Because movies, videos and televisions can slow down or freeze screen the details and key points of the movements, let students observe the movements carefully, deepen their understanding of the movements, reduce the difficulty of movement learning, and help them to master the movements faster.

The overall framework of the article is:

1. Design the teaching structure, improve the teaching means, make multimedia courseware, summarize the feedback of evaluation after class, and train the ability of multimedia application to realize the assistant teaching of physical education under the multimedia network environment.
2. Early preparation, survey results, the completion of multimedia network-assisted teaching research.
3. Through the experiment contrast, the proposed method has obviously improved the teaching effect.

2 Sports-Aided Teaching Method Based on Multimedia Network

Using multimedia network technology to improve teaching quality, expand teaching scale, promote teaching reform, increase teaching efficiency and stimulate students' interest in learning. The details are as follows (Fig. 1):

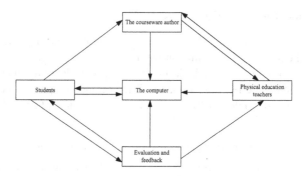

Fig. 1. Application of multimedia-assisted teaching

Sports activities are rich and colorful, and different levels have different actions and different difficulties. It is difficult for teachers to demonstrate every level or set of actions, and the action specifications are not necessarily high, and it is also impossible to remember every sports theoretical knowledge. There are difficulties in the process of demonstration and explanation for students. Multimedia assisted teaching has the advantages of teaching resource sharing, learning resource rich and colorful, rich forms of expression, not limited by time and space, multi-directional interaction and interaction, learning personalization, support for independent learning, creation of equal learning environment, teaching management automation, etc. Make full use of multimedia technology teaching means and can reasonably cooperate with traditional teaching methods to effectively improve the teaching quality and effect of sports.

2.1 Design of Teaching Structure

The overall structure of sports multimedia network teaching is determined by the teaching objectives, teaching contents and the nature of interaction. The essence of sports multimedia network courseware is the organization structure of multimedia information, which reflects the main framework and teaching function of sports multimedia network courseware. The overall structure of courseware is divided into two parts: teaching content and online interaction. Due to the introduction of online teaching methods, a large number of extended knowledge related to the main teaching content of the course can be linked into the teaching content, thus forming a specific teaching resource environment to support the personalized learning of students with different interests and hobbies. Due to the introduction of extended knowledge, the content of courseware is greatly enriched [4]. The content of Sports Multimedia Network Courseware certainly includes theoretical knowledge and practical technical knowledge. The contents of the online interactive part of Sports Multimedia Network Courseware include the overall learning objectives, assessment methods, class arrangement, learning progress and learning methods. Lecture points include teaching tasks, technical action points, difficulties, error prone and correction methods, practice methods, course discussion, teacher's Q&A, course announcement, assignment handling of each project [5].

2.2   Improvement of Teaching Means

Teaching courseware, pictures, videos and audio are combined to deliver to students, which makes the delivered information vivid and intuitive, gives students strong sensory stimulation, stimulates their desire and interest in learning, and makes them more truly experience the teaching situation. The specific comparison between multimedia courseware and single machine courseware is as follows (Table 1):

Table 1. Comparison between multimedia courseware and single machine courseware

	Multimedia courseware	Stand-alone courseware
Environmental conditions	It can be used in network environment It is no need to install and maintain courseware on the client side	The scope of use is limited It is necessary to install and maintain courseware on the client side
Interactivity	In addition to human-computer interaction, students and teachers can interact with each other and interact with others online	Limited to local human-computer interaction
Autonomy	Students have more autonomy and flexibility, and can choose their learning content freely and carry out exploratory learning	Students lack autonomy and flexibility
Resource sharing	Realize the sharing of teaching resources and link the network information resources	Can only use local resources, can not share teaching resources, small amount of information

In the teaching of physical education, visual method and language method are usually used. Because of the relationship between the technical level and age of physical education teachers, the demonstration action is not necessarily standard and standard. If only the language method is used for explanation, the effect may not be ideal. If we need to fully reflect the initiative and enthusiasm of students, and then give full play to the main role of students, we can solve these problems by using multimedia technology to repeat, change the playing speed, freeze frame playing sports courseware and so on. With the development and progress of information technology, physical education should change the teaching mode which is teacher centered, textbook centered and classroom centered, to the teaching mode which is teacher led and student-centered, and use modern education theory and information technology, so as to realize the theory and practice of teaching optimization [6].

2.3   Multimedia Courseware Production

In order to diversify the teaching content of physical education, the application of multimedia network courseware is introduced. In the process of making network courseware, we need to consider the writing and design of text, graphics, pictures, sound, animation, video and other materials, as well as the link between these different materials. Script includes text script, sound script, graphic design, animation design, video design and function design. The function design of Sports Multimedia Network Courseware includes the selection of interface and level, the determination of selection button and function button, the display mode of course content, the link mode of different types of materials, the design content of navigation mode, the determination of file structure of courseware, etc. The purpose is to use multimedia network means to complete the auxiliary teaching of specific content as much as possible. Generally, according to the requirements of the overall structure, the interface is designed as a three-level structure, namely, the main interface, content selection interface and content explanation interface [7]. The home page divides the content selection button into two groups, that is, the teaching content group and the online interaction group. In order to reduce the number of interface switching and improve the running speed of course-ware, not only the selection buttons of each section but also the switching buttons of each chapter are set in the content selection interface. Around a certain teaching content, there are many kinds of teaching methods, such as text introduction, graphics, pictures and video clips, animation explanation. In addition, you can also provide other hypertext link buttons, such as "enjoy" links with other websites. The function buttons of the interface comprehensively consider the various needs of students, and appro-priately increase the dynamic effect and interest of the buttons. Before making courseware collectively, we should collect, arrange and edit all kinds of materials used in the courseware, and try to avoid grasping materials temporarily in the process of courseware making. Material preparation includes two processes, one is the selection of original material, the other is the processing of material. The acquisition of original materials can be realized by image acquisition through scanner, recording and video recording by software, and can also be obtained from the Internet and various material discs. Many original materials can not be directly used in the courseware. They are modified according to the requirements of sports courseware or the purpose of teaching, which is mainly realized by media editing software. In order to improve the efficiency of courseware making, we should make use of the available material resources as much as possible [8].

In the script of courseware, according to the teaching design, the process of making courseware should be expressed in written form with words and figures in advance to guide courseware makers to collect materials and integrate courseware. The textbook is the blueprint of courseware development. A deep understanding of the knowledge structure and content system of the textbook is conducive to the compilation of courseware manuscript. Flash mx 2004 software is used to make the animation file with the extension of SWF. The human body is divided into head, trunk, left upper arm, left small arm, right upper arm, right small arm, left thigh, left calf, right thigh, right calf and other parts, and these graphic elements are established. Because in the middle of doing the movement, we need to turn the body and see different sides of the body, so we need to do the head, trunk, calf and other components. That is to say, according to the needs of the whole action, components are built to make the animation vivid. Build a complete "human". According to our action needs, set the shoulder joint, elbow joint, skeleton joint and knee joint of this person to be movable when making graphic components. Its sports interface is shown in the following figure (Fig. 2):

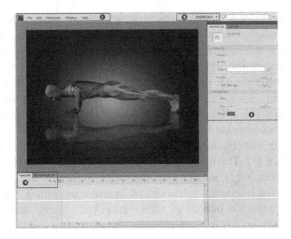

Fig. 2. Flash sports interface

Combined with the teaching content and teaching needs of sports, the network courseware of sports can be divided into teaching network courseware, testing network courseware and simulation network courseware. Teaching network courseware focuses on the teaching of sports technology and theoretical knowledge, which can provide students with self-learning. This kind of courseware simulates the steps and procedures of traditional classroom teaching or autonomous learning, usually provides the contents of teaching materials to students, teaches them various concepts and skills, and arranges exercises and assignments. Generally speaking, there should be action demonstration, explanation, action essentials, teaching methods, students' error prone and correction methods, theoretical knowledge, etc. The test-based network courseware focuses on providing repeated practice opportunities for the theoretical knowledge of sports, or evaluating the teaching effect after the completion of teaching activities. This kind of courseware should be the earliest and most extensive one. However, the simulation courseware is relatively complex and not widely used in sports, which needs the design and completion of computer professionals [9].

As a teacher in the 21st century, he should master the production technology of electronic teaching plan which integrates various media and technologies. Based on the sports multimedia network courseware, it can solve the problem that the time of theoretical explanation is tight and not systematic in the sports practice class, and the teaching content of sports can be completely and systematically taught on the Internet. In sports teaching, we should make full use of network courseware and electronic teaching plan, and use them reasonably with traditional teaching methods. We can use the courseware to explain the action essentials vividly, so as to achieve the purpose of assisting extracurricular exercise. We can use the conditions provided by the network to guide interested students to discuss and answer questions.

2.4   After-Class Evaluation Feedback

The end of classroom teaching does not mean the end of teaching activities. In the teaching process, teaching evaluation and feedback is also an essential link. The implementation of this link enables teachers and students to recognize their own shortcomings from different aspects, and to correct and improve in the subsequent teaching activities. Teachers use the link function of wechat group and QQ group to convey the learning situation of students to students in time after each PE class. After class, students should make full use of the broadcast function of mobile phone, computer or iPad and other media tools to watch carefully the sports action video displayed by themselves in class. Students can compare their own actions, which places are easy to forget, whether the steps are correct, whether the arm movements are coordinated, how the amplitude and strength of the actions are, what are the short-comings of their peers, how to avoid the same mistakes in the future study, and whether they can imitate the teaching video to further improve the details of the actions after mastering the correct actions. In view of these self-awareness and learning problems, teachers can regularly organize students to conduct self-evaluation and mutual evalu-ation in wechat group [10]. Finally, the teacher should correct the wrong actions of the comments, fully affirm the learning advantages of the students and praise the students who have made great progress in completing the actions, and privately communicate with the students who have poor mastery of the technical actions or poor learning attitude.

2.5   Multimedia Application Ability Training

Change the teaching concept of physical education teachers, keep learning advanced teaching methods and means, give full play to the role of multimedia network teaching, and realize the modernization of physical education. We should pay attention to the training of modern educational technology for sports teachers, give more opportunities for teachers to continue learning about computer application technology and hold modern teaching competitions regularly. To increase teachers' motivation of learning computer and promote the improvement of information quality of physical education teachers. Strengthen the communication among provinces, regions and schools, learn from each other, reduce or avoid low-level repeated practice, improve the starting point of practice, broaden the vision of making courseware; encourage the communication and cooperation of teachers in different disciplines to make high-quality teaching courseware. The teaching decision-making departments and schools at all levels should pay enough attention to the practical teaching of physical education and sports, have a long-term development perspective, invest a certain amount of education funds in the development of multimedia assisted teaching, equip with high-level hardware equip-ment, software materials, and reasonably arrange the use of existing electric class-rooms. If possible, improve the multimedia equipment in the venue to facilitate teaching or training. In addition, as a teacher, we should not only constantly study and innovate teaching methods, but also use heuristic teaching to let students participate in and play their main role. Students and teachers should pool their ideas and ideas and integrate into teaching together to create a harmonious environment and sufficient conditions for teaching. Teachers should not use multimedia in order to reflect the so-called modern teaching, but should not only retain the advantages of traditional teaching, but also play the advantages of multimedia teaching. Teachers have accumulated a lot of valuable teaching experience in long-term teaching activities. On the basis of traditional teaching experience, with the help of multimedia assistance, we should grasp the problem of "degree" and "quantity" between traditional teaching and multimedia assisted teaching, so as to make teaching more effective with less effort.

3 Investigation on Multimedia Network Assisted Instruction

In order to verify the effectiveness of the multimedia network-based teaching method, a questionnaire was designed.

3.1 Early Preparation

According to the stratified random sampling method of social survey, 20 schools are randomly selected from the first tier cities, which basically represent different economic level areas and various types of schools to send out questionnaires and collect data. Before the questionnaire design, according to the content and purpose of this study, we consulted many books about social investigation, which made full theoretical preparation for the questionnaire design of this paper. Before the formal investigation, after several modifications, 50 experts of scientific research in the field of were invited physical education and sports teaching management to carry out the consultation and test of the validity of the questionnaire. The validity of the questionnaire meets the requirements of this study. The details are as follows (Table 2):

Table 2. Validity test results of questionnaire

Inspection result	Overall design evaluation		Overall content evaluation		Overall evaluation of structure	
	Number of people	%	Number of people	%	Number of people	%
Very appropriate	48	96	47	94	48	96
More appropriate	2	4	3	6	2	4
Ordinary	0	0	0	0	0	0
Inappropriate	0	0	0	0	0	0
Very inappropriate	0	0	0	0	0	0

The data in the above table can show that the questionnaire has good reliability, stability and high consistency, and the reliability meets the requirements of this study. The specific procedures of this questionnaire are as follows ① According to the purpose of the study, the type, factors and level of the study were determined; ② Determine data characteristics and sample size; ③ The corresponding type of statistical data should be judged correctly, and the correct statistical value should be calculated according to the appropriate conditions of statistical method.

The questionnaire is mainly distributed and collected face to face by myself and the entrusted teacher or students. After the questionnaire is recovered, the questionnaire is eliminated, and the distribution and recovery are as follows (Table 3):

Table 3. Statistical table on the distribution and recovery of questionnaires

Questionnaire category	Teacher questionnaire	Student questionnaire
Distribute questionnaires	100	660
Collect the questionnaire	96	619
Valid questionnaires	96	600
Recycled probability	96%	93.8%
RETEST reliability	100%	96.9%

As shown in the above table, a total of 760 questionnaires were distributed to 20 randomly selected colleges and universities, of which 100 were distributed to teachers, 96 were recovered, the recovery rate was 96%; 96 were effective questionnaires, the efficiency rate was 100%; 660 were distributed to college students, 619 were recovered, the recovery rate was 93.8%. The Retest reliability was 96.9%.

3.2 Findings of Investigation

Through the above preparation process, complete the design, distribution, recovery and statistics of the questionnaire, the results are as follows (Table 4):

Table 4. Statistical table of teachers' attitude towards multimedia network assisted instruction in physical education teaching course

Teacher's attitude	Number of times	%
Very supportive	18	18.7
More support	24	25
Basic support	34	35.4
Grudging support	18	18.7
Nonsupport	2	2.1

The survey shows that 18.7% of PE teachers are very supportive of using multimedia to assist teaching in PE practice class, 25% of them are quite supportive, 35.4% of them are basic supportive, 18.7% of them are not very supportive, only 2.1% of them are not. From the data, we can draw the conclusion that most of the physical education teachers think it is necessary to use multimedia to assist teaching in physical education. They can give full play to the advantages of multimedia and make up for the shortcomings of traditional physical education teaching. Only a few physical education teachers who have fallen behind in the concept of education think it is unnecessary and do not support the use of multimedia assisted teaching (Table 5).

Table 5. Statistical table of students' attitude towards multimedia network assisted instruction in physical education teaching course

Student attitude	Number of times	%
Very supportive	201	33.5
More support	261	43.5
Basic support	57	9.5
Grudging support	74	12.3
Nonsupport	7	1.2

Sports multimedia assisted teaching is an irresistible trend of social development. The survey results show that 43.5% of the students are very supportive of the use of Multimedia Assisted Teaching in sports teaching, 43.5% of the students are relatively supportive, and only a few of the students are not. The above data shows that most of the students have a very positive attitude towards the use of Multimedia Assisted Teaching in physical education.

In order to better verify the effectiveness of the application of multimedia network in sports auxiliary teaching, a questionnaire survey was conducted on its teaching advantages. The results are as follows (Table 6):

Table 6. Statistical table of advantages of multimedia assisted instruction

Advantages of multimedia assisted teaching	Students		Teachers	
	Number of times	%	Number of times	%
Make the demonstration more accurate, intuitive and visual	313	52.2	70	72.9
It stimulates students' interest in learning and improves their enthusiasm for learning	400	66.7	72	75
Resource sharing is achieved	224	37.3	14	14.6
Create a classroom atmosphere	420	70	54	56.3

The results show that multimedia assisted instruction has advantages over traditional teaching methods in Teachers' demonstration, students' interest and classroom atmosphere. Multimedia assisted instruction makes teachers' demonstration more accurate, intuitive and vivid, and enables students to form action image more accurately and quickly in teaching, which is conducive to action learning. Compared with the traditional teaching method, it stimulates the students' interest in learning and improves their enthusiasm for learning. Teachers and students believe that multimedia assisted teaching has the advantages of resource sharing and knowledge broadening. These advantages of multimedia assisted teaching are all impossible in the traditional teaching method, and the exertion of these advantages can increase the teaching effect and improve the teaching quality.

4 Conclusions

Multimedia network application is a new interdisciplinary subject formed by computer science, pedagogy and so on. It has its own gradually formed theory and application category, and has its own advantages. It is feasible to develop multimedia network software and make full use of multimedia network to assist sports teaching, which can improve the effect of sports teaching. Due to the limited time, although this paper proposes a multimedia teaching method of physical education course, it lacks the verification of practical teaching based on theoretical suggestions, which is also my future research direction.

References

1. Li, J., Wu, X.: Research on business English teaching mode based on multimedia network assisted teaching. J. Gannan Med. Univ. **37**(05), 731–734 (2017)
2. Wu, Y.: Based on the network multimedia college English teaching of middle school students' autonomous learning ability. J. Shandong Agric. Adm. Coll. **34**(4), 66–67 (2017)
3. Duan, Z., Zhang, M., Guo, J., et al.: Construction and Application of online open course of 《process equipment design》 based on superstar network teaching platform. Shandong Chem. Ind. **47**(22), 146–147+149 (2018)
4. Zhang, D., Zhu, X.: Experimental teaching of computer software courses based on SPOC and multimedia network classroom. Exp. Technol. Manag. **34**(08), 195–198+205 (2017)
5. Song, C.: On web-based research-oriented teaching mode in listening course of English majors. J. Hubei Corresp. Univ. **30**(05), 137–138+163 (2017)
6. Wang, J.: On the application of multimedia assisted teaching in college teaching. Zhiye Jishu **16**(06), 42+53 (2017)
7. Zhang, W.: Research on the construction of network assisted teaching platform of college physical education. Microcomput. Appl. **34**(12), 111–114 (2018)
8. Xie, C.: Design of computer-aided dance teaching resource management system. Mod. Electron. Tech. **41**(16), 100–103 (2018)
9. Jie, D.E.N.G.: The application of interactive teaching method in English translation teaching—illustrated by online peer revision. J. Heilongjiang Coll. Educ. **37**(11), 142–144 (2018)
10. Hu, J.: The effect of different proportions of multimedia-assisted teaching time on learning effectiveness of students' physical education—taking pilates course as an example. J. Anhui Norm. Univ. (Nat. Sci.) **41**(06), 609–612 (2018)

Improving Accuracy of Mobile Robot Localization by Tightly Fusing LiDAR and DR data

Yuan Xu[1][✉][iD], Yuriy S. Shmaliy[2][iD], Tao Shen[1], Shuhui Bi[1][iD],
and Hang Guo[3][iD]

[1] School of Electrical Engineering, University of Jinan, Jinan 250022, China
xy_abric@126.com, {cse_st,cse_bish}@ujn.edu.cn
[2] Department of Electronics Engineering, Universidad de Guanajuato,
36885 Salamanca, Mexico
shmaliy@ugto.mx
[3] Institute of Space Science and Technology, Nanchang University,
Nanchang 330031, China
hguo@ncu.edu.cn

Abstract. In this paper, a tightly-coupled light detection and ranging (LiDAR)/dead reckoning (DR) navigation system with uncertain sampling time is designed for mobile robot localization. The Kalman filter (KF) is used as the main data fusion filter, where the state vector is composed of the position error, velocity error, yaw, and sampling time. The observation is provided of the difference between the LiDAR-derived and DR-derived distances measured from the corner feature points (CFPs) to the mobile robot. A real test experiment has been conducted to verify a good performance of the proposed method and show that it allows for a higher accuracy compared to the traditional LiDAR/DR integration.

Keywords: Light detection and ranging (LiDAR) · Dead Reckoning (DR) · Tightly integration · Uncertain sampling period

1 Introduction

Nowadays, mobile robots are widely exploited in indoor closed spaces Borenstein (2007). In view of the latter and as a basic issue for a robot to accomplish the task, the robot accurate localization has become a hot research topic Jiang (2015).

Many localization techniques were developed with time for indoor localization. For example, an effective weighted path length and support vector regression algorithm for radio frequency identification (RFID) is introduced in He (2018). An bluetooth-based accurate method is proposed for mobile robots and tested in Raghavan (2010). A WiFi-based solution is present to solve the localization problem for robots operating in unknown environments Benjamin (2015).

Y.-D. Zhang et al. (Eds.): ICMTEL 2020, LNICST 327, pp. 116–123, 2020.
https://doi.org/10.1007/978-3-030-51103-6_10

Although the received signal strength indication (RSSI)-based technology can also be used for indoor localization, it commonly provides an insufficient accuracy. In order to improve the localization accuracy, the ultra wide band (UWB) technology is developed for indoor needs Fan (2017). For instance, the UWB-based indoor robot localization system is designed Xu (2018). But, even though the UWB-based approach is able to improve the localization accuracy, it requires a pre-arranged additional equipment. In order to avoid an extra equipment, the beacon-free positioning approaches have been proposed. The famous example is an inertial navigation system (INS) Cai (2018). However, the INS-based schemes are not suitable for long-time operation due to induced time-drifts Wu (2019), Cao (2019). In this regard, the light detection and ranging (LiDAR) approach allows getting a higher positioning resolution in indoor environments Chang (2019), Akshay (2019) that is an obvious advantage.

In this paper, an tightly-coupled light detection and ranging (LiDAR)/dead reckoning (DR) navigation system is designed for mobile robot localization with uncertain sampling time. In this model, the Kalman filter (KF) is used as the main data fusion estimator. The state vector is composed of the position error, velocity error, yaw, and the sampling time. Observation is provided of the difference between the LiDAR-derived and DR-derived distances measured from the corner feature points (CFPs) to the mobile robot. A real test experiment is conducted to verify the performance and show that the system designed has a higher accuracy compared to the traditional LiDAR/DR integration schemes.

2   Tightly LiDAR/DR-Integrated Robot Localization Model

A block diagram of the tightly LiDAR/DR-integrated indoor mobile robot localization system designed is this paper is given in Fig. 1. In this scheme, the LiDAR provides measurements of the distance $\rho_i^L, i \in [1, n]$ between the mobile robot and the corner feature points (CFPs), where n is the numbers of the CFPs. The DR measures distances $\rho_i^D, i \in [1, n]$ between the mobile robot and the CFPs. The differences $\delta\rho^i = \rho_i^L - \rho_i^D, i \in [1, n]$, between the measured distances are used as measurements for the EKF. The vector output $\delta\mathbf{Po^D}$ of the EKF is used to compensate the vector drift $\mathbf{Po^D}$ in the DR.

Referring to Fig. 1, the state equation is written as

$$
\underbrace{\begin{bmatrix} \delta Pe_t \\ \delta Pn_t \\ \delta V_t \\ \varphi_t \\ \Delta T_t \end{bmatrix}}_{\mathbf{X}_{t|t-1}} = \underbrace{\begin{bmatrix} \delta Pe_{t-1} + \Delta T_{t-1}\delta V_{t-1}\sin(\varphi_{t-1}) \\ \delta Pn_{t-1} + \Delta T_{t-1}\delta V_{tt-1}\cos(\varphi_{t-1}) \\ \delta V_{t-1} \\ \varphi_{t-1} \\ \Delta T_{t-1} \end{bmatrix}}_{a(\mathbf{X}_{t-1})} + \omega_{t-1}, \tag{1}
$$

where t is the discrete time index, $(\delta Pe_{tt}, \delta Pn_{tt})$ is the DR-based position error, δV_{tt} is the DR-based velocity error, ϕ_{tt} is the mobile robot's yaw, the sampling time ΔT_t is considered as one of the states, and $\omega_t \sim \mathcal{N}(0, \mathbf{Q})$ is the system zero mean white Gaussian noise with the covariance $\mathbf{Q}$.

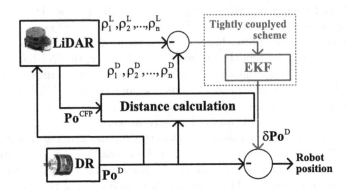

Fig. 1. Block diagram of the tightly LiDAR/DR-integrated indoor mobile robot localization system.

Based on the state equation, the distance ρ_i^D between the ith CFP and the mobile robot can be written as

$$\left(\rho_i^D\right)^2 = \left(Pe^D - Pe_i\right)^2 + \left(Pn^D - Pn_i\right)^2, i \in [1, n], \tag{2}$$

where (Pe_i, Pn_i) is the position of the ith CFP and $\left(Pe^D, Pn^D\right)$ is the DR-based position. A distance d_i^R between the ith CFP and the mobile robot can thus be specified by

$$\left(\rho_i^R\right)^2 = \left(Pe^R - Pe_i\right)^2 + \left(Pn^R - Pn_i\right)^2, i \in [1, n], \tag{3}$$

where $\left(Pe^R, Pn^R\right)$ is the real robot position. Because of $Pe^R = Pe^D - \delta Pe$ and $Pn^R = Pn^D - \delta Pn$, we have

$$\begin{aligned}&\left(\rho_i^D\right)^2 - \left(\rho_i^R\right)^2 \\&= \left(Pe^D + Pe^R - 2Pe_i\right)\delta Pe + \left(Pn^D + Pn^R - 2Pn_i\right)\delta Pn.\end{aligned} \tag{4}$$

Moreover, ΔT_t is used as one the elements of the observation vector that, referring to the above, is written as

$$
\underbrace{\begin{bmatrix} \delta\rho_{1,t} \\ \delta\rho_{2,t} \\ \vdots \\ \delta\rho_{g,t} \\ \Delta T_n \end{bmatrix}}_{y_n} = \begin{bmatrix} \left(\rho_1^D\right)^2 - \left(\rho_1^L\right)^2 \\ \left(\rho_2^D\right)^2 - \left(\rho_2^L\right)^2 \\ \vdots \\ \left(\rho_g^D\right)^2 - \left(\rho_n^L\right)^2 \\ \Delta T_n \end{bmatrix}
$$

$$
= \underbrace{\begin{bmatrix} 2\left(Pe^D - Pe_1\right)\delta Pe + 2\left(Pn^D - Pn_1\right)\delta y - \left(\delta Pe^2 + \delta Pn^2\right) \\ 2\left(Pe^D - Pe_2\right)\delta Pe + 2\left(Pn^D - Pn_2\right)\delta y - \left(\delta Pe^2 + \delta Pn^2\right) \\ \vdots \\ 2\left(Pe^D - Pe_n\right)\delta Pe + 2\left(Pn^D - Pn_n\right)\delta y - \left(\delta Pe^2 + \delta Pn^2\right) \\ \Delta T_t \end{bmatrix}}_{h(\mathbf{X}_{t|t-1})} + \nu_t,
$$

$$(5)$$

where $\nu_t \sim N(0, \mathbf{R})$ is the zero mean white Gaussian measurement noise with the covariance $\mathbf{R}$.

3   Experiment

In this section, a real test measurement is conducted to verify the performance of the localization system designed. First, we will present the design of the LiDAR/DR integrated integration system. Then, the performance of the proposed method will be discussed.

Experimental investigations have been provided in the No. 1 teaching building of the University of Jinan, China. In this test, we employ one LiDAR, one DR, one mobile, and one computer. The LiDAR, DR, and a computer are maintained on a mobile robot. The LiDAR measures ranges $\rho_i^L, i \in [1, n]$, from the robot to the CFPs. The DR measures the velocity of the mobile robot and provides ranges $\rho_i^D, i \in [1, n]$ from the robot to the CFPs. A computer is used to collect the LiDAR- and DR-based ranges via the RS232. In this work, we set $\Delta T_t = 0.75\,\text{s}$, which is also included to the observation vector listed in Eq. (5). Therefore, we suppose that ΔT_t is uncertain as it often is in digital systems and model it with

$$\Delta T_{tt} = 0.75 + \delta T, \tag{6}$$

where $\delta T \sim (0, 0.25)$ represents an uncertainty in the sampling time.

In Fig. 2 we sketch the reference path and the paths derived from DR and LiDAR. The reference trajectory and the ones derived by the LiDAR, LiDAR/DR tightly integration, and LiDAR/DR tightly integration with uncertain sampling period are shown in Fig. 3 and the position errors in east and north directions are sketched in Fig. 4. From these figures, one can deduce that the path estimated by the LiDAR/DR tightly integration has a large error due

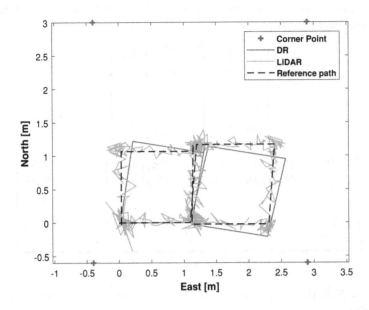

Fig. 2. The reference path and the paths derived from the DR and LiDAR.

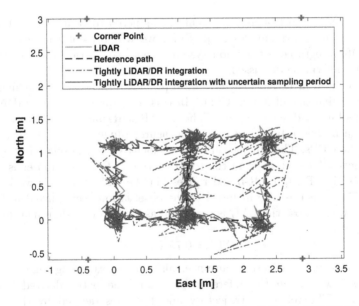

Fig. 3. The reference trajectory and ones derived by the LiDAR, LiDAR/DR tightly integration, and LiDAR/DR tightly integration schemes with uncertain sampling time.

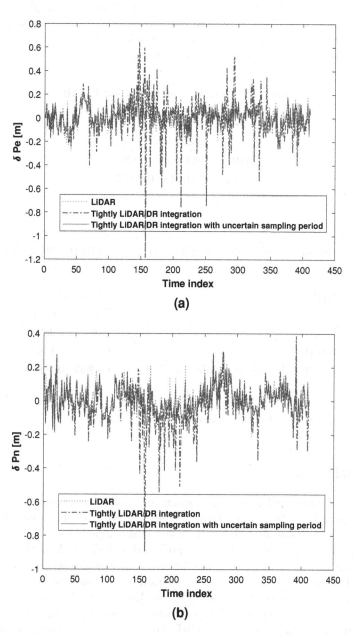

Fig. 4. Position errors estimated by the LiDAR, LiDAR/DR integration, and LiDAR/DR integration with uncertain sampling time: (a) east direction (b) north direction.

to the uncertain sampling time. Compared to the traditional tightly integration, the proposed LiDAR/DR tightly integration with uncertain sampling time has a stable performance. The root mean square error (RMSE) produced by the DR and LiDAR in the east and north directions are listed in Table 1. For the uncertain sampling time, the RMSEs are listed in Table 2.

Table 1. RMSE produced by DR and LiDAR

Method	RMSE (m)	
	East	North
DR	0.1186	0.1128
LiDAR	**0.0917**	**0.0896**

Table 2. RMSE produced nu LiDAR, LiDAR/DR, and LiDAR/DR with uncertain sampling time

Method	RMSE (m)	
	East	North
LiDAR	0.0917	0.0896
LiDAR/DR	0.1793	0.1290
LiDAR/DR with uncertain sampling time	**0.0832**	**0.0850**

Observing these tables, one concludes that the proposed localization scheme has the highest accuracy.

4   Conclusion

The tightly LiDAR/DR-integrated system with uncertain sampling time has been designed for indoor robot localization. The position error, velocity error, yaw, and uncertain sampling time were selected as components of the state vector of the tightly integrated model. The proposed model employs the difference between the LiDAR- and DR-derived distances measured from the CFPs to the mobile robot and combined in the observation vector. Real test measurements conducted in No. 1 teaching building of the University of Jinan, China, have confirmed a highest accuracy of the solution proposed among other available schemes, including the traditional LiDAR/DR integration. An overall conclusion that can be made is that the performance of the proposed tightly model is better than that demonstrated by traditional methods.

Acknowledgment. This work was supported by the Shandong Key R&D Program under Grants 2019GGXI04026 and 2019GNC106093.

References

Jiang, W., Li, Y., Rizos, C.: Precise indoor positioning and attitude determination using terrestrial ranging signals. J. Navig. **68**(2), 274–290 (2015)

He, X., Wu, M., Li, P., et al.: An RFID indoor positioning algorithm based on support vector regression. Sensors **18**(5), 1504 (2018)

Raghavan, A.N., Ananthapadmanaban, H., Sivamurugan, M.S., et al.: Accurate mobile robot localization in indoor environments using bluetooth. In: IEEE International Conference on Robotics and Automation (2010)

Benjamin, B., Erinc, G., Carpin, S.: Real-time WiFi localization of heterogeneous robot teams using an online random forest. Auton. Robots **39**(2), 155–167 (2015). https://doi.org/10.1007/s10514-015-9432-5

Fan, Q., Sun, B., Sun, Y., et al.: Data fusion for indoor mobile robot positioning based on tightly coupled INS/UWB. J. Navig. **70**(5), 1079–1097 (2017)

Xu, Y., Shmaliy, Y.S., Ahn, C.K., et al.: Robust and accurate UWB-based indoor robot localisation using integrated EKF/EFIR filtering. IET Radar Sonar Navig. **12**(7), 750–756 (2018)

Cai, X., Hsu, H., Chai, H., et al.: Multi-antenna GNSS and INS integrated position and attitude determination without base station for land vehicles. J. Navig. **72**(2), 342–358 (2019)

Cao, H., Zhang, Y., Han, Z., et al.: Pole-zero-temperature compensation circuit design and experiment for dual-mass MEMS gyroscope bandwidth expansion. IEEE/ASME Trans. Mechatron. **24**(2), 677–688 (2019)

Wu, Z., Wang, W.: INS/magnetometer integrated positioning based on neural network for bridging long-time GPS outages. GPS Solut. **23**(3), 1–11 (2019). https://doi.org/10.1007/s10291-019-0877-4

Akshay, S., Gao, G.X.: Adaptive covariance estimation of LiDAR-based positioning errors for UAVs. Navig.-J. Inst. Navig. **66**(2), 463–476 (2019)

Chang, L., Niu, X., Liu, T., et al.: GNSS/INS/LiDAR-SLAM integrated navigation system based on graph optimization. Remote Sens. **11**(9), 1009 (2019)

Borenstein, L., Johann, O.: Non-GPS navigation for security personnel and first responders. J. Navig. **60**(3), 391–407 (2007)

Xu, Y., Shmaliy, Y.S., Li, Y., Chen, X., Guo, H.: Indoor INS/LiDAR-based robot localization with improved robustness using cascaded FIR filter. IEEE Access **7**(1), 34189–34197 (2019)

Design of Human-Computer Interactive Fire Extinguishing Training System Based on Virtual Reality Technology

Xue-yong Cui$^{(\boxtimes)}$ and Jun-qin Diao

Anhui Xinhua University Academy of Arts, Hefei 230601, China
ht69250@163.com

Abstract. In order to improve the ability of human-computer interactive fire extinguishing training, a visual simulation model of man-machine interactive fire extinguishing training based on virtual reality technology is proposed. Taking the large-scale emergency scene fire drill as the research object, the virtual scene image reconstruction model of human-computer interaction fire fighting training is constructed, the boundary feature detection and particle tracking filter processing are carried out for the scene image of human-computer interaction fire fighting training, the dynamic structure of the scene image of human-computer interaction fire fighting training is reorganized with fuzzy edge feature extraction method, and the virtual reality simulation model of human-computer interactive fire fighting training scene image is established. The multi-person cooperative control method is used to simulate the virtual reality in the process of human-computer interactive fire fighting training, and the 3D simulation image of virtual scene fire drill in large-scale emergency scene is followed and rendered, and the optimization design of virtual reality VR simulation model of human-computer interactive fire fighting training is realized. The test results show that the virtual scene simulation of human-computer interaction fire fighting training using this method has good cooperation, high image fusion performance and strong reconstruction ability of fire fighting training scene simulation.

Keywords: Virtual reality technology · Human-computer interaction · Fire fighting · Training system

1 Introduction

With the development of the virtual reality technology, the virtual scene reconstruction is carried out in combination with the image processing technology, the virtual simulation method is adopted, the model of the environment system is built, and the cost of the drill is reduced while the cost of the drill is reduced, and the efficiency of the training exercise is improved. The three-dimensional solid model design is carried out by adopting the image information processing technology [1], a human-computer interaction fire-fighting training system is established in a virtual scene simulation environment, a three-dimensional visual analysis method is adopted, the human-computer interaction fire-fighting training optimization design is carried out, The model of

Y.-D. Zhang et al. (Eds.): ICMTEL 2020, LNICST 327, pp. 124–136, 2020.
https://doi.org/10.1007/978-3-030-51103-6_11

human-computer interaction fire-fighting training system based on virtual reality is studied, and it is of great significance in the optimization of emergency drill training [2].

A virtual scene three-dimensional simulation model of a large-scale emergency scene fire drill training is constructed, a three-dimensional visual reconstruction technology is adopted, and the optimization structure of a large-scale emergency scene fire drill training system is carried out [3]. The invention adopts the digital information processing technology to carry out a large-scale emergency scene fire drill training virtual scene three-dimensional simulation design, and combines the VR framework of the virtual reality to carry out a large-scale emergency scene fire drill training virtual scene three-dimensional simulation design, Three DStudio MAX and Multigen Creator are adopted for the design of the virtual scene model of the emergency drill of the main transformer of the fire main transformer [4]. The virtual scene reconstruction of the emergency drill of the main transformer of the fire is carried out by using the texture mapping technology, and the better visual reconstruction efficiency is achieved. But the above-mentioned method is not good in human-computer interaction and fire-fighting training, and the human-computer interaction capability is not strong [5]. In view of the above problems, this paper presents a visual simulation model of human-computer interaction fire-fighting training based on virtual reality technology. taking a large-scale emergency scene fire drill as a research object, building a virtual scene image reconstruction model for human-computer interaction fire-fighting training, and performing boundary feature detection and particle tracking filtering processing on a visual image of the human-computer interaction fire-extinguishing training, The multi-person cooperative control method is adopted to carry out the virtual reality simulation in the human-computer interaction fire-fighting training process, and the tracking and rendering of a large-scale emergency scene fire drill virtual scene three-dimensional simulation image is carried out, and the optimization design of the virtual reality VR simulation model for human-computer interaction fire-fighting training is realized. Finally, the simulation experiment is carried out to show the superiority of the method in improving the virtual reality simulation and visual simulation ability of man-machine interactive fire-fighting training [6].

2   Virtual Scene Model and Feature Extraction of Man-Machine Interactive Fire-Fighting Training

2.1   Virtual Scene Model of Man-Machine Interactive Fire-Fighting Training

The three-dimensional simulation software such as MAYA, 3DStudio MAX, SoftImage and LightWave 3D is used to design the solid model of the virtual reality emergency training model. the three-dimensional visual simulation and the characteristic reconstruction of the human-computer interaction fire-extinguishing training are carried out, the AMCC is the main control chip, the overall design framework of the human-computer interaction fire-fighting training system is carried out [7], the SPI interface is constructed, and the peripheral equipment of the virtual reality simulation model for human-computer interaction and fire-fighting training is constructed. In SD mode, the bus control is carried out to obtain the modular design of the virtual reality simulation model, and the general structure of the virtual reality system with human-computer interaction and fire-fighting training is shown in Fig. 1.

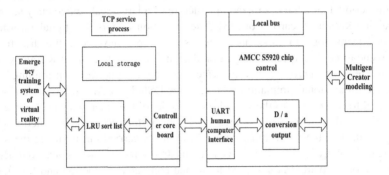

Fig. 1. Overall structure of virtual reality system for human-computer interactive fire fighting training

The Haar-like rectangular feature template matching method is used to segment the virtual visual image block of human-computer interactive fire extinguishing training. Combined with the dynamic weight registration method, the scene image reconstruction of human-computer interactive fire extinguishing training is carried out, the detection ability of scene reconstruction of human-computer interactive fire extinguishing training is improved, the overall framework of human-computer interactive fire extinguishing training system is carried out in square subregions, and the static visual model is established [8]. The distributed reconstruction model of human-computer interaction fire extinguishing training is obtained. In the Jakobi matrix $J(x, y, \sigma)$, the imaging area model of human-computer interaction fire extinguishing training can be expressed as follows:

$$J(x, y, \sigma) = \begin{pmatrix} \dfrac{\partial P}{\partial x} \\ \dfrac{\partial P}{\partial y} \end{pmatrix} = \begin{pmatrix} 1 & 0 & L_x(x, y, \sigma) \\ 0 & 1 & L_y(x, y, \sigma) \end{pmatrix} \tag{1}$$

Wherein, $L(x, y, \sigma) = G(x, y, \sigma) * I(x, y)$, $G(x, y, \sigma)$ is the seed point of the visual image of the human-computer fire extinguishing training, and $I(x, y)$ is the gray scale feature of the fusion of the visual image area of the human-computer fire extinguishing training, $\vec{N}$ represents matrix vector, ∂ is the adjustment coefficient. According to the gray scale histogram model [9], the visual reconstruction of the human-computer interactive fire extinguishing training is carried out, and the square template matching matrix is expressed as follows:

$$M = \begin{pmatrix} \frac{\partial^2 P}{\partial x^2} \vec{N} & \frac{\partial^2 P}{\partial x \partial y} \vec{N} \\ \frac{\partial^2 P}{\partial x \partial y} \vec{N} & \frac{\partial^2 P}{\partial y^2} \vec{N} \end{pmatrix} = \begin{pmatrix} (0, 0, L_{xx}(x, y, \sigma)) \cdot \vec{N} & (0, 0, L_{xy}(x, y, \sigma)) \cdot \vec{N} \\ \left(0, 0, L_{xy}(x, y, \sigma) \cdot \vec{N}\right) & \left(0, 0, L_{yy}(x, y, \sigma) \cdot \vec{N}\right) \end{pmatrix}$$
$$= \begin{pmatrix} L_{xx}(x, y, \sigma) & L_{xy}(x, y, \sigma) \\ L_{xy}(x, y, \sigma) & L_{yy}(x, y, \sigma) \end{pmatrix} \tag{2}$$

A visual reconstruction model of human-computer interaction fire-fighting training is constructed by a continuous 3D reconstruction method, and a boundary characteristic detection method is adopted to carry out the feature extraction of the visual image of the human-computer interaction fire-extinguishing training, and the virtual scene model design of the human-computer interaction fire-extinguishing training is realized.

2.2 Visual Image Feature Extraction of Human-Computer Interactive Fire Fighting Training

On the basis of constructing the virtual visual image reconstruction model of human-computer interactive fire extinguishing training, the fuzzy feature extraction of the visual image of human-computer interactive fire extinguishing training is carried out, such as the boundary feature quantity of each visual image is fuzzy mining, and the matching feature quantity of the constructed template is as follows:

$$d_j^a(s, t) = sqrt\left[\sum_{i=1}^{n} (s_j^a[i] - t[i])^2\right] \tag{3}$$

Where j is the current Haaris feature set, a[i] is the covariance value of the virtual view image pixel sequence of the current man-machine interaction fire-fighting training, and $d_j^a(s, t)$ represents the distance between the j-th characteristic values of the first pixel sample. calculating the boundary feature quantity of the virtual visual image of the human-computer interaction fire-extinguishing training to obtain an average value $d_j^a(s, t)$ of the edge contour feature distribution, and taking the $\mu(j)$ as a characteristic distribution threshold value of the virtual reality imaging, and the following:

$$\mu(j) = \frac{1}{k} \sum_{a=1}^{k} d_j^a(s, t) \tag{4}$$

Wherein, k is the total number of samples for scene reconstruction of human-computer interactive fire fighting training [10]. The fuzzy feature reconstruction of the visual imaging edge feature j of human-computer interactive fire extinguishing training is carried out. The fuzzy boundary coefficient C (x, y) and x, y represent the i and j matrices respectively in the virtual space. The number of pixel points in the virtual parameter panel and the main interface is n, where $\mu(i)$ represents the first matrix sample weight:

$$\mu(i) = \frac{1}{n} \sum_{k=1}^{n} p_k(i) \qquad (5)$$

The virtual image acquisition of human-computer interactive fire extinguishing training is carried out by using optical sensor, and a two-dimensional random variable is outputted to find out the dynamic weight coefficient of the visual image of human-computer interactive fire extinguishing training:

$$C(i,j) = \frac{1}{n-1} \sum_{k=1}^{n} ((p_k(i) - \mu(i)(p_k(j) - \mu(j)) \qquad (6)$$

In that whole virtual reality visual distribution system, the visual reconstruction of man-machine interaction fire-extinguishing training is carried out, and the covariance of the sub-region and the human-machine interaction can be represented by an $d \times d$-dimension matrix to obtain an 8×8 grid with the $[(d^4 + d^2)/2] - d^2$ useful characteristics of a human-computer interaction fire-extinguishing training vista, The best matching block area of the three-dimensional simulation design image of the fire main transformer emergency drill training virtual scene is as follows:

$$\{\tau_k(t_i) : |\tau_k| \geq 4, i = 2, \ldots, |\tau_k| - 2, k = 1, \ldots, K\} \qquad (7)$$

The gray-scale pixel decomposition method is adopted to obtain the grid block function of the three-dimensional simulation design of the fire main transformer emergency drill training virtual scene:

$$F(B,X) = B * X = ax^2 + bxy + cy^2 + dx + ey + f \qquad (8)$$

Wherein, $B = [a, b, c, d, e, f]^T$, $X = [x^2, xy, y^2, x, y, 1]^T$, $F(B, X_i)$. The virtual scene reconstruction of the emergency drill of the main transformer is carried out under the framework of the virtual reality [11].

3   Optimization of Visual Simulation Model for Human - Machine Interactive Fire Extinguishing Training

3.1   Dynamic Structure Recombination of Visual Image of Man-Machine Interactive Fire-Fighting Training

The virtual scene image reconstruction model of human-machine interactive fire-extinguishing training is constructed [12], the boundary feature detection and particle tracking filter processing of the visual image of human-machine interactive fire-extinguishing training are carried out. Given the fire main transformer emergency drill training virtual scene 3D simulation design image pixel sequence sample (x_1, y_1), $(x_2, y_2), \ldots (x_N, y_N)$, $y_i = 1$ or 0 respectively represent the fire main transformer emergency drill training virtual scene model design edge pixel set, take the edge pixel point as the information localization center, extract the fire main transformer emergency drill training virtual scene 3D simulation $h_t(x)$, carry on the feature matching in the block to get the emergency training feature extraction: virtual description:

$$h(x) = \begin{cases} 1 & TH = \sum_{t=1}^{T} a_t h_t(x) - \frac{1}{2}\sum_{t=1}^{T} a_t \geq 0 \\ 0 & others \end{cases} \tag{9}$$

Wherein, $a_t = \log\frac{1}{\beta_t}$, which represents the state information of the emergency drill training of the fire main transformer, adopts a template matching method in the color space of the distributed man-machine interaction fire-extinguishing training, and the matching function of the virtual scene template of the emergency drill training of the fire main transformer is as follows:

$$\begin{bmatrix} x' \\ y' \\ 1 \end{bmatrix} = \begin{bmatrix} 1 & 0 & 0 \\ 0 & \frac{\delta * \sin\alpha}{\sin(\alpha+\theta)} & \frac{n}{2\cos\alpha} - \frac{\delta * nose * \sin\alpha}{\sin(\alpha+\theta)} \\ 0 & 0 & 1 \end{bmatrix} \begin{bmatrix} x \\ y \\ 1 \end{bmatrix} \tag{10}$$

In the formula, θ is the pixel point of pixel region segmentation, y represents the color difference drift of virtual reality, nose represents the transparency of virtual scene design in fire main transformer emergency drill training, and δ is the rendering intensity of visual surface [13].

The self-correlation matching feature of the distributed man-machine interactive fire-fighting training is obtained by combining the self-adaptive feature matching method:

$$R(x, y) = ax^2 + by^2 + cxy + dx + ey + f \tag{11}$$

Wherein

$$\begin{cases} \frac{\partial R}{\partial x} = 2ax + cy = 0 \\ \frac{\partial R}{\partial y} = 2by + cx + e = 0 \end{cases} \tag{12}$$

By using sparse linear feature decomposition method and RGB color reconstruction [14], c represents the number of color components, R represented by three primary colors, the homomorphic feature transformation of human-computer interactive fire fighting training is as follows:

$$l(X, Y) = (2u_x u_y)/(u_x^2 + u_y^2) \tag{13}$$

Using the color contrast method, the scene reconstruction of man-machine inter-active fire extinguishing training is carried out. The virtual scene seed point matching expression of man-machine interactive fire extinguishing training is as follows:

$$H = \sum_{r=1}^{t} \sum_{p=1}^{k_1} (x_{ir} - x'_{irp})(x_{ir} - x'_{irp})^T \tag{14}$$

Based on the edge pixel point matching formula, the region segmentation model of distributed human-computer interaction fire fighting training scene image is obtained as follows:

$$G = tr(W_i^T [\sum_{r=1}^{t} \sum_{q=1}^{k_2} (x_{ir} - x_{irq})(x_{ir} - x_{irq})^T] W_i)$$
$$= tr(W_i^T H_2 W_i) \tag{15}$$

Wherein

$$H_2 = \sum_{r=1}^{t} \sum_{q=1}^{k_2} (x_{ir} - x_{irq})(x_{ir} - x_{irq})^T \tag{16}$$

In the different three-dimensional visual reconstruction model, the dynamic struc-tural reorganization of the visual image of human-computer interaction fire-fighting training is carried out [15].

3.2   Virtual Reality VR Simulation Design of Man-Machine Interactive Fire-Fighting Training

In combination with a fuzzy edge feature extraction method, a dynamic structure recombination of a human-machine interaction fire-extinguishing training visual image is carried out, a virtual reality simulation model of a man-machine interaction fire-extinguishing training visual image is established, and a process for designing a virtual reality simulation model algorithm for man-machine interaction fire-extinguishing training is established [16]:

1) Input the training sample $S_1 = \{(x_1, y_1) \cdots (x_n, y_n)\}$ of the virtual scene image pixel sequence of human-computer interaction fire fighting training, in which the edge pixel feature matching points on the grid model corresponding to $y_i = \{0, 1\}$ are obtained. Under the known principal transformer emergency training, the statistical eigenvalues of the virtual samples of human-computer interaction fire fighting training are obtained. At m feature matching points, Qm pixel regions are divided into feature points, and a total of n training samples are obtained.

2) Initialization sample weight: the dynamic weight segmentation method is used to carry out multi-person cooperative control, and the initialization weight is $\omega_{t,1} = 1/2m$; for the vector initialization characteristic value of $\omega_{t,0} = 1/2q$, fire main transformer emergency drill training.

3) The virtual scene image samples of man-machine interactive fire-extinguishing training are cut into 8×8 grids, and the multi-person cooperative control method is used to take two different sub-areas for state feature matching. if 8×8, the $S_{t+1} = S_t$, human-machine interactive fire-extinguishing training process is stopped; when the water and electricity fee is full, the iterative process of man-machine interactive fire-extinguishing training jumps to step (2).

4) The matching point weight of multiplayer collaborative virtual reality: $\omega_{t+1} = \omega_{t,i}\beta_t^{1-e_i}$, according to the emergency mode of human-computer interactive fire fighting training system. Initialize $e_i = 0$, get $e_i = 1$, $\beta_t = \varepsilon_t/(1 - \varepsilon_t)$;

5) According to the maximum distance p between the virtual scene image point cloud of the human-computer interaction fire-fighting training, cutting off the edge contour imaging error, and performing virtual reconstruction by adopting a plurality of cooperative methods;

6) Set t = t + 1, if t is not greater than the threshold, return to step 3.

7) the virtual reality simulation model of the human-computer interaction fire-extinguishing training visual image is established, and the output optimization control of the virtual visual image of the human-computer interaction fire-extinguishing training is realized, and the imaging output is obtained as follows:

$$h(x) = \begin{cases} 1, & \sum_{t=1}^{T} \alpha_t h(x) \geq \frac{1}{2}\sum_{t=1}^{T} \alpha_t \\ 0, & otherwise \end{cases} \tag{17}$$

Wherein $\alpha_t = \ln\frac{1-\varepsilon_t}{\varepsilon_t}$.

8) for $t = 1$, ...,T. he main transformer emergency target function $\varepsilon_j =$

$\sum_{j=1}^{n} \omega_{t,j}|h_j(xi) - y_i|$ with multi-people cooperative virtual reality is the smallest, and the G of each feature is sorted according to the size. And the control frame is continuously sent, so as to realize the optimization of the multi-people collaborative virtual reality emergency training and output the final result.

The implementation process is shown in Fig. 2.

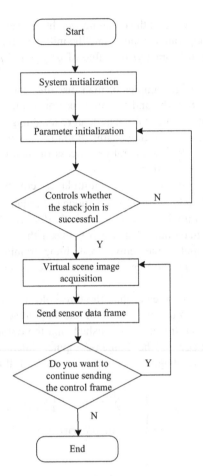

Fig. 2. Model development design process

4   Simulation Experiment and Result Analysis

In order to test the application performance of this method in the visual simulation of human-computer interaction fire fighting training, the experimental test and analysis are carried out. The development environment CPU is Inter Pentium 43000 MHz, memory 1.5 GB, the sample set of visual image of man-machine interaction fire extinguishing training image is 800, the size of 3D visual block area is 256 × 256 × 224, fuzzy feature matching coefficient 1.26, the feature resolution of virtual scene image imaging of human-computer interaction fire extinguishing training is 300 * 500, and the parameter writing interface is shown in Fig. 3.

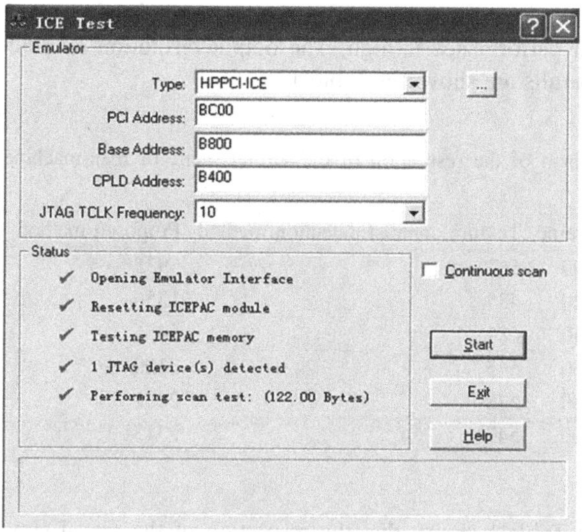

Fig. 3. Parameter write interface

The multi-person cooperative control method is used to simulate the virtual reality in the process of human-computer interactive fire fighting training, and the 3D simulation image of virtual fire drill in large-scale emergency scene is followed and rendered. The visual simulation output of human-computer interactive fire fighting training is shown in Fig. 4.

Fig. 4. Visual simulation output of man-machine interactive fire-fighting training

The analysis of Fig. 4 shows that the performance of visual simulation is good, and the image fusion performance is high. The output resolution of different systems is tested, and the results are shown in Table 1.

Table 1. Comparison of the resolution of the virtual scene of man-machine interactive fire-fighting training

Frame	Texture contrast detection method	Proposed method
100	567	935
120	435	1356
140	689	1456
160	545	2100
180	578	2241
200	547	2313

The analysis Table 1 shows that the resolution of the visual simulation of man-machine interactive fire-fighting training is better, and the parameters such as the mean square error and the output signal-to-noise ratio of the human-computer interaction fire-fighting training are tested by the method, and the results of the comparison are shown in Table 2.

Table 2. Performance Comparison Test

Method	MSE	Time overhead/ms	Output SNR/dB
Proposed method	0.032	26.4	43
Traditional manual design method	0.412	105.6	21

The results of the analysis in Table 2 show that the virtual visual simulation of human-computer interaction and fire-fighting training has good coordination, high image fusion performance, strong visual simulation and reconstruction capability of fire-fighting training, and good output performance.

5   Conclusions

In this paper, the image information processing technology is used to design the 3D solid model, and the human-computer interactive fire extinguishing training system is established in the virtual visual simulation environment. In this paper, a visual simulation model of human-computer interactive fire extinguishing training based on virtual reality technology is proposed. The virtual reality simulation model of human-computer interaction fire extinguishing training scene image is established by using continuous 3D reconstruction method. The virtual reality simulation in the process of human-computer interaction fire fighting training is carried out by using multi-person cooperative control method, and the tracking and rendering of virtual scene three-dimensional simulation image of large-scale emergency scene fire drill is carried out, and the optimization design of virtual reality VR simulation model of human-computer interaction fire fighting training is realized. It is found that the cooperation of virtual scene simulation with human-computer interaction fire fighting training is better, and the tracking and rendering ability of 3D simulation image is strong, and it has a good ability to reconstruct fire fighting training scene simulation. This method has good application value in visual simulation of machine cross fire fighting training.

6   Fund Projects

Key projects of Humanities and social sciences of Anhui Provincial Department of Education SK2018A0640.

Key projects of Humanities and social sciences of Anhui Provincial Department of Education SK2019A0745.

Provincial Quality Engineering in 2018 3Dsmax2018mooc599.

References

1. Xin, W., Yun, Z., Chen, N., et al.: Image saliency detection via adaptive fusion of local and global sparse representation. J. Comput. Appl. **38**(3), 866–872 (2018)
2. Cheng, M.M., Mitra, N.J., Huang, X., et al.: Global contrast based salient region detection. IEEE Trans. Pattern Anal. Mach. Intell. **37**(3), 569–582 (2015)
3. Jung, H.S., Kim, R.-C., Lee, S.-U.: A hierarchical synchronization technique based on the EREC for robust transmission of H.263 bit stream. IEEE Trans. Circ. Syst. Video Technol. **10**(3), 433–438 2000
4. Kim, W., Kim, C.: Spatiotemporal saliency detection using textural contrast and its applications. IEEE Trans. Circ. Syst. Video Technol. **24**(4), 646–659 (2014)
5. Yan, Q., Xu, L., Shi, J., et al.: Hierarchical saliency detection. In: CVPR 2013 Proceedings of the 2013 IEEE Conference on Computer Vision and Pattern Recognition. IEEE Computer Society, vol. 10, no. 153, pp. 1155–1162 (2013)
6. Wang, X., Ning, C., Xu, L.: Spatiotemporal saliency model for small moving object detection in infrared videos. Infrared Phys. Technol. **69**(5), 111–117 (2015)
7. Radenović, F., Tolias, G., Chum, O.: CNN image retrieval learns from BoW: unsupervised fine-tuning with hard examples. In: Leibe, B., Matas, J., Sebe, N., Welling, M. (eds.) ECCV 2016. LNCS, vol. 9905, pp. 3–20. Springer, Cham (2016). https://doi.org/10.1007/978-3-319-46448-0_1
8. Azizpour, H., Razavian, A.S., Sullivan, J.: From generic to specific deep representations for visual recognition. In: Proceedings of the 2015 International Conference on Computer Vision and Pattern Recognition Workshops, vol. 6, no. 23, 36–45. IEEE Computer Society, Washington, DC (2015)
9. Razavian, A.S., Sullivan, J., Carlsson, S.: Visual instance retrieval with deep convolutional networks. ITE Trans. Media Technol. Appl. **4**(3), 251–258 (2016)
10. Wolf, L., Hassner, T., Taigman, Y.: Effective unconstrained face recognition by combining multiple descriptors and learned background statistics. IEEE Trans. Pattern Anal. Mach. Intell. **33**(10), 1978–1990 (2011)

11. Li, G., Li, H., Shang, F., et al.: Noise image segmentation model with local intensity difference. J. Comput. Appl. **38**(3), 842–847 (2018)
12. Niu, S., Chen, Q., Sisternes, L.D., et al.: Robust noise region-based active contour model via local similarity factor for image segmentation. Pattern Recogn. **61**(9), 104–119 (2016)
13. Jie-yu, Z., Hong-ping, Z., Shu, C.: Face recognition based on weighted local binary pattern with adaptive threshold. J. Electron. Inf. Technol. **36**(6), 1327–1333 (2014)
14. Naseem, I., Togneri, R., Bennamoun, M.: Linear regression for face recognition. IEEE Trans. Pattern Anal. Mach. Intell. **32**(11), 2106–2112 (2010)
15. Wu, J.G., Shao, T., Liu, Z.Y.: RGB-D saliency detection based on integration feature of color and depth saliency map. J. Electron. Inf. Technol. **39**(9), 2148–2154 (2017)
16. Carlson, N.A., Porter, J.R.: On the cardinality of Hausdorff spaces and H-closed spaces. Topol. Appl. **160**(1), 137–142 (2017)

Virtual Interactive Planning Model of Landscape Architecture in Settlement Area Based on Situational Awareness

Jun-qin Diao and Xue-yong Cui[(✉)]

Anhui Xinhua University Academy of Arts, Hefei 230601, China
lsj6090032@163.com, ht69250@163.com

Abstract. In order to realize the visual effect optimization design of landscape virtual interactive planning in the resettlement area, the optimal design method of landscape virtual interactive planning in the resettlement area based on situational awareness is proposed. The feature sampling model of landscape virtual interactive planning optimization is established, the virtual reality simulation in landscape virtual interactive planning design is carried out by MPI visual simulation tool, the virtual interactive planning feature construction of landscape virtual interactive planning is carried out in Vega Prime software, and the virtual interactive planning information sampling model and block information fusion model of landscape virtual interactive planning are established. Create, edit and run virtual interactive planning optimization program of landscape in resettlement area, combine with cross-compiling method to simulate virtual interactive planning information of landscape in resettlement area, create 3D visual environment of virtual interactive planning of landscape in resettlement area in real-time interaction, and realize virtual interactive planning optimization design of landscape in resettlement area in virtual reality simulation environment. The simulation results show that this method can effectively realize the visual optimization design of virtual interactive planning of landscape in resettlement area, improve the visual feature expression effect of virtual interactive planning of landscape in resettlement area, and has good application value in virtual interactive planning design of landscape in resettlement area.

Keywords: Situational awareness · Resettlement area landscape · Virtual interaction · Planning model

1 Introduction

As an expression carrier of cultural connotation, landscape design is closely related to culture itself. Modern landscape pays more attention to professionalism and high quality, and its cultural connotation is richer and has strong emotional color. Through different design methods, different cultures, such as personal philosophy, political ideal, literature revision, art aesthetics and so on, are integrated into the landscape architecture [1]. As a special form of artistic expression, landscape design mainly converts the cultural connotation of garden landscape into the ideology of users, which has an important educational function. Garden art is an integral part of social and cultural

Y.-D. Zhang et al. (Eds.): ICMTEL 2020, LNICST 327, pp. 137–148, 2020.
https://doi.org/10.1007/978-3-030-51103-6_12

attributes, and landscape designers interpret their own understanding of the connotation of garden culture through a certain form of landscape art. The cultural inheritance of landscape does not mean the communication between things, but expresses the design creativity of human landscape inheritance with the help of garden art, and the user interprets the creative idea of designer from the point of view of cultural connotation [2]. The inheritance of landscape culture is continuous in time and space, not limited by time and region, and cannot replace other forms of language and culture inheritance. Landscape has the three-dimensional expression of historical and cultural inheritance. The cultural inheritance in landscape design also has certain inheritance and transcendence. Transcendental cultural inheritance is embodied in the fact that according to the rules of human survival and development, we should learn the selective trade-off and innovation by absorbing the cultural essence inherited from the original cultural heritage [3].

With the development of 3D virtual visual simulation technology, the virtual interactive planning and optimization design of landscape in resettlement area is carried out by using 3D virtual reality technology, and the simulation model of virtual interactive planning of landscape in resettlement area is established [4]. Combined with advanced visual simulation software and image processing program, the virtual interactive planning and optimization of landscape in resettlement area is carried out, and the effect of virtual interactive planning of landscape in resettlement area is improved. The optimization design of virtual interactive planning of landscape in resettlement area is based on the advanced visual simulation tool of virtual interactive planning of landscape in resettlement area [5]. Combined with the database of visual simulation, the expert system is constructed, and the optimization construction model of virtual interactive planning effect of landscape in resettlement area is established, so as to improve the simulation ability of virtual interactive planning of landscape in resettlement area [6]. The research on the optimization design method of virtual interactive planning of garden landscape in settlement area is of great significance in advertising design. In the virtual reality simulation scene, the advanced visual simulation function is effectively combined with the image processing program, the optimal design model of the virtual interactive planning of the landscape in the resettlement area is established, the virtual interactive planning of the landscape in the resettlement area is optimized by using the Vega Prime editor, and the scene map system (Vega Scene Graph, VSG) of the virtual interactive planning of the landscape in the resettlement area is designed The visual information analysis model of virtual interactive planning of landscape in resettlement area is established. Combined with the application software, the visual optimization design of virtual interactive planning of landscape in resettlement area is carried out. in this paper, the optimization design method of virtual interactive planning of landscape in resettlement area based on situational perception is proposed. The sampling model of virtual interactive planning feature information of landscape in resettlement area is established [7], and the virtual interactive planning information of landscape in resettlement area is simulated with cross compilation method, and the visual optimization of virtual interactive planning of landscape in resettlement area is realized by combining visual simulation software. Finally, the simulation test and analysis are carried out, which shows the superior performance of this method in improving the optimization ability of virtual interactive planning of landscape in resettlement area [8].

2   Overall Framework of the Planning and Optimization Design System for the Virtual Space Planning of the Garden Landscape in the Resettlement Area

In order to realize that optimal design of the virtual environment planning of the garden landscape in the resettlement area, a virtual space planning optimization design of the garden landscape of the resettlement area is carried out in combination with the Vega Prime bottom software framework agreement [9]. In this paper, the design of the program embedded in the process of the planning and optimization of the landscape in the resettlement area is set up by using the OpenGL program, and the three-dimensional rendering of the virtual space planning drawing of the garden landscape in the resettlement area is carried out by using the OpenGL, and the VP class library developed by the MultiGen Co., Ltd. is adopted. The VSG class library software is used for the integration sampling of the virtual landscape planning information of the garden landscape in the resettlement area, and the urban landscape virtual spatial planning optimization control model is established in the three-dimensional space, and the virtual spatial planning optimization of the landscape in the resettlement area is carried out in the three-dimensional space, So as to improve the planning and design effect of the landscape of the garden landscape in the resettlement area [10]. Under the technical support of virtual reality, the planning and optimization design of the landscape of the landscape of the resettlement area is carried out, and the overall framework of the planning and optimization design system of the garden landscape of the resettlement area is shown in Fig. 1.

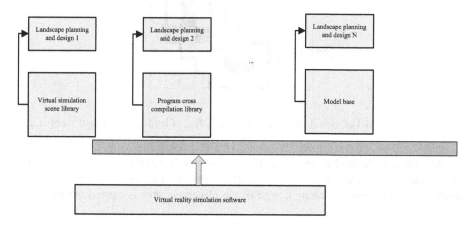

Fig. 1. General framework of the planning and optimization design system of the landscape of the garden landscape in the resettlement area

According to the above analysis, using the MPI visual simulation tool to carry out the virtual reality simulation in the virtual environment planning design of the garden landscape in the resettlement area, the feature structure is constructed in the Vega Prime software for the virtual environment planning of the landscape of the placement area [11].

3   Graphic Processing of Virtual Interactive Planning Optimization of Garden Landscape in Settlement Area

On the basis of the overall structure of the virtual interactive planning and design of the landscape in the resettlement area, the modeling structure of the virtual interactive planning and optimization design system of the landscape in the resettlement area is carried out. Combined with the image processing software, the virtual interactive planning and optimization design of the landscape in the resettlement area is carried out [9], and the sampling model of the feature information of the virtual interactive planning of the landscape in the resettlement area is established. The spatial sampling in the virtual interactive planning design of landscape in resettlement area is carried out by using block feature matching technology, and the block information fusion model of virtual interactive planning effect of landscape in resettlement area is established, as shown in Fig. 2.

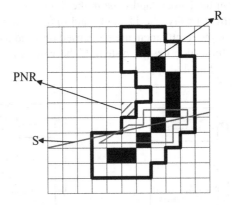

Fig. 2. Block model of virtual interactive planning effect of landscape in resettlement area

According to the landscape virtual space planning effect expression blocking model of the placement area shown in Fig. 2, the visual characteristic rendering of the landscape virtual space planning design of the placement area is carried out by using the method of the blocking information fusion [12], and the image expression of the landscape virtual space planning image of the placement area is as follows:

$$c(x,y) = \sum_W \left[I(x_i, y_i) - I(x_i + \Delta x, y_i + \Delta y) \right]^2 \tag{1}$$

Wherein, $(\Delta x, \Delta y)^T$ is the displacement of virtual interactive planning distribution of landscape in resettlement area, and (x_i, y_i) is the point in window W. By using the method of multi-scale feature decomposition, the virtual interactive planning and rendering of landscape in resettlement area is carried out, and the image block feature matching model is established. The approximate value of virtual interactive planning displacement of landscape in resettlement area is obtained as:

$$I(x_i + \Delta x, y_i + \Delta y) \approx I(x_i, y_i) + \left[I_x(x_i, y_i) \quad I_y(x_i, y_i) \right] \begin{bmatrix} \Delta x \\ \Delta y \end{bmatrix} \tag{2}$$

The $\nabla I = \begin{bmatrix} I_x & I_y \end{bmatrix}^T$ is the spatial gradient of the layout information distribution of the landscape of the landscape in the resettlement area. The visual database is used for the integration of the virtual space planning information of the landscape of the placement area and the space undersampling technique, and the visual optimization control in the planning and design process of the garden landscape of the resettlement area is carried out, and the control function is as follows:

$$c(x,y) = \begin{bmatrix} \Delta x & \Delta y \end{bmatrix} \begin{bmatrix} \sum_W I_x^2 & \sum_W I_x I_y \\ \sum_W I_x I_y & \sum_W I_y^2 \end{bmatrix} \begin{bmatrix} \Delta x \\ \Delta y \end{bmatrix} \tag{3}$$

In the virtual space planning image $I(x, y)$ of the garden landscape of the resettlement area, the point scanning technique is adopted to obtain the Hessian matrix of the detected characteristic point x at the scale.

$$H = \begin{bmatrix} L_{xx}(x, \sigma) & L_{xy}(x, \sigma) \\ L_{xy}(x, \sigma) & L_{yy}(x, \sigma) \end{bmatrix} \tag{4}$$

Where: $L_{xx}(x, \sigma)$ is the spatial distribution joint function of the landscape virtual interactive programming image in the resettlement area, and the autocorrelation function of L_{xy} and L_{yy}. the optimal scanning of the landscape virtual interactive planning of the resettlement area is carried out in the three-dimensional neighborhood of $3 \times 3 \times 3$, as shown in Fig. 3.

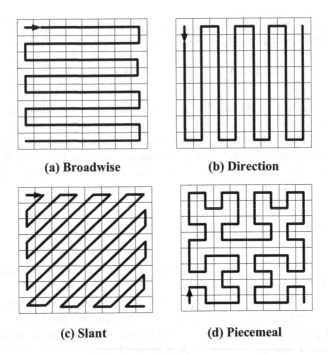

(a) Broadwise   (b) Direction

(c) Slant   (d) Piecemeal

Fig. 3. Optimization scanning of Virtual Interactive Planning of Landscape in settlement area

4   Optimization Design and Implementation of Virtual Interactive Planning of Landscape in Settlement Area

The virtual interactive planning and design of landscape in resettlement area is carried out in Vega Prime, the analysis model of virtual interactive planning of landscape in resettlement area is established, the graphic rendering model of virtual interactive planning of landscape in resettlement area is established in Lynx Prime [13], and the virtual visual simulation design of virtual interactive planning of landscape in resettlement area is carried out according to the virtual effect. Firstly, the plane distribution structure model of virtual interactive planning and design of landscape in resettlement area is constructed as shown in Fig. 4.

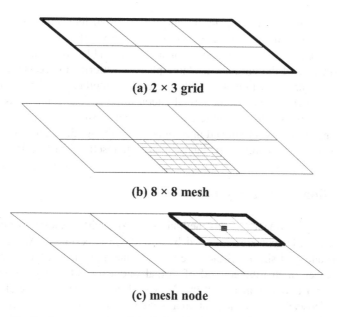

(a) 2 × 3 grid

(b) 8 × 8 mesh

(c) mesh node

Fig. 4. Plane distribution structure model of virtual interactive planning and design of garden landscape in settlement area

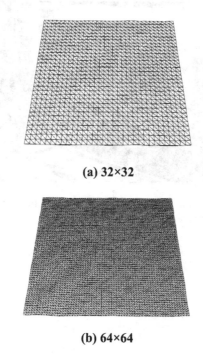

(a) 32×32

(b) 64×64

Fig. 5. 3D rendering model of virtual interactive planning of landscape in settlement area

According to the plane template design of the virtual interactive planning of the landscape in the resettlement area, the virtual interactive planning optimization program of the landscape in the resettlement area is created [14], edited and run in the 3D simulation scene, and the virtual interactive planning information of the landscape in the resettlement area is simulated with the cross-compilation method. The three-dimensional rendering model of the virtual interactive planning of the landscape in the resettlement area is established in the complex model as shown in Fig. 5.

The grid resolution is determined in conjunction with Fig. 5, and the optimal design of the virtual space planning of the landscape of the resettlement area is realized.

5   Simulation Test Analysis

The application performance of this method in realizing virtual interactive planning of landscape in resettlement area is tested by simulation experiment. According to the design scope and grid size of virtual interactive planning of landscape in resettlement area, the number of distributed grid of virtual interactive planning of landscape in resettlement area is determined, and the 3D visual environment of virtual interactive planning of landscape in resettlement area is created. The virtual interactive planning and optimization design of garden landscape in the virtual reality simulation environment is realized, and the virtual visual simulation results are shown in Fig. 6.

Fig. 6. Virtual scene simulation results of the virtual space planning design of the garden landscape in the resettlement area

According to the virtual scene simulation result of Fig. 6, the optimal visual design of the virtual space planning of the landscape of the resettlement area is carried out, and the gray scale is obtained as shown in Fig. 7.

Fig. 7. Grayscale map of garden landscape in settlement area

The 3D rendering of the virtual space planning pattern of the garden landscape in the resettlement area is carried out by using OpenGL, and the virtual spatial planning optimization control model of the garden landscape in the resettlement area is established, and the virtual spatial planning optimization of the landscape landscape in the resettlement area is carried out in the three-dimensional space, so that the visual optimization result is shown in Fig. 8.

Figure 8 shows that the virtual interactive planning and optimization design of landscape in resettlement area can be effectively realized by using this method, and the spatial visual effect is good. Taking the output signal-to-noise ratio as the index, the comparison results are shown in Table 1. The analysis Table 1 shows that the output signal-to-noise ratio (SNR) of virtual interactive planning and design of landscape in resettlement area is high, which shows that the quality of the design is better.

(a) The optimization results of this method

(b) Optimization results of literature [3] method

(c) Optimization results of literature [4] method

Fig. 8. Optimization effect of virtual interactive planning of landscape in resettlement area

Table 1. Comparison of output signal-to-noise ratio (dB)

Iterations	Proposed method	Reference [3]	Reference [4]
100	34.3	20.3	12.4
200	38.4	23.6	14.3
300	41.3	28.1	18.4
400	48.5	30.4	20.1

6   Conclusions

In this paper, a virtual interactive planning optimization design method for landscape architecture in settlement area based on situational perception is proposed. The sampling model of virtual interactive planning optimization feature information of landscape in resettlement area is established, the 3D rendering of virtual interactive planning graphics of landscape in resettlement area is carried out by using OpenGL, the optimization control model of virtual interactive planning of landscape in resettlement area is established, the virtual interactive planning and optimization of landscape in resettlement area is carried out in three dimensional space, and the effect of virtual interactive planning of landscape in resettlement area is improved. The analysis shows that this method can effectively realize the visual optimization design of virtual interactive planning of landscape in resettlement area, improve the visual feature expression effect of virtual interactive planning of landscape in resettlement area, and the output signal-to-noise ratio of virtual interactive planning image of landscape in resettlement area is high, which indicates that the design quality is good. This method has good application value in landscape planning and design of resettlement area.

7   Fund Projects

2019 top talents in Colleges and Universities GXYQZD2019089.

Key projects of Humanities and social sciences of Anhui Provincial Department of Education SK2018A0640.

Provincial Quality Engineering in 2018 3Dsmax2018mooc599.

References

1. Wang, Y.: Graphic design of graphic advertising based on information communication. J. Anhui Univ. Technol. (Social Science Edition) **34**(04), 37–39 (2017)
2. Bai, Y.: Solving ramsey number algorithm based on set theory. J. Jilin Univ. Sci. Ed. **57**(03), 647–652 (2019)
3. Jin, P.: Research on virtual design method of indoor landscape based on 3D vision. Modern Electron. Technol. **40**(24), 112–114 (2017)
4. Liu, N., Han, J.: DHSNet: deep hierarchical saliency network for salient object detection. In: Proceedings of the 2016 IEEE Conference on Computer Vision and Pattern Recognition, vol. 32, no.06, pp. 678–686. IEEE Computer Society, Washington, DC (2016)
5. Kim, W., Kim, C.: Spatiotemporal saliency detection using textural contrast and its applications. IEEE Trans. Circ. Syst. Video Technol. **24**(04), 646–659 (2014)
6. Zhang, F., Zhong, B.-J.: Image retrieval based on interested objects. Tien Tzu Hsueh Pao/Acta Electronica Sinica **46**(08), 1915–1923 (2018)
7. Jeon, H.-J., Kang, C.-S., Gi, W.Y., et al.: The influence of lithium content in xLi [Li1/3Mn2/3] O2·(1-x) Li [NiaCobMn (1-a + b)] O2 cathode materials prepared by co-precipitation method. J. Electroceram. **30**(03), 1–7 (2013)

8. Celikoglu, H.B., Sanchez-Medina, J.: Special section editorial on \"high performance computing in simulation and optimization of dynamic transportation networks\" [Guest Editorial]. IEEE Intell. Transport. Syst. Mag. **10**(01), 5–7 (2018)

9. Yang, L.: Technique for image de-noising based on non-subsampled shearlet transform and improved intuitionistic fuzzy entropy. Optik-Int. J. Light Electron Opt. **126**(04), 446–453 (2015)

10. Lim, B., Son, S., Kim, H., et al.: Enhanced deep residual networks for single image super-resolution. In: CVPRW 2017 Proceedings of the 2017 IEEE Conference on Computer Vision and Pattern Recognition Workshops, vol. 151, no. 03, pp. 1132–1140. IEEE Computer Society, Washington, DC (2017)

11. Yang, L.: Multi-focus image fusion method based on NSST and II CM. In: Proceedings of the 2017 International Conference on Emerging Internetworking, Data and Web Technologies, vol. 16, no. 01, pp. 679–689, Springer, Berlin (2017)

12. Yu, M., Zhang, H.: HDR imaging based on low-rank matrix completion and total variation constraint. Comput. Eng. **45**(04), 262–266 (2019)

13. Li, B., Wang, C., Huang, D.S.: Supervised feature extraction based on orthogonal discriminant projection. Neurocomputing **73**(01), 191–196 (2009)

14. Hou, C., Nie, F., Li, X., et al.: Joint embedding learning and sparse regression: a framework for unsupervised feature selection. IEEE Trans. Cybern. **44**(06), 793–804 (2014)

Time Series Data Reconstruction Method Based on Probability Statistics and Machine Learning

Haiying Chen[1(✉)] and Yinghua Liu[2]

[1] Xianning Vocational Technical College, College of Humanities and Arts, Xianning, China
chenhaiying5454@163.com
[2] Wuhan Institute of Design and Sciences, Wuhan, China

Abstract. In order to improve the reconstruction ability of time series data under probability statistical model, a time series data reconstruction method based on machine learning is proposed. The time series data distribution structure model under probability statistical model is constructed. The spatial multi-sensor information sampling method is used to sample the time series data information flow under the probability statistical model, and the phase space reconstruction method is combined to reconstruct the time series data information structure under the probability statistical model. The probability statistical model is established to decompose the time series data, and the distributed grid computing method is used to extract the big data association features of the time series data under the probability statistical model. Combined with the adaptive weight learning method, the optimal control of the scheduling is carried out. The big data cross-domain scheduling of the time series data under the probabilistic statistical model is realized under the support vector machine learning mode. The simulation results show that the method has good adaptability to time series data cross-domain scheduling under the probability and statistics model, and the load balance of data output is strong.

Keywords: Probability statistics · Machine learning · Time series · Reconstruction

1 Introduction

With the development of cloud computing technology, a large number of block chain slicing resource data are distributed in cloud computing storage space, so it is necessary to construct time series data scheduling model under probability and statistics model, combine cloud computing and cloud storage technology to schedule and retrieve time series data across domains, and improve the ability of time series data cross-domain retrieval and query. The research of related time series data cross-domain scheduling method is of great significance in optimizing cloud storage space design and cloud resource information retrieval optimization [1].

In embedded environment, block chain slicing storage system design and resource optimization scheduling under probability and statistics model are carried out, cloud

Y.-D. Zhang et al. (Eds.): ICMTEL 2020, LNICST 327, pp. 149–159, 2020.
https://doi.org/10.1007/978-3-030-51103-6_13

computing storage structure model is established, and block chain slicing storage and cross-domain scheduling under probability and statistics model are carried out by using link structure reorganization method [2, 3]. In the traditional method, the clustering energy consumption scheduling algorithm is mainly used in the cloud resource scheduling algorithm. Combined with the priority list control method, the big data partition scheduling of time series data under the probability statistical model is carried out. In reference [4], a time series data scheduling method based on adaptive priority list control is proposed, combined with the optimal storage and distributed retrieval model of cloud resources, the block chain slicing storage scheduling is carried out. However, the reconstruction ability of time series data under probability statistical model is not good. In reference [5], a resource scheduling algorithm for block chain slicing storage system under the probability and statistical model of double threshold equilibrium control is proposed. The time axis of cloud resource scheduling is divided into uniformly distributed time windows, and the time series data scheduling method is combined with the block structure reorganization method, but the computational overhead of this method is large and the real-time performance is not good.

In order to solve the above problems, a time series data reconstruction method based on machine learning is proposed. Firstly, the time series data distribution structure model under the probability statistical model is constructed, the spatial multi-sensor information sampling method is used to sample the time series data information flow under the probability statistical model, and the phase space reconstruction method is used to reconstruct the time series data information structure under the probability statistical model [6]. Then the feature decomposition model of time series data under probability and statistics model is established, and the big data association feature extraction of time series data under probability statistical model is carried out by combining distributed grid computing method, and the optimal control of scheduling is carried out by combining adaptive weight learning method. Big data cross-domain scheduling of time series data under probability and statistics model is realized under support vector machine learning mode. Finally, the simulation results show the superior performance of this method in improving the reconstruction ability of time series data.

2   Time Series Data Sampling and Information Structure Reorganization

2.1   Time-Series Data Sampling Under Probabilistic Statistical Model

In order to realize big data fusion and cross-domain scheduling of time series data under probability statistical model, it is necessary to construct time series data distribution structure model under probability statistical model, and to sample time series data information flow under probability statistical model by using spatial multi-sensor information sampling method, in order to realize time series data scheduling under probability statistical model. Firstly, the kernel structure and resource storage structure model of block chain slicing storage system under probabilistic statistical model are analyzed, and the information flow model and time series analysis of time series data under probabilistic statistical model are carried out. Under the probabilistic statistical model, cloud computing grid space realizes memory management, process management and spatial distributed structure storage management through resource cross-domain scheduling [7]. The distribution structure model of time series data under probability and statistics model is shown in Fig. 1.

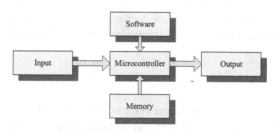

Fig. 1. Model of time series data distribution structure under probabilistic statistical model

In that whole probability statistic model, a block chain fragment storage system adopts a grid form, and various scattered time sequence data are reconstructed and linearly reformed, and through the analysis, a block chain fragment storage structure model of a block chain under a probability statistic model is obtained. in the method, a collaborative distribution method is adopted for designing a resource input interface [8], a time series data large data scheduling in a probability statistical model is adopted, and a cloud computing time series data large data characteristic distribution set of a resource searching module is set as follows:

$$P = \{p_1, p_2, \cdots p_m\}, m \in N \tag{1}$$

Wherein, the m represents the dimension of the data large data distribution space of the time series data under the probability statistical model, the p_m is a fuzzy degree function of the slice storage node, and the data fusion is carried out on the data large data stream of the time series data under the multiple probability statistical models, under the embedded platform, the fragment storage and the structure matching are carried out, and the time series data large data stream under the probability statistical model to be distributed is as follows:

$$flow_k = \{n_1, n_2, \cdots, n_q\}, q \in N \tag{2}$$

In the above formula, q represents the storage depth of the large data stream set of time series data under multiple probabilistic statistical models, n_q represents the data sequence of big data information flow of time series data under probabilistic statistical model, and N represents the total number of big data symbols of time series data under probabilistic statistical model. According to the above analysis, combined with the phase space reconstruction method, the time series data information structure is reorganized under the probability and statistical model [9].

2.2   Time Series Data Information Structure Reorganization

The characteristic decomposition model of time series data under probability statistical model is established, and the time series data structure under probability statistical model is reorganized with distributed grid computing method. The tasks in the client side of time series data scheduling under probability statistical model are submitted to the server, and a distribution set of time series data with N input time series data is obtained. the random number series components of time series data output are as follows:

$$x_{n+1} = 4x_n(1 - x_n) \ n = 1, 2, \cdots, NP \tag{3}$$

Wherein, NP is the distributed bandwidth of time series data in time period T, n represents the number of tasks in process management. The phase space reconstruction method is used to reconstruct the time series of time series data scheduling under probability and statistics model, and the phase space reconstruction equation is obtained as follows:

$$\begin{cases} x = (x_1, x_2, \ldots, x_n) \\ y = F(x) = (f_1(x), f_1(x), \ldots, f_m(x))^T \end{cases} \tag{4}$$

Wherein, the $x = (x_1, x_2, \ldots, x_n)$ is a set of interference data scheduled for the time series data under the probability statistical model; the $y = F(x)$ represents a delay function of the block chain fragment storage in the probability statistical model; and the type of the characteristic vector set n_i of the time series data scheduling is n_i, and the association rule mining method is adopted, the error of the inter-layer prediction of the time series data is $P(n_i) = \{p_k | pr_{kj} = 1, k = 1, 2, \cdots, m\}$, the priority attribute of the time series data scheduling under the probability statistical model is constructed by using the false nearest neighbor method, the reconstruction problem of the time series data under the probability statistical model is converted into a detection problem of a multivariate unknown parameter, The expression of the detected statistical feature quantity is as follows:

$$x_{\min,j} = \max \left\{ x_{\min,j}, x_{g,j} - \rho(x_{\max,j} - x_{\min,j}) \right\} \tag{5}$$

$$x_{\max,j} = \min \left\{ x_{\max,j}, x_{g,j} + \rho(x_{\max,j} - x_{\min,j}) \right\} \tag{6}$$

In the above formula, ρ is the transmission loss coefficient of time series data reconstruction, and the weight adjustment of the time series data scheduling is constructed in the interval $[x_{\min,j}, x_{\max,j}]$, and the bandwidth of the resource scheduling is SW. Based on the analysis, the information entropy of the time series data scheduling node in the probability statistical model is obtained:

$$H_i(x) = \sum_{k=1}^{K} p_k \ln \frac{1}{p_k} = -\sum_{k=1}^{K} p_k \ln p_k \tag{7}$$

The method for controlling the output of the chain-fragment storage resource to obtain the spatial distribution feature quantity meets the $\Phi : M \rightarrow R^{2d+1}$, so that the integration degree distribution of the time sequence data scheduling is as follows:

$$\Phi(z) = (h(z), h(\varphi_1(z)), \ldots, h(\varphi_{2d}(z)))^T \tag{8}$$

In the block chain slicing storage system based on probability and statistics model, the adaptive weighted control of time series data is carried out, and the ambiguity function is obtained as follows:

$$x_n = [x(0), x(1), \cdots, x(N-1)]^T \tag{9}$$

The third order statistical feature analysis method is used to decompose the time series data under the probability statistical model [10], and the eigenvalues are defined as follows:

$$cum(\lambda_1 x_1, \lambda_2 x_2, \cdots, \lambda_k x_k) = \left(\prod_{i=1}^{k} \lambda_i \right) cum(x_1, x_2, \cdots, x_k) \tag{10}$$

Among them, the k-order cumulant of time series data under probability statistical model is $c_{kx}(\tau_1, \tau_2, \cdots, \tau_{k-1})$, resource load balanced transmission information flow $\{x(n), x(n+\tau_1), \cdots, x(n+\tau_{k-1})\}$, according to the above analysis, the time series data information structure reorganization model is constructed, combined with subspace fusion method, the time series data reconstruction is carried out [11].

3   Time Series Data Reconstruction and Optimization

3.1   Association Feature Extraction

On the basis of sampling the information flow of time series data under probability statistical model by using spatial multi-sensor information sampling method, the feature decomposition model of time series data under probability statistical model is established, and the time series data structure of time series data under probability statistical model is reorganized with distributed grid computing method [12]. The optimal control model is obtained, and the output characteristic quantity of time series data cross-domain scheduling is described as follows:

$$x(t) = \text{Re}\{a_n(t)e^{-j2\pi f_c \tau_n(t)} s_l(t - \tau_n(t))e^{-j2\pi f_c t}\} \tag{11}$$

In the support vector machine learning mode, the load balancing characteristics of resource scheduling can be described as follows:

$$c(\tau, t) = \sum_n a_n(t)e^{-j2\pi f_c \tau_n(t)} \delta(t - \tau_n(t)) \tag{12}$$

In the above formula, $a_n(t)$ is the output information flow of block chain slicing scheduling on n channels, $\tau_n(t)$ is the delay on n transmission channels, f_c is the modulation frequency of block chain slicing scheduling under probabilistic statistical model, the adaptive weighting method is used to schedule the time series data in real time under probabilistic statistical model [13], and the attribute distribution quantification function of time series data in probabilistic statistical model is as follows:

$$S(i,j) = \frac{\sum_{u \in U_{ij}} (V_{u,i} - 3)(V_{u,j} - 3)}{\sqrt{\sum_{u \in U_{ij}} (V_{u,i} - \overline{V}_{\cdot i})^2} \sqrt{\sum_{u \in U_{ij}} (V_{u,j} - \overline{V}_j)^2}} \tag{13}$$

In which, the time series data information resource evaluation matrix of the probability statistical model is an optimized characteristic solution of the time series database distribution set of the $R = (r_{ij}, a_{ij})_{m \times n}$ and the probability statistical model:

$$\Phi = diag[e^{j\phi_1}, \cdots, e^{j\phi_P}] \tag{14}$$

According to the analysis, the sample set distribution model of the time series data under the probability statistical model is met:

$$T_{i,j}(t) = \frac{|p_{i,j}(t) - \Delta p(t)|}{p_{i,j}(t)} \tag{15}$$

Where, $U_{i,j}(t)$ is used to represent the association rule items of time series data under the probabilistic statistical model. according to the above analysis, the association features are extracted [14].

3.2   Weight Analysis and Time Series Data Reconstruction Output

Combined with adaptive weight learning method to optimize scheduling, the time series data of time series data in probabilistic statistical model is scheduled across domains under support vector machine learning mode. 4 tuples (E_i, E_j, d, t) is used to represent the main feature decision tree of time series data sharing scheduling under probabilistic statistical model. E_i, E_j is the bifurcation node of time series data in directed graph under probabilistic statistical model [15]. The differential fusion feature quantity of time series data sharing under probability and statistics model is obtained.

$$J_m(U, V) = \sum_{k=1}^{n} \sum_{i=1}^{c} \mu_{ik}^m (d_{ik})^2 \tag{16}$$

In the formula, m is a finite data set of time series data distribution under a probability statistical model, and $(d_{ik})^2$ is a similarity distribution map, and the decision-making independent variable of the time series data under the probability statistical model is as follows:

$$(d_{ik})^2 = \|x_k - V_i\|^2 \tag{17}$$

And

$$\sum_{i=1}^{c} \mu_{ik} = 1, k = 1, 2, \cdots, n \tag{18}$$

The optimal scheduling and mining of time series data under probabilistic statistical model is carried out [16], and the priority clustering model of time series data sharing scheduling under probabilistic statistical model is obtained as follows:

$$V_{u,i} = \frac{D_i^-}{D_i^+ + D_i^-}, \overline{V}_j = \frac{R_i^+}{R_i^+ + R_i^-} \tag{19}$$

Based on the analysis, the adaptive fusion clustering model of time series data under probabilistic statistical model is constructed, and the reconstruction of time series data under probabilistic statistical model is realized by using time series data fusion method [17–20].

4   Simulation Experiment and Result Analysis

In order to test the application performance of this method in time series data scheduling under probabilistic statistical model, Matlab is used to carry out simulation experiment, and embedded Linux technology is used to design the platform of time series data scheduling under probabilistic statistical model. The sampling time length of time series data under probabilistic statistical model is 60 s, the characteristic sampling

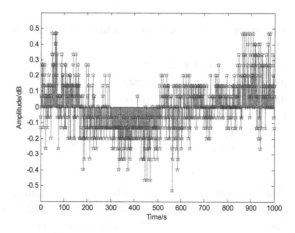

Fig. 2. Time domain waveform of time series data distribution under probability statistical model

frequency is 80 kHz, and the carrier frequency of block chain slicing storage is 12 kHz. A frequency component of 250 Hz is set between 400 and 600 sampling points for distributed adjustment of time series data, and the sampling of time series data information under probability and statistics model is shown in Fig. 2.

Taking the resource data of Fig. 2 as the research object, the time series data scheduling under the probability statistical model is carried out, and the reconstruction output is shown in Fig. 3.

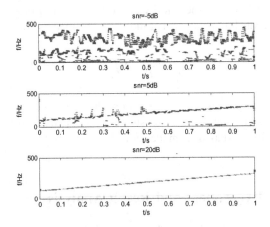

Fig. 3. Time series data to reconstruction output under probabilistic statistical model.

The analysis Fig. 3 shows that the proposed method can effectively realize the time series data scheduling under the probability statistical model, and test the equilibrium degree of the block chain storage resource scheduling under the probability statistical model. The comparative results are shown in Table 1.

Table 1. Equalization comparison of block chain reconstruction under probability and statistical model

Iterations	Proposed method	Reference [4]	Reference [5]
100	0.913	0.845	0.845
200	0.924	0.876	0.864
300	0.954	0.914	0.912
400	0.987	0.926	0.934
500	0.998	0.943	0.967

As shown in Table 1, compared with the traditional method, the proposed method has higher equilibrium of block chain and higher practical application. The analysis Table 1 shows that the proposed method has a better balance in the open source scheduling of the block chain time series data under the probability and statistics model.

In order to verify the complexity of the proposed algorithm and the running time of the algorithm, the experimental results are shown in Fig. 4.

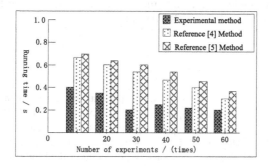

Fig. 4. Comparison of running time of different methods

As shown in Fig. 4, under the same conditions, the proposed method has the shortest running time, indicating that the proposed method has the lowest complexity and high operating efficiency.

5 Conclusions

In this paper, a time series data reconstruction method based on machine learning is proposed. The time series data distribution structure model under probability statistical model is constructed. The spatial multi-sensor information sampling method is used to sample the time series data information flow under the probability statistical model, and the phase space reconstruction method is combined to reconstruct the time series data information structure under the probability statistical model. The probability statistical model is established to decompose the time series data, and the distributed grid computing method is used to extract the big data association features of the time series data under the probability statistical model. Combined with the adaptive weight learning method, the optimal control of the scheduling is carried out. The big data cross-domain scheduling of the time series data under the probabilistic statistical model is realized under the support vector machine learning mode. The simulation results show that the method has good adaptability to time series data cross-domain scheduling under the probability and statistics model, and the load balance of data output is strong. The method has good application value in time series data reconstruction. However, due to the limited time, the efficiency of reconstruction of time series data needs to be improved, which is also my future research direction.

6 Acknowledgment

The 2018 annual scientific research project of Hubei Provincial Department of education. Based on modern statistical theory and machine learning theory, economic time series analysis is carried out (B2018371)

References

1. Bi, A., Dong, A., Wang, S.: A dynamic data stream clustering algorithm based on probability and exemplar. J. Comput. Res. Dev. **53**(5), 1029–1042 (2016)
2. Yang, J., Wei, C.: Testing serial correlation in partially linear additive models. Acta Mathematicae Applicatae Sinica, English Serie **35**(2), 401–411 (2019)
3. Zhang, X., He, Y.-H.: Modifid interpolatory projection method for weakly singular integral equation eigenvalue problems. Acta Mathematicae Applicatae Sinica English Serie **35**(2), 327–339 (2019)
4. Zhou, Z., Rahman Siddiquee, Md.M., Tajbakhsh, N., Liang, J.: UNet ++: a nested u-net architecture for medical image segmentation. In: Stoyanov, D., et al. (eds.) DLMIA/ML-CDS -2018. LNCS, vol. 11045, pp. 3–11. Springer, Cham (2018). https://doi.org/10.1007/978-3-030-00889-5_1
5. Sudre, C.H., Li, W., Vercauteren, T., Ourselin, S., Jorge Cardoso, M.: Generalised dice overlap as a deep learning loss function for highly unbalanced segmentations. In: Cardoso, M.J., et al. (eds.) DLMIA/ML-CDS -2017. LNCS, vol. 10553, pp. 240–248. Springer, Cham (2017). https://doi.org/10.1007/978-3-319-67558-9_28
6. Huang, G., Liu, Z., Laurens, V.D.M., et al.: Densely connected convolutional networks. In: CVPR 2017 Proceedings of the 2017 IEEE Conference on Computer Vision and Pattern Recognition, pp. 2261–2269. IEEE Computer Society, Washington, DC (2017)
7. Pan, C., Jia, Y., Cai, R., Yang, L.: Routing strategy for spatial information network based on MPLS. Comput. Eng. **45**(3), 85–90 (2019)
8. Lee, G.M., Lee, J.H.: On nonsmooth optimality theorems for robust multiobjective optimization problems. J. Nonlinear Convex Anal. **16**(10), 2039–2052 (2015)
9. Ma, M.Y., Chen, S.L., Zuo, Y.: Research on private set intersection cardinality protocol based on Goldwasser-Micali encryption system. Appl. Res. Comput. **35**(9), 2748–2751 (2018)
10. Patricia, N., Caputo, B.: Learning to learn, from transfer learning to domain adaptation: a unifying perspective. In: Proceedings of the IEEE Conference on Computer Vision and Pattern Recognition, pp. 1442–1449. Columbus, OH, USA (2014)
11. Sun, L., Guo, C.H.: Incremental affinity propagation clustering based on message passing. IEEE Trans. Knowl. Data Eng. **26**(11), 2731–2744 (2014)
12. Yu, Q., Tu, G., Li, N., Zhou, T.: Multi-hop multi-policy attributed-based fully homomorphic encryption scheme. J. Comput. Appl. **39**(8), 2326–2332 (2019)
13. Goyal, V., Pandey, O., Sahai, A., et al.: Attribute-based encryption for fine-grained access control of encrypted data. In: Proceedings of the 13th ACM Conference on Computer and Communications Security, pp. 89–98. ACM, New York (2006)
14. Mernik, M., Liu, S.H., Karaboga, M.D., et al.: On clarifying misconceptions when comparing variants of the Artificial Bee Colony Algorithm by offering a new implementation. Inf. Sci. **291**(10), 115–127 (2015)
15. Hsieh, T.J.: A bacterial gene recombination algorithm for solving constrained optimization problems. Appl. Math. Comput. **231**(15), 187–204 (2014)

16. Gentry, C., Sahai, A., Waters, B.: Homomorphic encryption from learning with errors: conceptually-simpler, asymptotically-faster, attribute-based. In: Canetti, R., Garay, J.A. (eds.) CRYPTO 2013. LNCS, vol. 8042, pp. 75–92. Springer, Heidelberg (2013). https://doi.org/10.1007/978-3-642-40041-4_5

17. Brakerski, Z., Cash, D., Tsabary, R., Wee, H.: Targeted homomorphic attribute-based encryption. In: Hirt, M., Smith, A. (eds.) TCC 2016. LNCS, vol. 9986, pp. 330–360. Springer, Heidelberg (2016). https://doi.org/10.1007/978-3-662-53644-5_13

18. Gentry, C., Peikert, C., Vaikuntanathan, V.: Trapdoors for hard lattices and new cryptographic constructions. In: Proceedings of the 40th Annual ACM Symposium on Theory of Computing, pp. 197–206. ACM, New York (2008)

19. Yan, X.X., Ye, Q., Liu, Y.: Attribute-based encryption scheme supporting privacy preserving and user revocation in the cloud environment. Netinfo Secur. 17(6), 14–21 (2017)

20. Zhou, Y.H., Shi, W.M., Yang, Y.G.: A quantum protocol for millionaire problem with continuous variables. Commun. Theoretical Phys. 61(4), 452–456 (2014)

Research on Intelligent Scheduling Optimization of Non-Full-Load Logistics Vehicle Based on the Monitor Image

Rui Li[1] and Haiying Chen[2(✉)]

[1] Software College, Nanyang Institute of Technology, Nanyang 473000, China
safasfas5465@163.com
[2] College of Humanities and Arts, Xianning Vocational Technical College,
Xianning, China
chenhaiying5454@163.com

Abstract. The traditional logistics vehicle scheduling method only estimates the total scheduling of batch vehicles, without considering the capacity limit of single logistics vehicle. It causes the problem of waste in vehicle transportation. Therefore, a vehicle scheduling method based on monitoring image under time constraints is proposed, using the time displayed in the monitoring image to constrain, the dynamic scheduling model is established by setting up the time window scheduling model to schedule the vehicle tasks within the time window conditions. According to the images obtained from the monitoring, combined with the need to divide several stages under the time constraints, the vehicles to ensure the logistics transportation can be scheduled according to the actual situation, make a highly optimal decision, achieve the maximum vehicle load rate, and ensure the smooth implementation of the dynamic strategy of the non full load logistics vehicle scheduling under the time constraints. Finally, the simulation test results show that the proposed method can improve the efficiency and rationality of logistics vehicle scheduling, the algorithm is stable and reliable, and has strong practicability.

Keywords: Integration of collection and delivery · Vehicle scheduling · Dynamic scheduling · Time window

1 Introduction

Non full load logistics vehicle scheduling is an important part of vehicle scheduling. The reasonable optimization of non full load logistics vehicle scheduling can improve the vehicle full load rate, which has great economic benefits and social value. It is widely valued by experts and scholars in related fields, and has made some progress [1]. In the traditional method of non full load logistics vehicle scheduling, the non full load logistics vehicle scheduling based on non-linear transformation path selection algorithm is mainly used, these non full load logistics vehicle scheduling methods only consider the distance and transportation speed in the process of cargo transportation, the implementation process is relatively simple, but the capacity limit of single transport vehicle is considered in the actual logistics vehicle scheduling process [2].

© ICST Institute for Computer Sciences, Social Informatics and Telecommunications Engineering 2020
Published by Springer Nature Switzerland AG 2020. All Rights Reserved
Y.-D. Zhang et al. (Eds.): ICMTEL 2020, LNICST 327, pp. 160–168, 2020.
https://doi.org/10.1007/978-3-030-51103-6_14

The overall capacity of the bulk vehicle is estimated and scheduled, resulting in a non-full-load waste in the transportation of the vehicle.

Aiming at the problems of traditional methods, a method of vehicle scheduling based on monitoring image is proposed, through the establishment of time window scheduling model and optimization of the model, time window constraints are obtained under the premise of image monitoring, and tasks are scheduled based on this condition. According to the needs of the division of a number of time constraints under the stage, in any stage of logistics transport vehicles can be scheduled according to the actual situation, make a high degree of optimal decision, to achieve the maximum vehicle load ratio. The simulation results show that the non full load logistics vehicle scheduling method can improve the efficiency and rationality of logistics vehicle scheduling, the algorithm is stable and reliable, and has strong practicability.

2 Non-Full Load Logistics Vehicle Scheduling Algorithm

Cargo collection and distribution are two important parts of logistics system [3], which play an important role in the whole logistics system, and vehicle scheduling information computing is also more and more important. The competition of logistics center urges users to pay attention to their own internal and also to strengthen the speed of software facilities, i.e. goods dispatching, so as to make rapid response to market demand. Only in the case of reasonable cost reduction, can we have a foothold in the fierce competition [4].

In the scheduling of non full load logistics vehicles, once the cargo transportation is required to arrive in a relatively tight time, the scheduling system shall be started to determine the logistics vehicle scheduling according to the current cargo and special needs [5]. Combined with the total vehicle capacity, the vehicle scheduling model of non full load logistics is established, to complete the vehicle scheduling problem constrained by the time window of non full load logistics vehicle scheduling, that is to say, the customer has time requirements for the distribution task, and the distribution activity should be carried out between the earliest time at and the latest time QT $(AT < t < QT)$ [6]. Time window is obtained by monitoring image, and time window is constrained. The problem of time window can be divided into hard time window and soft time window. For the hard time window problem, the vehicle must be delivered within the time window [AT, QT], while the customer does not accept the delivery outside the time window, that is, the solution obtained is infeasible. For the soft time window problem, if the vehicle arrives earlier than at, it needs to wait until at to deliver. At this time, it needs to bear a certain cost loss. If it arrives later than QT, it needs to bear a certain penalty cost. If it arrives too early or too late, the customer will not accept delivery, that is, it becomes a hard time window problem again. At this time, it needs an infinite penalty cost expressed by Q. The penalty cost can be expressed as follows:

$$f(t) = \begin{cases} Q \to +\infty, t \le e \\ v(AT - t), AT \le t \le QT \\ 0, AT \le t \le QT \\ k(t - QT), P \le t \le QT \\ A \to +\infty, t \ge r \end{cases} \quad (1)$$

Among them, $f(t)$ is the penalty cost caused by the time window constraint, e is the earliest acceptable waiting time, K is the cost coefficient generated when waiting; R is the latest acceptable time, V is the penalty cost coefficient to be borne by delayed delivery; P is the infinite penalty cost. Logistics distribution path optimization problem can be described as: a distribution center needs multiple vehicles to deliver goods to multiple customers in this area. The location of customers and distribution centers is determined, and each customer's demand for goods is determined [7]. Each distribution center has a variety of models, each model load is determined, and the maximum driving distance is determined. Each vehicle can serve multiple customers. Each customer can only have one vehicle to provide service. The vehicle starts from the distribution center and returns to the original distribution center after completing the distribution task. The distribution center shall deliver the goods to the customer within the time specified by the customer [8]. It is required to reasonably schedule vehicles to arrange the distribution path so that the objective function, i.e. the minimum distribution cost, can be optimized. The basic constraints of the problem are as follows:

(1) There are multiple distribution centers, each of which has a variety of models;
(2) The distribution center type is different, the weight is different, and the maximum travel distance is different;
(3) The distribution center is known to the customer's distribution needs and locations;
(4) The actual demand per customer does not exceed the maximum load of each model, and the total distance of a distribution mission is not greater than the maximum distance of a distribution vehicle;
(5) Each vehicle is only scheduled once, and each customer can only have one vehicle for distribution, and the vehicle will return to the original distribution center from the distribution center;
(6) The customer requires that the goods be delivered within a certain amount of time, taking into account the time window constraints.

From the above time window constraints, the mathematical model of multi-unloaded logistics vehicle distribution scheduling can be obtained as follows.:

$$\min G = \sum_{i=1} \sum f(t) + \sum_{i=1}^{j} X_{abcd}(Y + S) \le P \quad (2)$$

Among them, X is the distribution center set, I is each distribution center, a is the customer set, B is the vehicle set, C is the customer demand, D is the maximum load capacity of the vehicle in the distribution center, s is the distance from the distribution center to the customer, and Y is the transportation cost per unit mileage of goods.

3   The Design of the Vehicle Scheduling Model for Non-Full-Load Logistics

3.1   Design of the Non-Full-Load Logistics Vehicle Scheduling Process

The non full load logistics vehicle scheduling algorithm based on the monitoring image is divided into two stages. In the first stage, the adaptive genetic algorithm is applied to obtain the elite population as the initial solution, and in the second stage, the tabu search algorithm is applied to re optimize the initial solution to obtain the final solution to take account of customer satisfaction, while avoiding the "shock" phenomenon of scheduling [9, 10]. Considering the actual situation, the dispatching center will not dispatch vehicles far away from the dynamic demand point for service, so before searching all current paths, first select the candidate route set according to the location of the new demand point, using the buffer analysis function of GIS, combined with the demand of the customer point, so as to reduce the search time of logistics vehicle path [11, 12]. At the same time, through comprehensive consideration of customer satisfaction, vehicle driving distance, vehicle waiting time and other factors, the Logistics comprehensive cost increase caused by the addition of new customers is worth to be optimized. Combined with the above ideas, the non full load logistics vehicle scheduling process under the constraints of time window is designed, as shown in Fig. 1.

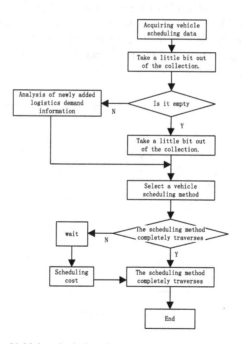

Fig. 1. Vehicle scheduling flow under time window constraint

3.2   Optimization of Vehicle Scheduling for Non-Full Load Logistics Based on Shortest Path

In the process of logistics and distribution, a traffic network structure is composed of each city point connected by roads. In the process of transportation, the shortest path problem is the minimum path from the logistics and distribution center to each city point. In general, the weight of the shortest path is represented by the distance between the source point and the target point, vehicle travel time and cost. In order to solve the shortest path problem in logistics network, it is necessary to find the lowest transportation cost, the shortest driving time and the shortest driving distance from the distribution center to the target point.

The input contains a weighted digraph g and a source vertex s in *G*. *V* is the set of all vertices in *G*. Every edge in a graph is an ordered pair of elements formed by two vertices. (U, V) indicates that there is a path from vertex *u* to *v*. With the set of all edges of *E*, w (*U, V*) is the nonnegative cost value from vertex u to vertex v. The cost of an edge can be thought of as the distance between two vertices. The cost value of a path between any two points is the sum of the cost values of all sides of the path. We know that there are vertices s and *T* in *V*, and find the lowest cost path from *s* to *t*. The driving cost between two points is equal to the driving time plus the distance between two points plus the driving cost. The distance between two customer points in the traffic network is known, and the driving time of vehicles can be roughly estimated. However, the amount of vehicle driving can only be determined according to the actual situation and experience. With the weight between two points, the problem is transformed into the shortest path problem. Therefore, the traffic network structure with weight is designed as shown in Fig. 2: v_0 in the figure represents the distribution center. The arc with arrow on the figure represents the road, and the value on the arc represents the driving cost of the vehicle.

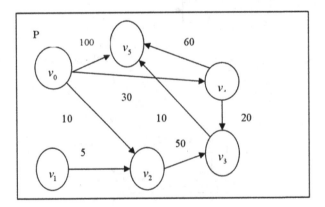

Fig. 2. Traffic network

Using the above network to carry out vehicle traffic, on the way of vehicle delivery, the dispatching center receives the demand from new customers. The dispatching

center records the customer information, goods information and service time requirements of the new demand according to the principle of first come first serve. According
to the interval time and the situation of new customers, judge whether to conduct
dynamic scheduling, as shown in Fig. 3. The judgment of scheduling time is to avoid
two kinds of situations: too long delay and too short delay. Too long delay will lead to
untimely customer service and affect the customer service level of logistics enterprises;
too short delay will lead to scheduling "oscillation", that is to say, too frequent
scheduling scheme changes will affect the implementation effect of the scheduling
scheme. The time interval threshold and new customer threshold are set by the dispatching center personnel according to their experience. In this paper, the threshold of
scheduling time is 30 min, and the threshold of new customers is 5.

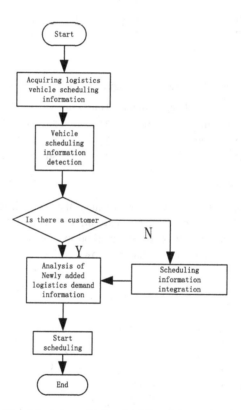

Fig. 3. Non-full-load logistics vehicle scheduling information detection process

4   Experimental Results and Analysis

In order to verify the effectiveness of the algorithm, it is assumed that there are three
types of non full load logistics vehicles under the time constraints obtained from the
monitoring image, a total of 300 vehicles, including 100 vehicles of a, B and C, each
with a full load time of 10 min and a cost of 20 tons. The vehicle attributes of each

vehicle (including load, fuel consumption, freight, etc.) are shown in Table 1. The simulation environment of this paper is: Windows 7 operating system, using the simulation software MATLAB 7.0, combined with the method of random data generation. In order to facilitate the comparative analysis of data, three kinds of transportation tasks with the same vehicle distribution are simulated, and the scheduling time, stability and scheduling frequency are taken as the basis to evaluate the effectiveness of this method (Table 2).

Table 1. Dispatching logistics vehicle attribute

Models	Load	Load	No load	Additional	Speed
A	8	32	0.94	0.22	80
B	14	46	0.81	0.16	86
C	20	60	0.93	0.14	70

Table 2. Dispatching logistics vehicle task data

Task	Weight/kg	Volume	Time/min	A	B	C
A	460	200	24	50	40	35
B	420	280	24	50	40	35
C	400	360	24	50	40	35

Due to the complex application environment of the non full load logistics vehicle scheduling method, the logistics vehicle scheduling time is tested first. The purpose of this paper is to test the effectiveness of this method in scheduling efficiency compared with traditional methods, and the results are shown in Fig. 4.

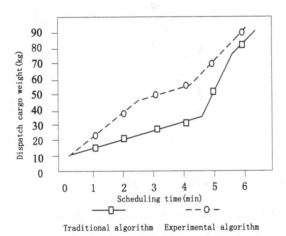

Fig. 4. Comparison and test results of dispatching efficiency of non-full-load vehicles

From the analysis of Fig. 4, it can be seen that there is not much difference between the two methods in the process of clustering under the condition of the same cargo volume and the same scheduling volume. However, with the continuous increase of the number of scheduled cargo, the method in this paper is used for the non full load logistics vehicle scheduling. The required scheduling time is far less than that of traditional methods, so we can see that, with the passage of time, the scheduling results of this algorithm are much higher than that of traditional algorithms. The purpose of the comparison is to test the stability of the two methods under the condition of large cargo volume. A total of 400000 tons of logistics resources are set for scheduling, and tested by cbkal software. The results are shown in the Fig. 5.

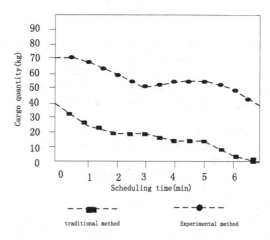

Fig. 5. Comparative Test results of Stability of vehicle scheduling methods for non-full load Logistics

Analyzing the experimental data in the Fig. 5, we find that under the same scheduling time, the stability rate of this method is obviously higher than that of the traditional method when scheduling non full load logistics vehicles. With the extension of scheduling time, its advantages are more obvious. On the whole level, the average scheduling stability of the non full load vehicle scheduling optimization method with time window can reach more than 50%. However, the average scheduling stability of the traditional method is about 13%, which proves that the stability of this method is much higher than that of the traditional method, and the scheduling results are stable and reliable.

5   Conclusions

In order to solve the problem of vehicle scheduling in logistics distribution, the time window algorithm is used to optimize the logistics distribution system, and the vehicle scheduling optimization algorithm under the constraints of time window is established, and the comparison and analysis of the non full load logistics vehicle scheduling situation is carried out, and a new model is established to realize the optimal design of non full load logistics vehicle scheduling. Finally, through the square array experiment, it is proved that the vehicle scheduling optimization method with time window is of high value.

References

1. Yang, G.: Research on optimization method of vehicle scheduling for non-full load logistics. Comput. Simul. **34**(03), 147–150 (2017)
2. Zheng, Z., Long, J.Y., Gao, X.Q.: Production scheduling problems of steelmaking-continuous casting process in dynamic production environment. J. Iron Steel Res. (Int.) **24** (6), 586–594 (2017)
3. Gili, S., Marra, W.G., D'Ascenzo, F., et al.: Comparative safety and efficacy of statins for primary prevention in human immunodeficiency virus-positive patients: a systematic review and meta-analysis. Eur. Heart J. **37**(48), 3600–3610 (2016)
4. Pan, C., Zhang, J., Qin, W.: Real-time OHT dispatching mechanism for the interbay automated material handling system with shortcuts and bypasses. Chin. J. Mech. Eng. **30**(3), 663–675 (2017)
5. Chamana, M., Chowdhury, B.H., Jahanbakhsh, F.: Distributed control of voltage regulating devices in the presence of high PV penetration to mitigate ramp-rate issues. IEEE Trans. Smart Grid **77**(99), 1 (2016)
6. Yu, Y.L., Li, W., Sheng, D.R., et al.: A hybrid short-term load forecasting method based on improved ensemble empirical mode decomposition and back propagation neural network. J. Zhejiang Universityence A **17**(2), 101–114 (2016)
7. Chen, L., Ming, Y.E., Jiang, Y.E., et al.: Study on ecological operation and influence of power generation of Longtan-Yantan cascade reservoirs on Hongshui River. J. Hydroel. Eng. **35**(2), 45–53 (2016)
8. Niu, W., Feng, Z., Cheng, C., et al.: Parallel multi-objective optimal operation of cascaded hydropower system. J. Hydraulic Eng. **48**(01), 104–112 (2017)
9. Sediqi, M.M., et al.: An optimization approach for unit commitment of a power system integrated with renewable energy sources: a case study of Afghanistan. Energy Power Eng. Engl. Vers. **8**, 528–536 (2017)
10. Tao, D., Lin, Z., Wang, B.: Load feedback-based resource scheduling and dynamic migration-based data locality for virtual hadoop clusters in OpenStack-based clouds. Tsinghua Sci. Technol. **22**(2), 149–159 (2017)
11. Shi, J., Lee, W.-J., Liu, X.: Generation scheduling optimization of wind-energy storage system based on wind power output fluctuation features. IEEE Trans. Ind. Appl. **16**(99), 1–7 (2017)
12. Gutiérrez-Mena, J.T., Gutiérrez, C.A., Luna-Rivera, J.M., et al.: A novel geometrical model for non-stationary MIMO vehicle-to-vehicle channels. IETE Tech. Rev. **7**, 1–12 (2017)

Research on Short-Term Load Forecasting Based on PCA-GM

Hai-Hong Bian[1,2(✉)], Qian Wang[1], and Linlin Tian[3]

[1] Nanjing Institute of Technology, Nanjing 211167, Jingsu, China
llq201801@163.com
[2] Jiangsu Collaborative Innovation Center for Smart Distribution Network,
Nanjing 211167, Jingsu, China
[3] School of Information and Control, Shenyang Institute of Technology,
Shenyang 113122, China

Abstract. In this paper, a short-term load forecasting model based on PCA dimensionality reduction technology and grey theory is proposed. After the correlation analysis between meteorological factors and load indicators, the data is carried out by combining PCA dimensionality reduction technology and grey theoretical load forecasting model. In this paper, the validity of the load data verification model in a western region is selected. The analysis of the example shows that compared with the general gray prediction model GM (1, 1), the accuracy of the model prediction result is much higher, which proves the model. Effectiveness and practicality.

Keywords: Short-term · Load forecasting · PCA-GM

1 Introduction

Under the current situation of increasingly tight international power supply situation, accurately predicting the power load and rationally arranging the power system production and power supply plan are of great significance to the safe and reliable operation of the power system, the sustained development of the national economy and the normal life of the residents [1]. There are many methods for power load forecasting, such as gray forecasting, regression analysis and forecasting, traditional trend analysis and exponential smoothing. These algorithms have their own advantages and disadvantages: (1) From the applicable conditions, regression analysis and trend analysis are devoted to the research and description of statistical laws, and are suitable for large-sample and consistent predictions of past, present and future development models; exponential smoothing The method is to use the principle of inertia to extrapolate the growth trend, while the gray model method seeks the law by sorting the original data, which is suitable for analysis and prediction under the condition of poor information; (2) from the data form adopted, gray The theoretical model uses the generated numerical sequence modeling. The regression analysis method and the trend extrapolation method are all based on the original data modeling. The exponential smoothing method directly predicts the future value by exponentially weighting the original data; (3) from the applicable time In terms of classification, regression analysis and trend

© ICST Institute for Computer Sciences, Social Informatics and Telecommunications Engineering 2020
Published by Springer Nature Switzerland AG 2020. All Rights Reserved
Y.-D. Zhang et al. (Eds.): ICMTEL 2020, LNICST 327, pp. 169–177, 2020.
https://doi.org/10.1007/978-3-030-51103-6_15

extrapolation are more suitable for medium and long-term predictions. For short-term predictions, it is more suitable for grey theory models. Although the traditional grey theoretical model is more efficient and simpler than other predictive models, it does not predominate when using long data series predictions. The data columns are too long, the system is subject to many disturbances, and the instability factors increase. On the contrary, the accuracy of the model is reduced, and the credibility of the prediction result is also reduced.

Because of the above problems, this paper improves the traditional grey theory algorithm, and considers the complexity and multi-dimensionality of meteorological factors that affect the load. A short-term load forecasting model based on PCA dimensionality reduction and grey theory is proposed. A correlation analysis and PCA dimension reduction method are used to analyze the degree of correlation between meteorological factors and load indicators, and the factors with low correlation degree are eliminated, and the multidimensional variables are reduced to one dimension by PCA dimensionality reduction technology, simplify the prediction model and reduce the amount of calculation.

2   Short-Term Load Forecasting Under Grey Theory

2.1   GM (1, 1) Model

The grey system theory is based on the concept of associative space, smooth discrete function and other concepts to define gray derivatives and gray differential equations [2]. By processing the original data, the system changes the law, and then the discrete data series is used to build the dynamic model of the differential equation. This is the intrinsic gray [3]. The basic model of the system, and the model is approximate, non-unique, so this model is called the gray model, denoted as GM (Grey Model).

The gray model in the general sense is GM (u, h), which means that u-order differential equations are established for h variables. The model used for prediction is generally GM (u, 1), and the most practical application is GM (1, 1), so only the GM (1, 1) model needs to be established.

The above standard "(0)" indicates the original sequence [4], and the superscript "(1)" indicates the cumulative generation sequence. The main steps of modeling and predicting GM (1, 1) are as follows:

(1) Let the original sequence be

$$X^{(0)} = \left[x^{(0)}(1), x^{(0)}(2), x^{(0)}(3), \cdots x^{(0)}(n) \right] \tag{1}$$

(2) Accumulating the original sequence (AGO) generation

$$x^{(1)}(j) = \sum_{i=1}^{j} x^{(0)}(i), j = 1, 2 \cdots n \tag{2}$$

Can get

$$x^{(1)}(i) - x^{(1)}(i-1) = x^{(0)}(i), i = 1, 2, 3 \cdots n \tag{3}$$

(3) Establish the corresponding differential equation as

$$\frac{dx^{(1)}}{dt} + mx^{(1)} = n \tag{4}$$

Where, m is the development coefficient and n is the ash dosage [5]. The effective interval of m is that the matrix composed of m and n is the grey parameter $\hat{m} = (m \quad n)$. only m and n are required, then $x^{(1)}$ can be obtained, and then the prediction value of $x^{(0)}$ can be obtained.

(4) Build the mean of the accumulated generated data B and the constant term vector Y_n

$$B = \begin{bmatrix} -\frac{1}{2}(x^{(1)}(1) + x^{(1)}(2)) & 1 \\ -\frac{1}{2}(x^{(1)}(2) + x^{(1)}(3)) & 1 \\ \vdots & \vdots \\ -\frac{1}{2}(x^{(1)}(n-1) + x^{(1)}(n)) & 1 \end{bmatrix}, Y_n = \begin{bmatrix} x^{(0)}(2) \\ x^{(0)}(3) \\ \vdots \\ x^{(0)}(n) \end{bmatrix} \tag{5}$$

(5) Solving the ash parameter $\hat{m}$ by the least squares method, then

$$\hat{m} = (B^T B)^{-1} B^T Y_n$$

(6) Substituting the ash parameter into the formula (1.4) and solving it,

$$\hat{x}^{(1)}(i) = [x^{(0)}(1) - \frac{n}{m}]e^{-m(i-1)} + \frac{n}{m}, i \geq 1 \tag{6}$$

(7) The above results are reduced and reduced to obtain the predicted value

$$\hat{x}^{(0)}(i) = \hat{x}^{(1)}(i) - \hat{x}^{(1)}(i-1), i \geq 2 \tag{7}$$

The establishment of the GM (1, 1) model generally requires a series of tests. Once the established GM (1, 1) model fails the test by the three methods of residual, correlation and posterior difference, the model must be tested. Corrected. Compared with the traditional methods of statistical analysis through a large number of samples, the superiority of gray system modeling is reflected in two aspects [6]. One is that the former requires a large amount of raw data, and its accuracy can be guaranteed, while the latter does not have such demanding requirements [7]. The second is that the latter generally uses a certain way to generate and process the original data (such as accumulation generation and subtraction generation), and organizes the disordered raw data into regular generation data, thereby weakening the randomness of the original random sequence. A smooth discrete function is obtained, which is further modeled based on these generated data. Therefore, the gray model has the characteristics of less information required for modeling and higher modeling accuracy.

2.2 Analysis of the Influence Degree of Meteorological Factors on Load

There are many factors that affect short-term load forecasting. However, we are unable to assess the extent to which these factors affect the load. Differences in load structure and climate will also change the form of the load. Therefore, short-term load is difficult to predict to a certain extent, which is the main solution to short-term load forecasting. The results show that meteorological factors are the most important factors affecting the load. Temperature is considered to be one of the main influencing factors [8]. In addition to the highest temperature of the day, the effects of average temperature and minimum temperature on the load should also be considered.

The influence of meteorological factors on power load forecasting is very complicated, and there are often interactions between different meteorological factors. This requires further analysis of the coupling strength between multiple meteorological factors and load forecasting [9]. The larger the correlation coefficient, the higher the correlation between the two; the lower the correlation coefficient, the lower the correlation.

The formula for calculating the correlation coefficient is as follows:

$$R = \frac{\sum\limits_{i=1}^{n}(x_i - \bar{x})(y_i - \bar{y})}{\sqrt{\sum\limits_{i=1}^{n}(x_i - \bar{x})^2 \cdot \sum\limits_{i=1}^{n}(y_i - \bar{y})^2}} \tag{8}$$

In the above formula: R is the correlation coefficient; $\bar{x}$ and $\bar{y}$ are the average values of meteorological indicators and load indicators [10], respectively. Among them, the larger R is, the higher the correlation degree is. When $|R| > 0.8$, x is highly correlated with y; when $|R| < 0.8$, x is associated with y low; when $|R| = 1$, x and y are completely related.

The correlation data between the highest temperature, average temperature, minimum temperature, and rainfall data of a certain week in July 2018 in the western part of July 2018 was calculated and correlated with the load data of the day. Four correlation coefficients were obtained, which were denoted by $R_1 \sim R_4$, such as Table 1 shows.

Table 1. Correlation coefficient result table

R_1	R_2	R_3	R_4
77.4%	81.0%	39.8%	11.0%

From the correlation coefficient, the correlation between rainfall and load data is small; due to the small amount of rainfall in summer, the relationship between rainfall and load in summer is also very small. Therefore, when constructing short-term load forecasting using the grey theory method, it is only necessary to consider the influence of the highest temperature, average temperature and minimum temperature on the load data.

3 Improved PCM-Based GM Model

PCA (Principal Component Analysis) is a common dimensionality reduction method. It focuses on mitigating dimensional disasters, minimizing information loss while compressing data, and understanding simple multidimensional data. When understanding feature extraction and processing data, the problems involving high-dimensional feature vectors tend to fall into dimensional disasters. As the data set dimension increases, the number of samples required for algorithm learning increases exponentially. When processing a large number of highest temperature, daily average temperature, minimum temperature and load data after correlation analysis [11], the data is sparsely high due to the higher dimension, and the same data set is explored in the high dimensional vector space. It is more difficult to explore sparse data sets.

Principal component analysis, also known as the Karhunin-Lough transform, is a technique for exploring high-dimensional data. PCA is commonly used for the exploration and visualization of high-dimensional data sets, and can also be used for data compression, data pre-processing, and so on. PCA can synthesize highly-dimensional variables that may be correlated into linearly independent low-dimensional variables, which are called principal components. The new low-dimensional datasets retain the variables of the original data as much as possible, and can effectively extract the most valuable data. Save a lot of time to achieve uniformity and coordination of load forecasting results.

There are two common methods for achieving PCA dimensionality reduction. One is based on the eigenvalue decomposition covariance matrix, and the other is based on the SVD decomposition covariance matrix. This article selects the latter, the specific steps are:

(1) Input: data set $X = \{x_1, x_2, x_3 \cdots x_n\}$, need to be reduced to k-dimensional;
(2) The average value is the average of each feature minus its respective average value;
(3) Calculate the covariance matrix $\frac{1}{n}XX^T$;
(4) Calculating the eigenvalues and eigenvectors of the covariance matrix $\frac{1}{n}XX^T$ by SVD;
(5) Sorting the eigenvalues from small to large, selecting the largest k, and then composing the corresponding k eigenvectors as the row vector to form the eigenvector matrix P;
(6) Convert the data into a new space constructed by k feature vectors, which is $Y = PX$.

In the PCA dimensionality reduction, the largest k eigenvectors of the sample covariance matrix need to be found, so that the matrix composed of the largest k eigenvectors can be used for low dimensional projection dimensionality reduction. In the process of using MATLAB to achieve PCA dimensionality reduction, a combination of SPE (square prediction estimation, also known as Q statistic) and T2 (principal component analysis) is used to monitor the fault. The Q statistic indicates the degree of deviation of the test value from the principal at this time. If the Q statistic is too large, it indicates that an abnormal situation has occurred in the process, so it is usually used to display the abnormal situation; the T2 statistic indicates the principal component analysis. Statistics. It can be seen from Fig. 1 and Fig. 2 that the Q statistic and the T2 statistic are basically within the threshold, indicating that the dimension reduction process is operating normally.

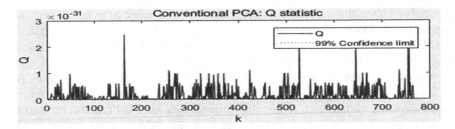

Fig. 1. Q Statistics

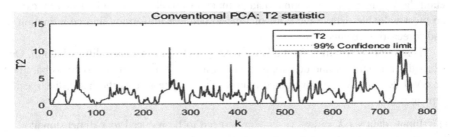

Fig. 2. T2 Statistics

4   Experiment and Result Analysis

In order to verify the accuracy of the load forecasting model based on PCA dimensionality reduction and grey theory proposed in this paper, the load data from July 27 to August 2, 2018 in the western region was taken as a sample, and in August 2018. On the 3rd day, as the forecast date, the 96 point load of the day was predicted. The specific process is shown in Fig. 3:

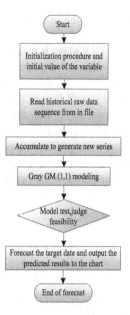

Fig. 3. Flow chart based on PCA and grey theory load forecasting

In this example, after the analysis of the correlation degree of meteorological factors, the influence of rainfall on the load index is removed, and only the three factors affecting the daily maximum temperature, the daily average temperature and the daily minimum temperature are considered; The maximum temperature, the daily average temperature and the daily minimum temperature are used for dimensionality reduction. Finally, the gray system theory is used to predict the power system load value. The simulation results are shown in Fig. 4 and Fig. 5.

Figure 4 shows the comparison between the predicted 96 points and the actual 96 points. The red line is the predicted value and the blue line is the actual value.

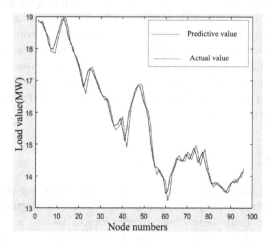

Fig. 4. Comparison of predicted and actual values (Color figure online)

It can be seen from the analysis of the above figure that the change range of the actual value is 18.8 MW–14.2 MW, and the change range of the predicted value of the method in this paper is 18.8 MW–14.3 MW. It can be seen from the observation that the predicted value and the actual value of the research method in this paper have a high degree of fit, which shows that the prediction accuracy of the method in this paper is high, and it can realize the accurate prediction of the short-term load. The reason why this method has high prediction accuracy is that the improved PCA GM based prediction model considers the complexity and multi-dimensional of climate factors, improves the accuracy of load forecasting, overcomes the shortcomings of traditional prediction methods, so it has high prediction accuracy.

Figure 5 clearly shows the error between the predicted value and the actual value.

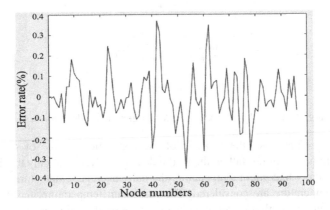

Fig. 5. Error between predicted and actual values

It can be seen from the analysis of Fig. 5 that the prediction error of this method varies from −0.36% to 0.37%, the prediction error is small, and the change amplitude of the prediction error is small, which further proves that this method can realize the accurate prediction of short-term load, and verifies the comparison results of Fig. 4. The reason is that by calculating the correlation between meteorological factors and load index, the influencing factors with high correlation are selected. On this basis, the multi-dimensional meteorological factors are reduced to one dimension, and the GM (1, 1) grey theoretical model is used for prediction, which effectively solves the problems existing in the traditional methods, reduces the prediction error and improves the prediction accuracy.

5   Conclusion

(1) By calculating the degree of correlation between meteorological factors and load indicators, the influencing factors with high correlation degree are selected. On this basis, the multi-dimensional meteorological factors are reduced to one-dimensional, and the GM (1, 1) gray theoretical model is used for prediction.

The error between the predicted value and the actual value obtained after bringing in the sample is only within $\pm 0.04\%$, which is closer to the actual situation, which proves that the model is reasonable.

(2) Considering the complexity and multidimensionality of climatic factors, the improved PCA-GM-based prediction model improves the accuracy of load forecasting and overcomes the shortcomings of traditional forecasting methods.

(3) Compared with the traditional GM (1, 1), the result is more accurate, and the calculation model is simplified, the calculation amount is reduced, the prediction accuracy is improved, and the validity of the model is proved.

In the load forecasting process, only four influencing factors of maximum temperature, average temperature, minimum temperature and precipitation are considered. In the actual situation, there are still many influencing factors to be explored. Uncertainty and controllability in load forecasting The variables are more complicated and there are still many problems to be further studied.

Fund Projects. The project was supported by the Open Research Fund of Jiangsu Collaborative Innovation Center for Smart Distribution Network, Nanjing Institute of Technology (No. XTCX 201807).

2019 Jiangsu Province Graduate Practice Innovation Plan (SJCX19_0519).

References

1. Gao, C., Li, Q., Su, W., et al.: Temperature correction model considering the effect of accumulated temperature in short-term load forecasting. Trans. China Electrotech. Soc. **30**(4), 242–248 (2015)
2. Liu, B., Ma, J., Li, X.: A topic representation model of "Feature Dimensionality Reduction" text complex network. Data Analysis Knowl. Disc. **1**(11), 53–61 (2017)
3. Li, Q.: New characteristics of load characteristics analysis and load forecasting in smart grid (2014)
4. Zhang, Q.: Short-term load forecasting considering temperature accumulation effect (2014)
5. Wang, W., Wang, B., Yu, H., et al.: Maximum load forecasting of power grid based on load decomposition and grey theory. Zhejiang Electric Power **26**(8), 48–53 (2018)
6. Cheng, D., Liu, J., Guo, W., et al.: Power load forecasting of central china power grid based on accumulated temperature effect. Meteorol. Sci. Technol. **46**(269(04)), 186–193 (2018)
7. Zhang, H., Li, W., Xiang, C., et al.: Short-term load forecasting based on temperature cumulative effect and grey correlation degree. Electric Autom. **26**(3), 12–19 (2019)
8. Cheng, Z.: Research on short-term load combination forecasting method considering accumulated temperature effect (2010)
9. Zhang, Q., Wang, Y., Lu, Y.: Study on summer daily maximum load forecasting considering accumulated temperature effect. Power Demand Side Manag. **26**(6), 1–5 (2013)
10. Renyuan, Z.: Analysis of the cumulative effect of temperature in Shanghai electricity load forecasting. Modern Electric Power **35**(2), 38–42 (2018)
11. He, W., Ye, P., He, N., et al.: Review of power system saturation load prediction research. J. Shenyang Inst. Technol. (Natural Sci. Ed.) **13**(4), 340–346 (2017)

Research on Preprocessing Algorithm of Two-Camera Face Recognition Attendance Image Based on Artificial Intelligence

Lin-lin Tian[1], Wan-li Teng[1], and Hai-hong Bian[2,3(✉)]

[1] School of Information and Control, Shenyang Institute of Technology,
Shenyang 113122, China
tianlinlin1258@tom.com
[2] Nanjing Institute of Technology, Nanjing 211167, China
llq201801@163.com
[3] Jiangsu Collaborative Innovation Center for Smart Distribution Network,
Nanjing 211167, China

Abstract. The traditional double-camera face recognition attendance image preprocessing algorithm can not distinguish the target from the complex background. In order to solve this problem, an artificial intelligence based double-camera face recognition attendance image preprocessing algorithm is proposed. First, artificial intelligence technology is used to extract the features of face recognition attendance image, and then spatial denoising algorithm is used to remove the noise of face recognition attendance image. On this basis, multi-channel texture weighting algorithm is used to realize the double-camera face recognition attendance image preprocessing. Therefore, a double-camera face recognition image preprocessing algorithm based on artificial intelligence is completed. In the experiment, the infrared image of the face is tested to see whether the evaluation factors obtained by the two algorithms can distinguish the target from the complex background. Experimental results show that the algorithm has a short computing time and can distinguish targets in complex background in a short time.

Keywords: Two camera face recognition · Image preprocessing · Artificial intelligence technology · Noise reduction processing

1 Introduction

The initial face recognition method is based on geometric features. The core idea of this kind of method is: first detect and extract the main face parts such as eyes, nose and mouth, and then integrate the geometric distribution relationship and mutual parameter ratio of these parts for face recognition [1]. Typical algorithms include semi-automatic face recognition attendance image preprocessing algorithm and profile algorithm. Face recognition based on geometric features of attendance image preprocessing algorithm, light and attitude interference resistance is weak, and geometric features of the face information is less than the overall face features. Therefore, currently, the commonly used methods are to extract the features of the whole face image, then conduct the

Y.-D. Zhang et al. (Eds.): ICMTEL 2020, LNICST 327, pp. 178–191, 2020.
https://doi.org/10.1007/978-3-030-51103-6_16

operations of reducing and classifying, and finally get the face recognition results. In order to enhance the interference resistance of illumination and attitude, a face recognition algorithm based on artificial intelligence was proposed.

The double-camera face recognition attendance image preprocessing algorithm based on artificial intelligence has the support of perfect statistical theory. They treat face samples as random vectors and use statistical methods to extract and classify faces. On this basis, extract the features of each face image, string the gray value of all pixels into a high-dimensional vector, and then remove the noise of double-camera face feature recognition attendance image through principal component analysis, so as to realize attendance image preprocessing. Based on artificial intelligence dual camera face recognition attendance image preprocessing algorithm specific implementation process is as follows.

2   Artificial Intelligence Dual Camera Face Feature Recognition Attendance Image Feature Extraction

Using the principle of artificial intelligence technology to extract double camera face feature recognition attendance image features. Artificial intelligence technology has the function of data classification and dimensionality reduction. Its essence is to create a coordinate system in the data space and project the data into the new coordinate space respectively. It can effectively remove the correlation degree between different types of data vectors, effectively realize data classification, and at the same time, it can also remove some coordinate system composed of data redundancy, so as to compress the data and reduce the dimension of feature space.

Complete artificial intelligence technology double camera face recognition attendance image feature extraction steps are as follows:

The first step is to normalize the face database and obtain the feature matrix X based on the data. Each column of the matrix represents an feature vector.

The second step is to find the autocorrelation matrix $R = E[X^T X]$ of the characteristic matrix X.

The third step is to solve the eigenequation and obtain the eigenvalue and eigenvector from the correlation matrix R. The eigenvalue obtained constitutes a diagonal autocorrelation matrix, which is a diagonal matrix, so each element is the autocorrelation coefficient of the transformation matrix. The eigenvector should be linearly independent and normalized [2].

The fourth step is to obtain the transformation matrix by treating the eigenvectors as column vectors. $\phi = (\phi_1, \phi_2, \ldots, \phi_n)$.

The fifth step is to obtain the transformation features by calculating $\alpha = \phi^T X$. Since the autocorrelation matrix of the new features is a diagonal matrix, it is linearly independent [3].

The sixth step is to select the features with larger eigenvalues for the application of classification. This is because the eigenvalue is a measure of variance, and the larger its value is, the more relevant the data is, and the larger the value range is [4].

Step 7: take 5 different scales $v = (0, 1, 2, 3, 4)$ and 8 different directions $\mu = (0, 1, 2, 3, 4)$ to get 6 different Gabor nuclei. Double-camera face recognition attendance image Gabor features, as shown in Fig. 1:

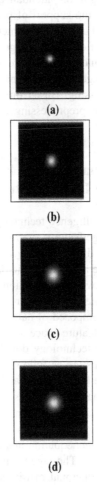

Fig. 1. Gabor features of double-camera face recognition attendance image

Complete Gabor features of human faces were obtained by using the Gabor nucleus.

3 Dual Camera Face Feature Recognition Attendance Image Noise Removal

The quality degradation of double-camera face feature recognition attendance image is often affected by some random errors, which is called noise [5]. Noise may occur in the process of generating, transmitting, converting, receiving and processing video images. For example, when a camera is used to obtain a video image, both the light level and the temperature change of the sensor are the main factors that produce image noise. In the transmission process, the image is interfered by the atmospheric environment, electromagnetic multiple number and other transmission channels, which will also introduce image noise [6]. Noise will reduce the image quality, which will make the subsequent image analysis and understanding difficult, and seriously affect the user's visual experience. For this reason, the noise reduction processing of face recognition attendance image with dual camera is proposed.

According to the design requirements of video image preprocessing system, a fast and efficient spatial noise reduction algorithm is most suitable. A fast adaptive gaussian noise reduction algorithm is proposed, which is described as follows:

For pixel $\mu_{|i,j\rangle}$ to be processed, take its surrounding filtering neighborhood $R_{|i,j\rangle}$, as shown in Fig. 2:

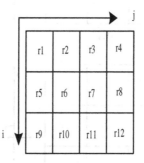

Fig. 2. The domain template for the pixels to be processed

Determine whether the pixel to be processed is polluted by gaussian noise through standard deviation $std - dev$, and define the formula:

$$std - dev = \left| \mu_{|i,j\rangle} - w - mean \right| \tag{1}$$

Where, $w - mean$ represents the grayscale weighted mean of pixels in the neighborhood set, and $\mu_{|i,j\rangle}$ represents the weight of 4. Here, the weight is set as 2 to facilitate hardware calculation and processing [7].

At the same time, sobel operator is introduced to detect the edge texture information of the image in order to protect the detail texture of the image from being

excessively smooth during noise reduction [8]. And the final filtering mode is determined accordingly. The operator detection template of the four squares is shown in
Fig. 3.

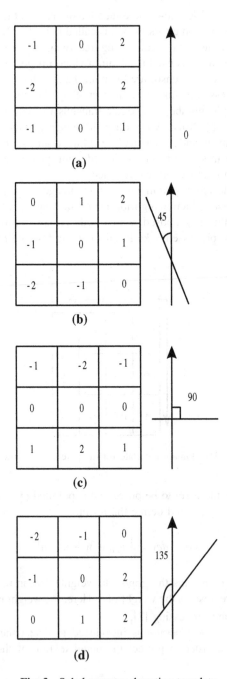

Fig. 3. Sobel operator detection template

The template is convolved with the neighborhood template of the pixel to be processed. After the calculation of standard deviation $std - dev$ and the representation of max-edge by the most human edge information, the paper first determines whether the processed pixels are polluted by noise by comparing $std - dev$ with the noise threshold NTH [9]. If $std - dev$ < NTH, it means that the pixel point is not polluted and its gray level is guaranteed to remain unchanged. If $std - dev$ > NTH, it means that the pixel point is judged as a noise point. For noise points, if the corresponding max-edge is greater than or equal to the edge judgment threshold NTH, it indicates that the noise point is located at the image details, and then the noise point is replaced and interpolated according to the edge direction.

The noise removal process of double-camera face feature recognition attendance image is as follows (Fig. 4):

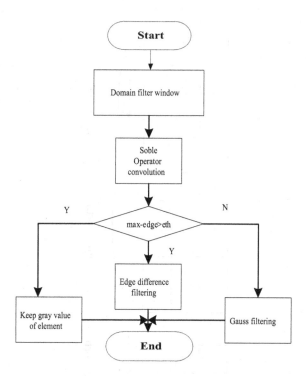

Fig. 4. Noise removal process of double-camera face feature recognition attendance image

In the concrete implementation, the solution of $std - dev$ and the convolution calculation of max-edge can be processed in parallel, so as to improve the processing speed of the whole algorithm process.

4 Dual Camera Face Feature Recognition Attendance Image Preprocessing

In order to further improve the suppression effect of noise reduction algorithm on high-concentration impulse noise, a fast and efficient algorithm based on decision theory was proposed [10]. The algorithm principle is as follows:

For neighborhood filtering window, first of all gray levels of pixels in the window to ascending, row from left to right, on this basis, the gray values of all pixels from top to bottom of sequence alignment, and finally, based on the results of the procession sorted, gray levels of pixels on the diagonal to ascending diagonal arrangement, the whole process as shown in the figure below (Fig. 5).

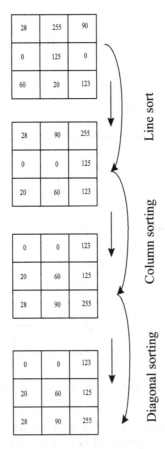

Fig. 5. Sequence flow of impulse noise suppression

It can be seen that the gray value of the pixel in the window is minimum value p_{min}, the gray value of the last pixel is maximum value p_{max}, and the gray value of the center point of the filter window is maximum value p_{med}. If the pixel $\mu_{|i,j\rangle}$ to be processed meets $p_{min} < \mu_{|i,j\rangle} < p_{max}$, it can be determined that $\mu_{|i,j\rangle}$ is not polluted by impulse noise and its gray value remains unchanged. Otherwise, $\mu_{|i,j\rangle}$ is the noise point. If $\mu_{|i,j\rangle}$ is polluted by noise, it satisfies $p_{min} < \mu_{|i,j\rangle} < p_{max}$, then the median p_{med} of the table neighborhood filter window is not polluted by noise, and the gray value of the pixel to be processed is replaced by this median. Finally, in the case of noise points, the gray value of the neighborhood pixel to the left of the median value is used to replace the pixel to be processed.

The artificial intelligence based two-camera face recognition attendance image prediction and processing algorithm also has a good processing effect for the image polluted by high concentration pulse noise, and only a few pulse noise points remain in the processed image.

Image preprocessing can improve the image signal-to-noise ratio transmitted to the main processor and reduce the amount of data stored and processed, so it is of great significance. According to the principle of target detection, this chapter will use multi-channel texture weighting algorithm to realize the double-camera face recognition attendance image preprocessing algorithm based on artificial intelligence, and analyze and compare these algorithms.

Image contrast and signal-to-noise ratio are two commonly used indicators to evaluate the excellent preprocessing, which are defined as follows:

$$contrast = (g_t - g_d) \tag{2}$$

Where, contrast represents the contrast before and after the image, and g_t is the average gray value of the target. g_d refers to the average grayscale size of the background.

Preprocessing time is another important parameter of preprocessing performance, which is defined as: $T = t_2 - t_1$, t_1 is the system time of image processing, t_2 is the time after image processing.

In image preprocessing, the performance parameter used to indicate the ability of spatial filter to suppress background clutter is absolute mean square error, which is defined as follows:

$$MSE = \frac{1}{m \times n} \sum (Y_{ij} - X_{ij}) \tag{3}$$

Among them, X_{ij} is the original initial image, Y_{ij} is the output image, and MSE is the smaller, the better the filter performance.

After obtaining the two commonly used indexes to evaluate the excellent preprocessing, in order to avoid the complex and time-consuming process of solving the
sparse representation coefficient of the image in the overcomplete dictionary, the image
after solving the column vectorization and the atom of the overcomplete dictionary
with the same vectorization were transformed into vector projection, and the algorithm
principle was simpler. After first use significant processing as a filter for image preprocessing module, by setting up a reasonable threshold to realize image binarization,
and extract the image block is in the original image as the interest area, and then USES
the two-dimensional gaussian model generation sample image, and then construct
super perfect target dictionary, with columns to quantify the interested region after
image block with a list of vectorization atom in turn do vector projection and calculation of the normalized projection coefficient, using the projection coefficient of
background and target on the atoms have bigger difference, can set a threshold to
distinguish. The ROI blocks larger than the threshold were used to enhance the image
gray scale compared with the original image, while those smaller than the threshold
were used to suppress the image gray scale. The final image could quickly detect the
target.

If the infrared image is $I(i,j)$, the significance image is $S(i,j)$:

$$A(f) = \alpha(fft[I(i,j)])$$
$$p(f) = \phi(fft[I(i,j)]) \tag{4}$$
$$L(f) = \log(A(f))$$

In the above formula, ϕ, α respectively represent the phase and modulus of complex Numbers, log represents convolution, *fft* represents Fourier transform and its
inverse transformation, and $A(f)$ represents mean.

According to the above formula, the calculation amount of the algorithm in this
paper is small and the real-time performance is high. Figure 1 shows the image containing the infrared small target and the result image of significant domain detection. In
contrast image 2, only one target is included, and the contrast between the target and
the background is very low. Also, after the detection of the significance domain, some
areas with sharp changes in the noise background are also detected as the significance
domain. In Fig. 6, in addition to the two dim targets, a dark interference is artificially
added to the upper right of the image, and the infrared small target is processed after the
processing of the significant domain.

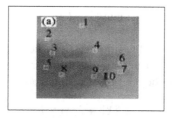

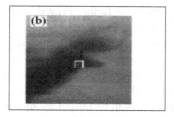

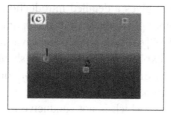

Fig. 6. Pre-processing results of infrared small target image and corresponding image significant domain

In order to obtain the optimal solution of infrared small target, it is necessary to know the optimization and the optimal solution of norm. Considering the matrix $A \in B$, an underdetermined linear system with an equation of $Ax = b$ is defined. If the system still has many unknown equations, case 1 is that the system has no solution. The other case is that if b is in the space extended by the matrix A column vectors, the equation has an infinite number of solutions. To avoid the case of no solution, the A matrix can be determined as full rank. If the signal is a single frame image, matrix A represents the corresponding compression or other domain transformation operations, and a relatively small amount of information b is obtained. If we want to get the original signal from information b, we should look for the more excellent solution sets in the infinite solution sets of x.

By defining constraints to narrow the scope of the solution, one of the methods is to adopt the normalized method, which estimates the characteristics of x by defining a $j(x)$ function. Generic optimal solution problem $\{p_j\}$ is defined as follows:

$$\{p_j\} : \min j(x) \, subect = Ax \tag{5}$$

In the above equation, p_j defines the least two norm problem for the most commonly used Euclidean mean square norm $j(x)$, $\{p_2\}$ problem and has a unique solution x. Lagrange factor is used to define Lagrange equation:

$$L = \|X\| + \lambda(Ax - b) \tag{6}$$

In Eq. (6), $\|X\|$ is the Lagrangian factor. Take the derivative of both sides of the above equation with respect to x.

The above $\{p_2\}$ problem is widely used in engineering field. The $\|X\|$ solution is simple and has closed form and uniqueness. Therefore, $\{p_2\}$ problem should be widely applied in signal and image processing field, but it does not mean that $\{p_2\}$ has the universality of solving the optimal solution in all problems. On the contrary, most of the time, researchers and engineers will be misled by the mathematical simplicity of the optimal solution, which is not conducive to the selection of the optimal $j(x)$.

Considering the optimal solution of other such norm and the problem of sparse tendency of norm, this paper gives a detailed introduction and further supplementary explanation. Considering other problems such as the optimal solution of $\{p_j\}$ norm of $p < 1$ and the sparsity trend of $\{p_2\}$ norm, this paper gives a more detailed introduction and further supplementary explanation. However, it should be pointed out that the norm problem of $\{p_2\}$ is very different from that of $\{p_j\}$ on the surface, but actually it is completely the opposite. In addition, the reverse recovery signal is computationally intensive and time consuming, which is not conducive to practical applications.

Therefore, a double-camera face recognition image preprocessing algorithm based on artificial intelligence is completed.

5 Experimental Analysis

In order to prove the performance and robustness of the proposed double-camera face recognition attendance image preprocessing algorithm based on artificial intelligence, we have done experiments on face recognition attendance image. At the same time, we compare the existing two better algorithms. One is the traditional algorithm, the other is based on artificial intelligence dual camera face recognition attendance image preprocessing algorithm (Fig. 7).

5.1   Experimental Procedures

Experimental process:

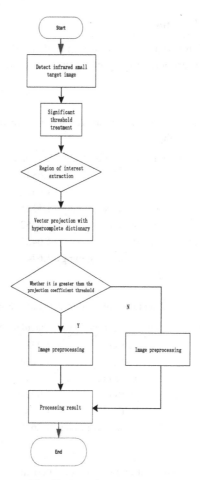

Fig. 7. Experimental process

First, the tested face recognition attendance image was processed in significant domain, and the region of interest of 8 * 8 size in the figure was extracted according to the binarization results. Through gaussian gray model GIM, atoms of 8 * 8 sizes are generated, 30 different target center positions are set, and 19404 infrared small target samples are generated. Therefore, an ultra-complete infrared small target dictionary of 64 × 19404 dimension is formed. Then, according to the vector projection method, the projection coefficient of each ROI in the overcomplete dictionary was calculated. The threshold was set as 0.79, and the threshold was judged according to the normalized vector projection coefficient. The gray value of ROI greater than the threshold was enhanced, and the enhancement factor was 1.2. The inhibitory factor less than gray inhibition was 0.37, and the inhibitory factor for background of non-interested region was 0.17. Finally, all regions are integrated to obtain a normalized gray distribution after image preprocessing algorithm.

5.2 The Experimental Results

The detection results of the traditional algorithm and the double-camera face recognition attendance image preprocessing algorithm based on artificial intelligence are as follows:

Table 1. Detection results of double-camera face recognition attendance image preprocessing algorithm based on artificial intelligence

Detection algorithm	Evaluation factor	Calculation results
Chart (a)	SCRG	1.6542
	BSF	1.6542
	Time consuming	1.5121
Chart (b)	SCRG	1.6452
	BSF	1.6421
	Time consuming	1.6745

Table 2. Traditional algorithm detection results

Detection algorithm	Evaluation factor	Calculation results
Chart (a)	SCRG	1.2316
	BSF	1.3264
	Time consuming	0.1326
Chart (b)	SCRG	1.0321
	BSF	1.2361
	Time consuming	1.2636

Face images contain strong clutter and noise. According to Table 1 and Table 2, the algorithm in this paper has a higher value than the traditional algorithm. This fact also proves that this method is more effective than other methods in such bad conditions. Although this algorithm will be more time-consuming than other methods, it is acceptable to some extent.

6 Conclusion

In view of the problems existing in the traditional algorithms, this paper proposes a double-camera face recognition image preprocessing algorithm based on artificial intelligence. First extract the face is the attendance image image features, and then remove the image noise, the realization of double camera face recognition attendance image preprocessing. Therefore, the research on the pre-processing algorithm of attendance image based on the artificial intelligence of dual camera face recognition is completed. Because this study only preprocesses the attendance image, but does not consider the problem of unclear recognition when employees wear glasses, so in the future, we will further study the two camera face recognition attendance system based on artificial intelligence.

7 Acknowdgments

National Natural Science Foundation of China (61603262, 61403071);

Liaoning Natural Science Foundation (20180550418);

Shenyang Institute of Technology i5 Intelligent Manufacturing Research Institute Foundation (i5201701).

References

1. Zheng, P.: Multi-level deep network fused for face recognition. Pattern Recog. Artif. Intell. **30**(5), 448–455 (2017)
2. Zhang, Y.: One sample per person face recognition based on deep autoencoder. Pattern Recog. Artif. Intell. **14**(6), 335–336 (2017)
3. Qi, M.: Face recognition algorithm based on improved deep networks. Acta Electronica Sinica **25**(3), 118–119 (2017)
4. Liu, S., Fu, W., Zhao, W.: A novel fusion method by static and moving facial capture. Math. Prob. Eng. (2013). https://doi.org/10.1155/2013/503924
5. Zhang, J., Liu, H.: Multimodal recognition of faces and fingerprints based on generalized canonical correlation analysis and robust probabilistic collaborative representation. J. Univ. Shanghai Sci. Technol. **40**(2), 158–165 (2018)
6. Zhang, P., Xu, X.: Computer-aided lung cancer diagnosis approaches based on deep learning. J. Comput.-Aided Des. Comput. Graph. **30**(1), 90–99 (2018)
7. Zheng, P., Li, B.: Local neighborhood based multi-manifold metric learning for face recognition. Appl. Res. Comput. **35**(4), 1250–1253 (2018)
8. Zhang, B., Wu, Q.: Method for recognizing gesture of traffic police based on DTW algorithm. Appl. Res. Comput. **34**(11), 3494–3499 (2017)
9. Cao, Y., Yuan, Y.: Image recognition method for building wall cracks based on feature distribution. J. Shenyang Univ. Technol. **40**(2), 235–240 (2018)
10. Qin, H.: Case study on application of face recognition technology in the library. J. Acad. Libraries **12**(2), 135–140 (2018)

Research on Military Intelligence Value Evaluation Method Based on Big Data Analysis

Li-li Xu[1] and Feng Jin[2(✉)]

[1] 95795 Troop, Guilin, Guangxi, China
xulili128@2980.com
[2] Information and Communication College,
National University of Defense Technology, Guilin, China
jingfeng250052@sohu.com

Abstract. The conventional methods of military intelligence assessment could not comprehensively analyze the value of military intelligence. To this end, a military intelligence value assessment method based on big data analysis was proposed. Big data analysis technology was introduced to determine data value density k; overall system architecture was established; operation mode was optimized, relevant analysis technology was formulated; military intelligence value evaluation was achieved. Experimental data showed that the application of big data analysis technology could comprehensively analyze the value of military intelligence.

Keywords: Big data · Military intelligence analysis · System structure

1 Introduction

As the development of human society has entered the wave of the information age, data has become an important strategic resource for promoting economic and social development, scientific and technological progress and new military changes. Big Data, or massive data, is a collection of data with large-scale, diverse types, time-sensitive processing, and low reliability of data sources. Big data has penetrated into all areas of human social life, and it has brought unprecedented changes and challenges to the analysis and research work of military intelligence. On the one hand, with the rapid development of military informationization, a large amount of intelligence data can be easily obtained, and data overload phenomenon occurs; On the other hand, the data obtained is all-encompassing, the content is complicated, the reality is mixed, the authenticity is difficult to distinguish, and there is a shortage of effective intelligence information [1]. How to mine and extract accurate and high-value intelligence information from these overloaded data has become an urgent problem to be solved. Literature [2] proposes the construction method of patent value evaluation system based on big data analysis. Starting from the big data analysis of evaluation indicators, the evaluation model of "total sub sub sub sub sub" three-tier structure is constructed, That is to say, "law, technology and economy" as the first level indicators, "patent protection scope, patent stability, patent technology quality, patent technology applicability, market, competition, applicant factor" as the second level indicators, and the quantifiable 81 subdivision indicators as the third level indicator

Y.-D. Zhang et al. (Eds.): ICMTEL 2020, LNICST 327, pp. 192–200, 2020.
https://doi.org/10.1007/978-3-030-51103-6_17

system, and the calculation method of patent value degree based on this system is described. However, the data density of the technique is low, and the time required to intercept the same data is longer.

In view of the above problems, in order to comprehensively evaluate the value of military intelligence, we need to use big data analysis technology to calculate the data density k. Then the overall system architecture is established, the operation mode is optimized, and finally relevant analysis techniques are developed.

2 Design of Military Intelligence Value Evaluation Method Based on Big Data Analysis

The use of big data analytics to assess the value of events is primarily determined by the data value density k.

Data value density k:

$$k = \frac{\sum UjQ_I}{\sum Uj} \tag{1}$$

i is the size of a single data, and U is the size of the total amount of data in the unit interval. The large amount of data often means extremely high value, but because of the large scale and variety, the value density of data is very low, which also increases the difficulty of value mining. At present, big data processing related technologies and tools have made great progress, forming a relatively complete big data processing ecosystem. It includes large-scale distributed computing technology, in-memory computing technology, stream processing technology, etc. Corresponding software products such as Big Data Batch Computing Framework Hadoop, Rapid Analysis Framework Spark, Stream Processing Framework Storm, etc. As shown in Fig. 1

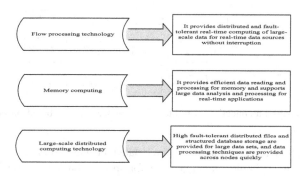

Fig. 1. Big data processing related technology

In the early big data analysis, people didn't know the value of the data, so most people took the statistical analysis method. As people's understanding of the value of data continues to deepen, the methods of data analysis are gradually moving toward predictive and normative directions. Big data technology on the one hand makes people's data sources more efficient and diversified; On the other hand, while the data source is growing rapidly, it also makes people more efficient and efficient in data processing. Together with the application of visual data analysis technology, enterprise decision making is more justified.

2.1   Establish Overall System Architecture Based on Big Data Analysis

The overall architecture design of the system first needs to meet the analysis and application requirements of massive heterogeneous military intelligence data, and adopts big data processing technology to provide business applications such as collection, reorganization analysis, comprehensive processing, and intelligence services for military intelligence data. This provides intelligence data and service support for strategic decision-making and operational command; Secondly, in order to solve the problem of resource sharing and collaborative application among multiple nodes in the intelligence system, the openness and flexibility of the system architecture should be fully considered to achieve loose coupling between data and platform, platform and application [3]. Finally, it is necessary to fully comply with or refer to relevant military system standards and technical specifications to achieve seamless connection with existing intelligence systems and improve the utilization efficiency of existing equipment. The overall structure of military intelligence analysis and service system based on big data can be divided into three aspects, including application layer, service layer and resource layer as shown in Fig. 2.

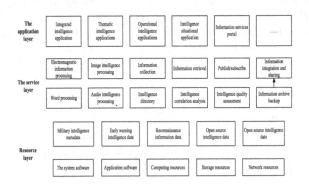

Fig. 2.   Overall architecture of military intelligence analysis based on big data

The application layer is based on the related business services running on the service layer. It is an implementation layer for specific business applications, and is mainly used to support human-computer interaction operations. The application layer includes software modules such as integrated intelligence applications, thematic intelligence applications, combat intelligence applications, intelligence situation applications, and intelligence service portals.

The service layer relies on the information sharing and collaborative environment provided by the application layer to provide information services for each service domain while implementing specific service processing functions. The service layer includes services such as electromagnetic information processing, image information processing, text information processing, audio information processing, information collection, information catalog, information retrieval, intelligence correlation analysis, intelligence release/subscription, intelligence quality assessment, intelligence integration sharing, and information archive backup provide shared service support for military intelligence analysis business applications [4].

The resource layer is the lowest level of the overall architecture, providing a physical support environment for system operation, including infrastructure and data resources. The infrastructure part provides the necessary hardware resources such as computing, storage, and network, as well as basic software resources such as operating system and application software, for system construction and operation, and provides resource scheduling management, configuration management and status monitoring and other functions; The data resource part of the system integrates various data resources into one, providing unified data resource collection, integration and logical description capabilities, mainly composed of military intelligence meta data and various business intelligence data.

The overall architecture design of the system adopts layered and service-oriented thinking. By providing standardized service interfaces, service components and service access methods, the system architecture is open and flexible, and the service components are reused in various business applications. The mature big data processing technology framework is used to build distributed platform support, and natural language understanding, graphic image analysis, data mining and other technologies are used to build a business support environment to meet the needs of various military intelligence analysis services; By implementing the process and componentization of various types of intelligence analysis services, and complying with relevant system interface specifications, the integration capability with existing equipment is fully guaranteed.

2.2   Develop Relevant Analytical Techniques

The key technologies of military intelligence analysis and service system based on big data mainly include information big data computing analysis framework, multiple source heterogeneous intelligence data processing, distributed wide-area intelligence application services, etc. [5].

Intelligence big data calculation and analysis framework technology. Facing the storage, management and analysis requirements of massive heterogeneous intelligence big data, it provides a massive data processing platform that is easy to develop and use, and provides support for real-time, dynamic and accurate analysis and processing of intelligence big data. Through the use of distributed networked architecture, the load balancing and anti-destruction between computing nodes are realized, and the dynamic allocation of intelligence analysis tasks and dynamic resource management are guaranteed. Hadoop + Spark + Storm provides a typical framework for intelligence big data analytics processing, using high-cost, inexpensive machines to build HDFS distributed file systems to provide high-throughput intelligence data storage and access. It uses the MapReduce parallel programming model to provide parallel computing of large-scale intelligence data. Based on Spark and Storm big data processing engine, the real-time intelligence data stream is processed quickly and interactively to support the high-real-time and large-capacity processing requirements of battlefield intelligence.

Multiple source heterogeneous intelligence data processing technology. Due to the existence of many sources of information resources, large differences in types, and mixed content, there is a great deal of uncertainty in intelligence analysis. How to extract and refine the truth from the big data of intelligence is a key challenge for intelligence big data analysis. The multiple source heterogeneous intelligence information comprehensive processing analyzes multiple intelligence such as images, texts, videos, audio, and electromagnetics according to different sources and formats, and extracts time, space, goals, and themes of intelligence big data. Furthermore, through the distribution characteristics of elements, the intelligence information from different sources are correlated and verified, and a multiple level and all-round unified intelligence situation is formed [6]. Ensure the accuracy and effectiveness of intelligence analysis results. In the intelligence processing mode, the cloud computing layered service model concept is adopted, and the computing and storage resources of each node in the system and various general intelligence analysis processing services are uniformly operated and managed. The distributed deployment, combination and use of loosely coupled intelligence service components through the network improves the real-time performance of intelligence processing.

Distributed wide-area intelligence application service technology. For wide-area distributed battlefield space and diverse sensor and operational target types [7]. For the information security needs of different combat missions, the service analysis technology is used to flexibly define the intelligence analysis and processing flow. The organization analyzes the intelligence analysis services deployed on each node to work together, provides a unified interface for intelligence data management and service application, and realizes the sharing of systematic intelligence analysis capabilities to ensure the overall processing capacity is maximized. Promote the transformation of intelligence processing and service models from the traditional "Task, Process, Use, and Distribute" (TPED) model to the modern "Task, Release, Process, and Use" (TPPU) model. It provides on-demand access and on-demand shared intelligence application services for various types of sensors and strike weapons distributed in a wide area, meeting the real-time processing of massive data and the operational requirements of sensitive operations and rapid strikes [8].

2.3 A Comprehensive Evaluation of the Method of Military Intelligence Value

Modern military warfare is an information-led war. The ability to acquire information and intelligence is the key to influencing the war process and winning and losing. Improving the ability to acquire information is especially important for accurately releasing and enhancing the combat capability of the military system. With the application of big data technology in the military field, the ability of commanding agencies to obtain information and intelligence has been greatly improved, mainly in the following two aspects: On the one hand, the efficiency of the military in handling information intelligence has greatly improved. In modern military warfare, the amount of information on battlefield information is extremely large, and most of them are unstructured data. If this information is processed in a conventional manner, it will not only take a long time, but also be extremely inefficient. However, with the help of big data technology, the military's processing speed of information and intelligence will exponentially jump, and its ability to acquire and process intelligence information in a unit of time has been greatly improved. On the other hand, more valuable information and intelligence. Under the constraints of information and intelligence, such as investigative methods and the battlefield environment, big data technology can quickly sort and sort information, and then analyze and feedback. The application of big data technology helps to mine more high-value military intelligence information about target objects in the war, thus making up for the lack of intelligence in reconnaissance and surveillance systems.

Military intelligence value assessment method based on big data analysis technology. First, we must understand the value density k of the data introduced, and determine its intelligence value according to the size of k. Then establish a system overall architecture, fully consider the openness and flexibility of the system architecture, to achieve the connection between data and platform, platform and application. Finally, we need to optimize the operation mode and develop relevant analysis techniques. The technology includes information big data computing and analysis framework technology, multiple source heterogeneous intelligence data comprehensive processing technology, and distributed wide-area intelligence application service technology. Through the above introduction, the military intelligence value assessment is finally achieved.

3 Experiment Analysis

In order to explore whether big data technology is comprehensive in assessing the value of military intelligence, a research experiment will be established here to compare the assessment of the value of military intelligence with literature [2] technology and the assessment of the value of military intelligence by big data technology. Comparing the curves and dense points of the data value density k, the abstract evaluation value is reflected in the form of data.

3.1 Comparison of the Value of Data Intercepted at the Same Time

Data value density means the value of data intelligence in a unit area. On the other hand, k is the speed of intercepting high-value intelligence, and comparing the value of data intercepted in the same time can compare the excellence of literature [2] technology with big data technology. Here, valuable information is recorded as Q_I, and the information acquired in the same period of time is recorded as U_I. The picture below shows the comparison (Fig. 3).

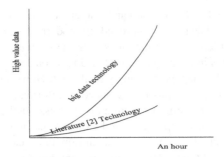

Fig. 3. Data value comparison

It is clear that the amount of high-value data acquired by big data technology is much larger than that of literature [2] technology in an hour.

$$k = \frac{\sum U_i Q_I}{\sum U_I} \qquad (2)$$

According to the data and substituted into the formula, calculate the k value, it can still be concluded that the k value of big data is larger than the literature [2] technology (Table 1).

Table 1. Data value density comparison

Amount of intelligence	Literature [2] technology k value	Big data analysis technique k value
20	0.3151	0.9846
40	0.3521	0.9765
60	0.3269	0.9854
80	0.3752	0.9903
100	0.3656	0.9712

3.2 The Time Required to Intercept the Same Value Data

Similarly, according to the results of the previous experiment, and based on this experiment, the same value data is intercepted. We compare the time required for big data technology and literature [2] technology, and the resulting data is plotted in Fig. 4, as shown below.

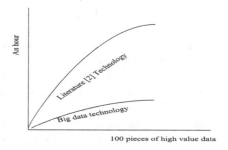

Fig. 4. Data time comparison

From the above picture, we can clearly see the time taken to obtain the same hundred high-value data. Obviously, big data technology is much faster than literature [2] technology, so that the excellent and advanced of big data technology can be compared.

$$U = \frac{\sum K_i Q_I}{\sum\limits_{i=1}^{m} K_I} \qquad (3)$$

4 Conclusion

This paper is based on the analysis of military intelligence value based on big data analysis. Big data analysis technology is to obtain some valuable information from a large amount of data, which has been applied in all aspects of production and life, and to improve the economic and social benefits of various industries. At present, the research on big data technology is also deepening, and relevant personnel can apply big data extensively and deeply in the military field to improve military management level and military decision-making level. Big data plays a very important role in the military field and has a very broad development prospect in the future. It needs to be highly valued.

The information age is a relatively long period of time relative to the industrial age and the agricultural era. There are significant differences in production factors and social drivers in different eras. The symbol of the information age is the widespread use of computers, the use of optical fiber communications, the Internet, and the use of integrated circuits. The emergence of big data analysis technology marks that the information age has entered a new stage, that is, the era of big data. Big data technology has been widely used in various industries of production and life. This paper mainly discusses the application of big data analysis technology in the military field. Big data technology will profoundly affect the military field and bring about changes in military thinking and thinking.

The arrival of the era of big data has changed the face of traditional military intelligence analysis. Based on advanced big data processing technology and tools, building a modern military intelligence analysis and service system has become a new solution. This paper conducts preliminary research and design on the military intelligence analysis architecture based on big data, and analyzes some of the key technologies. In the exploration of new military revolution, based on big data processing technology to solve the slow and complicated dilemma of modern military intelligence analysis, it is of great significance for improving information warfare intelligence analysis and operational command and decision. It is also the development direction of the current military intelligence revolution. However, this method does not consider the security of military intelligence data storage, and will focus on the military intelligence data storage to provide security protection for high-security data.

References

1. Guo, J., Huang, S.: Research on military intelligence analysis and service system architecture based on big data. J. China Electron. Sci. Res. Inst. **12**(4), 389–393 (2017)
2. Li, C.: Research on the construction of patent value evaluation system based on big data analysis. New Prod. Chin. New Technol. **8**(20), 3–6 (2016)
3. Peng, M., Li, H.: Research on patent analysis methods and models from the perspective of big data. Int. Theory Pract. **39**(7), 108–113 (2016)
4. Zhang, J., Hu, X., Zhang, H., et al.: Analysis and application of satellite image big data intelligence. Big Data **17**(5), 43–53 (2016)
5. Wu, X., Liu, X., Dong, J., et al.: Effects of big data on military intelligence and countermeasures. Sci. Technol. **9**(6), 206 (2018)
6. Wu, M., Hu, J., Zhang, H.: Big data analysis technology and its application in military field. Electron. Prod. **8**(14), 72–73 (2016)
7. Tao, L., Ling, Y.: Big data technology and its processing framework in intelligence compilation. Electron. Countermeas. **14**(1), 15–19 (2016)
8. Liu, S., Cheng, X., Fu, W., et al.: Numeric characteristics of generalized M-set with its asymptote. Appl. Math. Comput. **243**, 767–774 (2014)

Design and Simulation of Power Grid Energy Saving Control Model

Chao Song[1]($\boxtimes$) and Jia Xu[2]

[1] Dalian University of Science and Technology, Dalian, China
songchao0031@sina.com
[2] Dalian Jiaotong University, Dalian, China

Abstract. Aimed at the problem of volatile energy consumption of the traditional grid energy-saving control model, design a less volatile energy consumption of the power grid energy-saving control model, by building a basic calculation model, to meet energy-saving targets and related constraints, using the priorities and time calendar solving basic measurement model, in order to realize the energy-saving power generation dispatching; according to the results of energy-saving power generation scheduling, a power load stratification probability prediction method based on empirical mode decomposition and sparse Bayesian learning is used to establish a load forecasting model for load sampling. Based on the sampling results, the emission control cost and the reserve capacity cost, the nominal purchase cost of the non-renewable energy unit and the nominal purchase cost of the renewable energy unit are respectively constructed, and then the energy-saving control model of the power grid is constructed. In order to prove that the energy consumption fluctuation model of the power-saving control model is small, the model is Compared with the traditional grid energy-saving control model, the experimental results show that the energy consumption volatility of the model is less than that of the traditional power grid energy-saving control model, which reduces the nominal power purchase cost of renewable energy units and is more suitable for power grid energy-saving control.

Keywords: Power grid · Energy-saving control model · Basic measurement model · Energy-saving power generation dispatch · Minimum specific consumption

1 Introduction

In recent years, environmental protection and climate change have become the primary concerns of all countries in the world. Energy conservation, energy efficiency and the development of renewable energy have also become the basic energy policies of countries [1]. China has also proposed an ambitious goal of accelerating the construction of a resource-conserving and environment-friendly society, and proposed the reduction of energy consumption per unit of GDP during the 11th Five-Year Plan period, and the discharge of major pollutants during the 11th Five-Year Plan. A binding indicator of total reduction. As a basic industry of national economy, electric power is not only the creator of high-quality clean energy, but also a major energy consumer and

Y.-D. Zhang et al. (Eds.): ICMTEL 2020, LNICST 327, pp. 201–214, 2020.
https://doi.org/10.1007/978-3-030-51103-6_18

a major emitter of pollution, and plays an important role in the completion of energy-saving emission reduction targets [2]. Energy saving and consumption reduction of power industry run through the whole industrial chain from power production to power utilization, and the energy saving and consumption reduction potential of the electric power production link is the most significant. According to statistics, power generation energy accounts for more than 30% of the total energy consumption of the whole society, and coal consumption accounts for 50% of the total social coal consumption. Therefore, the energy saving and consumption reduction in the process of power production is of great significance to the completion of the indicators [3].

In the power production process, the power dispatching mode plays an important role in the allocation of power resources. At present, power generation dispatching in most provinces in China is based on the average capacity of the unit to allocate electricity generation hours [4]. This practice has led to the failure of the power generation capacity of clean energy units such as large thermal power units, hydro-power and nuclear power to fully utilize their environmental benefits. The high-energy, high-energy small thermal power units can generate more power, resulting in waste of energy resources and environmental pollution. Therefore, the General Office of the State Council and the Development and Reform Commission and other departments have issued a notice on the "Trial Implementation of Energy-Saving Power Generation Dispatching Measures": requiring energy-saving power generation dispatching, mandatory energy-saving emission reduction, and reduction of energy consumption and pollutant emissions [5]. Therefore, the General Office of the State Council and the Development and Reform Commission and other departments have issued a notice on the "Trial Implementation of Energy-Saving Power Generation Dispatching Measures": requiring energy-saving power generation dispatching, mandatory energy-saving emission reduction, and reduction of energy consumption and pollutant emissions. However, the energy-saving power generation scheduling mode may also bring severe challenges to the safe and stable operation of the power grid. The current energy-saving control model of the power grid has a large fluctuation of energy consumption, resulting in poor energy-saving control effect of the power grid, reducing the nominal power purchase cost of renewable energy units. Therefore, this paper designs a new energy-saving control model of the power grid.

2 Design Power Grid Energy Saving Control Model

2.1 Energy Conservation Power Generation Dispatch

(1) Build basic measurement model

By building a basic calculation model, the priority order and traversal were used to solve the problem before meeting the goal of energy conservation and relevant constraints, so as to realize the scheduling of energy-saving power generation [6]. In energy-saving generation scheduling, energy saving is the primary goal. Therefore, according to the current situation of energy-saving power generation scheduling, taking the total energy consumption of power grid as the objective function, and the basic calculation model is established as follows:

$$\min F = \sum_{t=1}^{T} \sum_{k=1}^{M} \left[s_{k,t} f_k(P_{k,t}) + Z_k s_{k,t}(1 - s_{k,t-1}) \right] \tag{1}$$

Where, F is the mathematical function of energy consumption of the system; $\min F$ represents the minimum objective function of total energy consumption of the power grid. M is the number of generator sets; T is the number of time periods; $s_{k,t}$ is the operation status of generator sets, when the value is 0, it means shutdown; when the value is 1, it means operation; $P_{k,t}$ is the active output of generator group k in t period; $f_k(P_{k,t})$ is the energy consumption function of the generating unit, which means the coal consumption when the output of the t period of the k unit is $P_{k,t}$; Z_k is the starting coal consumption of unit k.

The constraints of the basic calculation model generally meet the following constraints [7].

1) Active power balance constraint:

$$\sum_{i=1}^{t} P_{Li,t} = \sum_{k=1}^{M} s_{k,t} P_{k,t} \tag{2}$$

Where, $P_{Li,t}$ is the active load of node i at time t.

2) Rotational reserve capacity constraint:

$$\sum_{k=1}^{M} s_{k,t} P_{k,\text{tmax}} \geq \sum_{i=1}^{t} P_{Li,t} + R_t \tag{3}$$

Where, R_t is the reserve capacity of grid t at the moment; $P_{k,\text{tmax}}$ is the maximum output value of generator unit k in period t.

3) Unit output constraint:

$$P_{k,\text{tmin}} \leq P_{k,t} \leq P_{k,\text{tmax}} \tag{4}$$

Where, $P_{k,\text{tmin}}$ is the minimum value of the output of generator unit k in period t.

4) Unit minimum start stop time constraint:

$$\begin{cases} (b_{k,t-1} - T_k^{ON})(s_{k,t-1} - s_{k,t}) \geq 0 \\ (x_{k,t-1} - T_k^{OFF})(s_{k,t-1} - s_{k,t}) \geq 0 \end{cases} \tag{5}$$

Where, $b_{k,t-1}$ and $x_{k,t-1}$ are the continuous operation downtime in period t; T_k^{ON} and T_k^{OFF} are the minimum start and stop time of generator set k.

5) Speed restriction of climbing and downhill of generator set:

$$\begin{cases} P_{k,t} - P_{k,t-1} \leq R_{Hi}\Delta t \\ P_{k,t-1} - P_{k,t} \leq R_{Li}\Delta t \end{cases} \tag{6}$$

Where, R_{Hi} and R_{Li} are the climbing and downhill rates of generator unit k respectively.

6) Same constraint of output change trend:

$$\left(\sum_{i=1}^{t} P_{Li,t} - \sum_{i=1}^{t} P_{Li,t-1} \right) \left(P_{k,t} - P_{k,t-1} \right) \geq 0 \tag{7}$$

(2) Basic calculation model solution

The priority method combined with ergodic algorithm is used to solve the basic calculation model, so as to achieve the goal of minimum total energy consumption.

The priority method refers to the start-up sequence of the units in the system according to certain indicators, and then the corresponding unit operation or shutdown according to the size of the load [8]. This method also conforms to the basic idea of energy-saving power generation scheduling method, and is implemented strictly according to the sequence table of units.

1) Sequencing of generator sets

The energy consumption level of the unit is an important index of the unit sequencing, which reflects the energy consumption of the generator unit under normal operation. According to the sequence table, the energy conservation and emission reduction of the maximum energy conservation scheme can be realized. Because the same type of thermal power generating units need to be ranked from low to high strictly according to the energy consumption of each generating unit, and when the energy consumption of each generating unit is the same, it needs to be ranked according to the pollutant emission level of each generating unit, so the minimum specific consumption of the generating unit is used for evaluation.

The minimum specific consumption of the unit can be obtained according to the following formula:

$$\begin{cases} \mu_{kmin} = a_k P_{k,m} + b_k + \frac{c_k}{P_{k,m}} \\ P_{k,m} = \begin{cases} P_{kmin}, P_{k,m} \leq P_{kmin} \\ \sqrt{\frac{c_k}{a_k}}, P_{kmin} \leq P_{k,m} \leq P_{kmax} \\ P_{kmax}, P_{k,m} \geq P_{kmax} \end{cases} \end{cases} \tag{8}$$

Among them, μ_{kmin} is the minimum specific consumption of the unit; a_k, b_k and c_k are the maximum, medium and minimum energy consumption of the unit respectively; P_{kmin}, $P_{k,m}$ and P_{kmax} are the lowest, medium and highest emission levels respectively.

In the actual power grid dispatching process, the unit maintenance plan and the minimum start-up and shutdown time constraints can directly determine the status of the unit in some time periods, and these units can not participate in the unit sequencing in those time periods.

2) Determination of the number of generator sets

In the case of basic constraints, in order to ensure the output balance of adjacent periods, the minimum number of generating units is determined. Take the unit climbing as an example, the load value of the system in period t is P_t, and the number of generating units in period $t - 1$ is N. According to the climbing output limit of the unit, the maximum output that each generating unit can allocate can be obtained. At least, only $N + 1$ generating unit is required to meet the output in period t, that is, $N + 1$ generating unit is the number of basic generating units [9]. When the unit goes downhill, the corresponding number of generator units can be obtained in the same way.

After the priority method is used to sequence the generating units, the units in the first order are generally put into operation, and the units in the second order are generally not involved in the operation and are in shutdown state. However, due to the proximity of the sort index, it is not possible to determine the intermediate unit manually. Therefore, it is necessary to evaluate and calculate the combination mode of marginal units by ergodic method, so as to obtain the system energy consumption of each combination mode and finally determine the optimal combination mode of units. For a large number of units, cloud computing should be considered and advanced computing methods should be used to improve the depth of optimization. The priority algorithm can get the optimal strategy based on the current situation when solving the optimization problem, that is, to find the local optimal solution. The basic model solution flow is shown in Fig. 1.

The priority method and ergodic method are combined to determine the stop or start unit by using the basic constraints of the unit and the unit sequence table obtained by the sequencing index, and then the step-by-step relaxation constraints are used to search the state of the marginal unit to obtain the optimal unit combination scheme. By using this method, we can simplify the calculation and improve the speed of operation under the condition of ensuring the optimization depth.

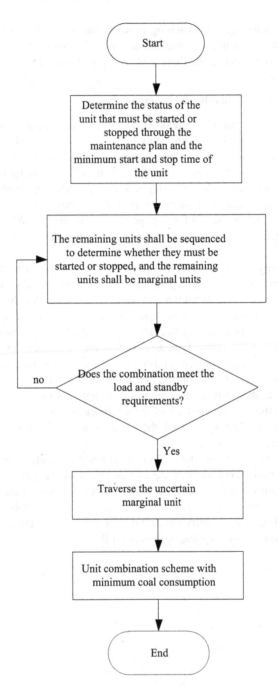

Fig. 1. Basic model solving process

2.2 Load Sampling

(1) Building load forecasting model

According to the results of energy-saving power generation scheduling, a hierarchical probability prediction method based on empirical mode decomposition and sparse Bayesian learning is adopted. Through the distributed collaborative network, a series of data such as the load size of each node in the power grid are analyzed, temperature and humidity, is extracted and eliminated by classification, and then the effective data is sent to Bayesian learning and training, so as to build Establish load forecasting model.

The specific prediction steps are as follows:

1) kernel principal component analysis (KPCA) is used to obtain the principal component characteristics of load samples.
2) using Mahalanobis distance weighting method to determine the similarity between the trained samples and the predicted samples, excluding the samples with low similarity;
3) empirical mode decomposition is used to process power load samples and extract high-frequency and low-frequency components from power signals.
4) SBL is used to predict the high frequency and low frequency load.
5) integrate the prediction results of different frequency bands to obtain the final power load prediction results.

Through the above steps, the user load demand can be predicted as a certain confidence interval "band", as shown in Fig. 2.

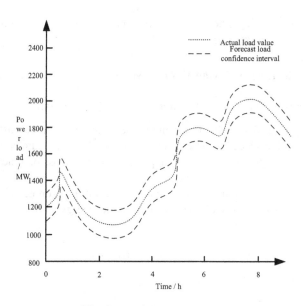

Fig. 2. Load forecast interval

(2) Load sampling

The load demand forecasting interval with confidence of ς is divided into r cells with different confidence, and each sub interval obeys a certain fuzzy membership function.

Assuming that the first l load forecasting interval is $[P_{L/\min}, P_{L/\max}]$, the interval is divided into r sub intervals, which are respectively expressed as $a_1, \ldots a_\vartheta, \ldots a_r$, where:

$$a_\vartheta = \left[\begin{array}{c} P_{L/\min} + \dfrac{(\vartheta-1)\left([P_{L/\max}-P_{L/\max}]\right)}{r}, \\ P_{L/\min} + \dfrac{\vartheta\left([P_{L/\max}-P_{L/\max}]\right)}{r} \end{array} \right] \tag{9}$$

The confidence of interval a_ϑ is set as ς_ϑ, and the load distribution in a_ϑ obeys the membership function.

As the load forecasting interval contains r sub interval, the load sampling value is determined by the sampling value and reliability of r sub interval. That is, the sampling value δ_{j_ϑ} $(j = 1, 2, \ldots, r)$ obtained by sampling j_ϑ times in the ϑ interval, then the sampling value of the uncertainty is:

$$\begin{aligned} \delta_j &= \frac{\varsigma_1 \delta_{j_1} + \varsigma_2 \delta_{j2} + \ldots + \varsigma_r \delta_{j\vartheta} + \ldots + \varsigma_r \delta_{jr}}{\varsigma_1 + \varsigma_2 + \ldots + \varsigma_r} \\ &= \frac{\displaystyle\sum_{j=1}^{r} \varsigma_\vartheta \delta_{jr}}{\displaystyle\sum_{j=1}^{r} \varsigma_\vartheta} \end{aligned} \tag{10}$$

In the actual sampling, the triangular membership function is used in the load distribution of each sub section.

2.3 Power Grid Energy Saving Control Model

(1) Nominal power purchase cost of non renewable energy unit

According to the results of load sampling, the emission control cost and reserve capacity cost are introduced to construct the nominal power purchase cost of non renewable energy units and the nominal power purchase cost of renewable energy units.

Considering energy saving factors based on unit environmental benefits, the nominal power purchase cost of the non renewable energy group is constructed as follows:

$$C_G = C_{GS} + \eta_{EPi} \times C_{EC} + C_{GT} \tag{11}$$

Where, C_G is the nominal power purchase cost of non renewable energy units; C_{GS} is the total fuel cost; η_{EPi} is the environmental penalty cost coefficient; C_{EC} is the total cost of emission control; C_{GT} is the operating cost of each unit at its maximum output.

(2) Nominal power purchase cost of renewable energy unit

Due to the randomness, intermittence and variability of renewable energy generation, in order to ensure the stability and safety of the grid, it is necessary to increase the reserve capacity of the grid. In order to reasonably reflect the power value of renewable energy, the loss to the grid caused by renewable energy should be reasonably reflected while considering its environmental benefits [10]. Therefore, reserve capacity penalty cost is introduced, and its expression is as follows:

$$C_{RPi} = \rho_{RPi} \cdot \left[Min\left(0, P_{wj} - P_{wja}\right)\right] \tag{12}$$

Among them, C_{RPi} represents the penalty cost of reserve capacity; ρ_{RPi} represents the penalty cost coefficient of reserve capacity for grid connection of renewable energy; P_{wj} represents the actual generation capacity of renewable energy generating units; P_{wja} represents the planned generation capacity of renewable energy generating units.

The nominal power purchase cost of renewable energy units can be expressed as follows:

$$C_R = \sum_{i-1}^{n} C_{PRi} + C_{RT} \tag{13}$$

Among them, C_R is the nominal power purchase cost of renewable energy units; C_{RT} is the power purchase cost of renewable energy units; n represents the number of renewable energy generating units; i is the number of renewable energy generating units.

(3) Construction of power grid energy saving control model

Since the output of renewable energy units is a random variable, the output of non renewable energy units, generation cost and up and down rotating reserve capacity are also random variables. Therefore, the mathematical expectation minimization of the nominal power purchase cost of power grid enterprises is taken as the objective function, and the constraints with uncertainty are expressed in the form of probability, so that they can meet a certain confidence level interval. Therefore, the power grid energy-saving control model is constructed as follows;

$$
\begin{cases}
MinE\left[\displaystyle\sum_{t=1}^{T}\sum_{i=1}^{N} U_{it}C_{Gi} + \sum_{t=1}^{T}\sum_{i=1}^{N} U_{it}C_{Ri}\right] \\[2mm]
\displaystyle\sum_{i=1}^{N} P_{Git} + \sum_{i=1}^{M} P_{Wit} = P_{Dt}, t \in T \\[2mm]
p_{Git}^{min} \le P_{Git} \le p_{Git}^{max} \\[2mm]
P\left\{\displaystyle\sum_{i=1}^{N}\left(p_{Git}^{max} - P_{Git}\right) \ge SR_u\right\} \ge \theta_1 \\[2mm]
P\left\{\displaystyle\sum_{i=1}^{N}\left(P_{Git} - p_{Git}^{min}\right) \ge SR_d\right\} \ge \theta_2 \\[2mm]
P\left\{S_{dGi} \le P_{Git} - P_{Gi(t-1)} \le S_{uGi}\right\} \ge \theta_3
\end{cases} \tag{14}
$$

Where, T is the number of cycle hours; N is the number of thermal power units; M is the number of renewable energy units; C_{Gi} is the nominal power purchase cost of thermal power unit i; C_{Ri} is the nominal power purchase cost of renewable energy unit i; P_{Dt} is the load demand of period; P_{Git} is the output of pyroelectric motor group i at time interval t, p_{Git}^{min} and p_{Git}^{max} are the upper and lower limits of its output respectively; SR_u and SR_d are the upper and lower rotation reserve requirements of power grid in period t respectively; S_{dGi} and S_{uGi} are the decline and rise rates of the active power output of thermal power unit i in period t; θ_1, θ_2 and θ_3 are confidence levels; P is thermal power unit; P_{Wit} is the power consumption of thermal power unit; $MinE$ is the minimum number of energy savings.

3   Simulation Experiment and Analysis

3.1   Experimental Process and Method

Taking the actual data of a certain region's power system as experimental data, the grid energy-saving control model designed in this paper is used to conduct grid energy-saving control experiments. The power supply structure of the power system is: thermal power proportion is 70.18%, hydropower is 28.24%, and gas and wind power is only 1.57%. The proportion of each capacity class of the thermal power unit is shown in Table 1.

Table 1. Proportion of each capacity class of thermal power units

Unit capacity (MW)	Number of tables	Capacity	Proportion
600	6	3600	10.03%
300 and 320 and 330	44	13660	38.06%
200 and 220	8	1460	4.07%
135 and below	96	6470	18.03%
Total	154	25190	70.18%

According to the thermal power unit capacity bin, the electricity price of the power system in the region is measured. According to the characteristics of the power system power distribution, the thermal power unit capacity is divided into 600 MW, 300 MW, 200 MW, 135 MW and below. The electricity price is only determined according to the coal consumption level of power generation. The standard coal price corresponding to the capacity level of each gear and the average coal price corresponding to the different calorific values of the four regions of the system are used as the basis for calculating the electricity price, and the electricity price of the thermal power units in the four regions is obtained. The valuation is shown in Table 2.

Table 2. Valuation of electricity price of thermal power units in four regions

Capacity class (MW)	A area	B area	C area	D area
600	99.1–105.4	81.3–83.6	85.5–89.3	85.2–91.8
300	105.6–109.1	86.2–88.3	90.8–94.2	90.2–97.6
200	108.2–119.7	95.6–97.3	100.2–104.4	100.1–107.9
135	130.3–177.9	117.6–151.3	119.0–158.3	117.1–161.0

Select the typical daily load curve of the power system in this region to control the grid energy, as shown in Fig. 3.

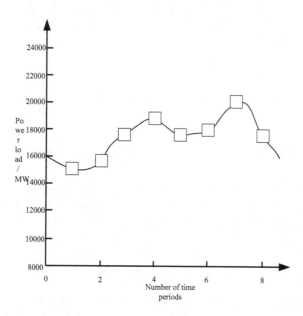

Fig. 3. Typical daily load curve of the power system in January

In order to ensure the effectiveness of the experiment, the traditional grid energy-saving control model is compared with the grid energy-saving control model designed in this paper. The energy volatility of each grid energy-saving control model is compared, and the energy volatility is smaller, which proves its grid energy-saving control. The performance is even better. The judgment of energy consumption volatility is the stability of the energy consumption fluctuation curve, and the more stable the energy consumption fluctuation curve, the smaller the energy consumption volatility.

3.2 Analysis of Results

The experimental results of the energy consumption volatility comparison between the traditional grid energy-saving control model and the grid energy-saving control model designed in this paper are shown in Fig. 4.

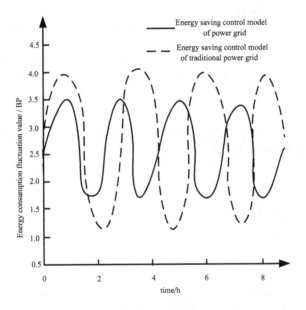

Fig. 4. Comparison of energy volatility experiment results

According to the energy consumption volatility comparison experiment results in Fig. 4, with the increasing cycle hours T, the energy consumption of both models is in a state of fluctuation. Compared with the traditional power grid energy-saving control model, the power grid energy-saving control model designed in this paper has a stronger stability of energy consumption fluctuation curve, that is, the energy consumption volatility is smaller and more stable, and it enhances the energy-saving control effect of the power grid.

In order to further verify the effectiveness of the model in this paper, taking the nominal power purchase cost C_R of the renewable energy unit as the experimental index, a comparative analysis is made on the nominal power purchase cost of the traditional power grid energy-saving control model and the renewable energy unit of the power grid energy-saving control model designed in this paper. The comparison results are shown in Fig. 5.

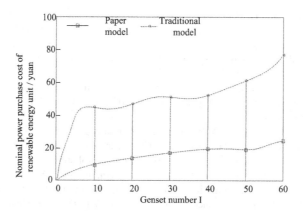

Fig. 5. Comparison of nominal purchase cost of the two models

According to Fig. 5, with the number of renewable energy generating unit I not increasing, the nominal power purchase cost of the renewable energy generating unit of the traditional power grid energy-saving control model and the power grid energy-saving control model designed in this paper is also increasing, but the nominal power purchase cost of the renewable energy generating unit of the power grid energy-saving control model designed in this paper is higher than that of the traditional power grid energy-saving control model The nominal power purchase cost of generating units is low.

4 Conclusion

Due to the large fluctuation of energy consumption in the traditional power grid energy-saving control model, the effect of power grid energy-saving control is poor, which reduces the nominal power purchase cost of renewable energy units. Therefore, this paper designs a power grid energy-saving control model with small fluctuation of energy consumption, which realizes the reduction of energy consumption fluctuation, reduces the nominal power purchase cost of renewable energy units, and provides energy-saving control tools for power grid of great significance.

References

1. Qi, Z., An, H., Duan, M., et al.: Energy-saving and loss-reduction control method for power grid considering the reactive power response capability of electric vehicle chargers. Proc. CSU-EPSA **29**(35), 129–134 (2017)
2. Liao, Z., Wang, Y., Wang, L.: Research on business model and development capability of distributed energy with power grid enterprise's participation. Control Theory Appl. **34**(38), 126–132 (2017)
3. Chen, M., Xiao, X.: Cooperative secondary control strategy of microgrids based on distributed internal model design. Trans. China Electrotech. Soc. **32**(10), 149–157 (2017)

4. Li, X., Li, Y., Cao, Y., et al.: Wide-area damping control strategy of interconnected power grid based on cyber physical system. Power Syst. Prot. Control **20**(21), 35–42 (2017)
5. Liang, X., Wang, J., Ke, Y., et al.: Research on grid dispatching of energy saving and environmental protection based on environmental comprehensive evaluation. Electr. Measur. Instrum. **54**(55), 124–128 (2017)
6. Wu, J., Zhao, F., Zhao, L., et al.: Reactive power optimization in distribution network considering residual capacity of photovoltaic inverter. Adv. Technol. Electr. Eng. Energy **36** (41), 38–43 (2017)
7. Lu, Z., Li, D., Lu, X., et al.: Multiple faults repair strategy under ice storm for distribution network with distributed generators. Trans. China Electrotech. Soc. **33**(32), 423–432 (2018)
8. Jiang, S., Zhao, Z., Zhang, X., et al.: Research on control strategy of PV air-conditioning system under grid harmonic condition. Refrig. Air-conditioning **18**(21), 12–16 (2018)
9. Zhao, W., Qi, L., Sun, X., et al.: Research on high-frequency stability of islanding microgrid adopting high efficient impedance analysis method. Acta Energiae Solaris Sinica **38**(45), 1166–1175 (2017)
10. Liu, S., Fu, W., Deng, H., et al.: Distributional fractal creating algorithm in parallel environment. Int. J. Distrib. Sensor Netw. (2013). https://doi.org/10.1155/2013/281707

Design and Improvement of Optimal Control Model for Wireless Sensor Network Nodes

Jia Xu[1,2] and Chao Song[2(✉)]

[1] Dalian Jiaotong University, Dalian, China
xujia00200@sohu.com
[2] Dalian University of Science and Technology, Dalian, China
songchao0031@sina.com

Abstract. Sensor network coverage is one of the basic problems in the Internet of Things. Coverage is one of the important indicators to measure the performance of sensor network nodes. Through the research on coverage problems, we can seek ways to improve the quality of sensor network services. A wireless sensor network node control effect is proposed. A distributed algorithm based on probability model is first constructed to optimize the probability perception algorithm. The above-mentioned two-dimensional space algorithm is extended to three-dimensional space, and the greedy heuristic algorithm is used to obtain the control solution to achieve the optimal control of the current wireless sensor network. In addition, the matlab simulation program is written, and the algorithm is compared with the simulation results of the average algorithm and the random algorithm. The simulation results of the proposed algorithm have significant advantages.

Keywords: Wireless sensing · Network control · Optimization control

1 Introduction

Wireless Sensor Networks (WSNs) are composed of many sensor nodes. The sensor nodes are placed in the area to be monitored according to certain methods through certain topologies. Through some suitable methods and their respective information exchanges, achieve the role of collaboratively perceiving physical world information, collecting and collating information of perceived objects within the network coverage area. The development of wireless sensor networks allows people to perceive the Earth's information more through sensor nodes, especially in radiation coverage areas, virus-infected areas or environments that humans cannot reach. Wireless networks can play an irreplaceable advantage [1]. WSNs are not the Internet of Things [2]. In fact, WSNs have been widely used before the emergence of IOT. The temperature control network consists of multiple temperature sensors and works by inspection. It has multiple sound, light and electricity. Security nets for mechanical and even image detection capabilities, etc. Therefore, although the sensor network in the traditional sense has developed very mature, compared with WSNs, the sensor nodes and the network architecture are completely different. In response to this situation, the academic community has proposed a larger IOT. The concept and include the WSNs

Y.-D. Zhang et al. (Eds.): ICMTEL 2020, LNICST 327, pp. 215–227, 2020.
https://doi.org/10.1007/978-3-030-51103-6_19

concept. WSNs has a relatively important coverage problem, which not only relates to the quality of service of the network, but also affects the research of other problems in the network [3, 4]. When studying the coverage problem, it is necessary to consider the characteristics of limited wireless sensor network resources and strong topological dynamics. Researchers have done a lot of research work on the coverage of wireless sensor networks for different application scenarios. Coverage is a measure of the detection of the target surveillance area, which directly affects the quality of the monitoring of the area to be tested. The research on wireless sensor network coverage problem can improve the service quality of the monitoring network in the area to be tested. At present, the existing wireless sensor network node control method can not complete the node, and ensure that the network can cover the entire area. In reference [5], a load balancing control method of channel boundary nodes in optical fiber networks is proposed. The task scheduling server is used to allocate node tasks, and the operation results are fed back to the relevant clients. When allocating the acquired blocking nodes, the priority quantification method of the nodes needs to be considered. In combination with the four relevant factors of the available memory of the optical network nodes, the main frequency of the CPU, the number of running nodes and the number of waiting nodes, the priority of the nodes is obtained and the accuracy of node allocation is optimized. At last, the node queue is defined, and the load balancing model of the channel boundary node is constructed. However, the positioning error of this method is large under unknown power, and the control accuracy of wireless sensor network nodes is low.

In response to the above problems, the optimal control model of the wireless sensor based on the probability model is proposed to improve the optimal control problem of the current network node [5].

2 Wireless Sensor Network Node Optimal Control Model

2.1 Optimized Probabilistic Perception Algorithm

The sensor node is determined by its own monitoring conditions. Only the information to be monitored within its sensing range may be "perceived". For the establishment of different wireless sensing network perception models, the sensing network can be better simulated because only a good sensing model can design a good simulation coverage algorithm [6].

The optimized probability-aware model is different from the traditional probability-aware model, which is closer to the physical world. The monitoring of the information to be measured is not a constant 0 or 1, but the distance between the sensor and the information to be measured, and the sensor. The physical properties and the number of neighbors of the sensor are determined by variables. As the distance between the target and the sensor changes, the probability that the information to be tested is monitored by a sensor changes exponentially.

For conventional areas, the design is adopted $P(s,p) = e^{-ad}$. In the form of setting the current target p to the sensing node s to a distance d, at this time, $P(s,p)$ denotes the probability that p is monitored by s, related to the distance between the two. If a represents the coefficient at which the current target to be measured is attenuated by the distance detected by the sensing node, refer to the following Fig. 1.

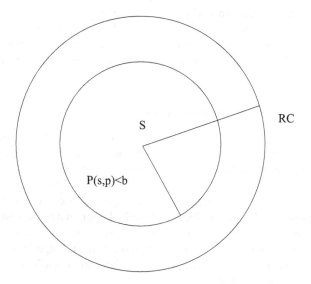

Fig. 1. Probability based sensor node monitoring model

Generally speaking, the greater the distance of the sensor node from the target to be measured, the smaller the probability that the target to be measured can be monitored. When the distance does not exceed a certain threshold, the probability of monitoring is not zero; otherwise, the sensor node is away from the target to be tested. The smaller the distance, the greater the probability that the target to be measured can be monitored [7].

Usually we define that the probability that the detection target at grid point j can be detected by the sensor node at grid point i is p:

$$p_{ij} = \begin{cases} e^{-ad} & e^{-ad} > b \\ 0 & e^{-ad} < b \end{cases} \tag{1}$$

p_{ij} indicates the target to be measured, and i represents the monitoring probability of the sensing node j. In general, the positive and negative distance of p_{ij} is the same without obstacles. If there are obstacles, the distance between the two is different. The above coverage model is obtained in the absence of neighboring nodes. If the target j to be tested is simultaneously covered by K sensor nodes, the K sensing regions are respectively represented as $R(s_1), R(s_2), \cdots R(S_K)$, then to the sensing overlap area $p(j)$:

$$p(j) = 1 - \prod_{i=1}^{k} (1 - p_{ij}) \tag{2}$$

Among them, p_{ij} is the probability defined above [8].

Since the sensor node senses the target to be measured by converting the excitation of the physical environment into an electrical signal, the quality of the excitation signal is closely related to the distance of the sensor node to the target to be measured. The signal is interfered by electromagnetic waves, and the distance increases. The strength of the signal will also change. It is considered that the monitoring of the target to be measured will be greatly attenuated as the transmission distance of the signal increases. Therefore, the path attenuation model is used to describe the sensor node's sensing ability [9, 10].

2.2   Dimensional Greedy Heuristic Algorithm

The above process optimizes the probability model of the wireless sensor network. According to the optimization results, a two-dimensional heuristic control algorithm is constructed based on the greedy algorithm. The greed method is a method that does not pursue the optimal solution and only obtains a more satisfactory solution. The greed method can generally get a satisfactory solution quickly because it saves the time it takes to find the optimal solution. The greed method often makes the best choice based on the current situation, without considering the various possible overall situations, so the greedy law does not require backtracking. The heuristic algorithm uses the same constraint equation or function at each step to direct the algorithm to proceed. The schematic is as follows (Fig. 2):

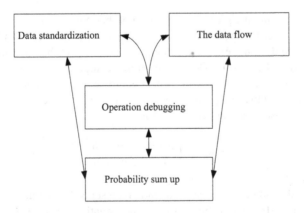

Fig. 2. Schematic diagram of the greedy algorithm

Since the two-dimensional space based on the probabilistic model sensor network will be affected by the inequality of obstacles and geographical environment, such as the presence of obstacles, the point j in the node sensing area will be monitored by node i with a probability of 0, that is, both. The distance is similar. Different cluster nodes and ordinary nodes have different status, which will make the monitoring probability different. The design is designed with ideal conditions.

The minimum probability of covering each point in the area to be tested, the minimum coverage by K sensor nodes (K-coverage), and the algorithm for deploying up to MS sensor nodes in the area: the monitoring area is dimensioned, the minimum number of sensor nodes are determined and the placement position in the grid is determined, so that each grid point realizes K-coverage with probability T, that is, each target grid point of the monitoring area is at least Monitored by K sensors, where K > 1, P(j) > T. Among them, MS is the maximum number of sensor nodes provided, K is the coverage, and T is the three basic conditions of the algorithm.

Some basic requirements for the design algorithm can be referred to the above probabilistic algorithm, such as each point is monitored by at least k sensors, etc. In addition, the design algorithm discusses the current sensor network node coverage, if for regional coverage problems, Refer to the above probability optimization algorithm to transform the grid point coverage problem into the coverage problem of the relevant area.

First, the two-dimensional to-be-monitored plane region is divided by an n × n grid (the ideal plane region is considered here, and other regions can be converted into similar regions), and the number of grid points on the plane is N = n2, which is known by the probability algorithm. The probability p_{ij} that the grid point. J is detected by the sensor node i in the area to be tested is called the monitoring matrix D:

$$D = \left[p_{ij} \right]_{N,N} \tag{3}$$

The monitoring matrix has n4 elements. According to the definition of missed detection probability and the monitoring probability matrix D, M = (n × n) can be defined as the missed detection probability matrix, where mij = 1 − pij, in the two-dimensional greedy heuristic algorithm. Initialize M = (1, 1, ... 1), where the elements of the matrix are all 1 means that all grid points are not detected. The position of the sensor nodes in the grid is determined by the iterative greedy heuristic algorithm. The algorithm iteratively deploys the position of one sensor node at each step, and refreshes the coverage probability p(j), j = 1, 2 .... N. When the required grid point reaches the requirement with the coverage degree K and the probability T, or the total number of sensor nodes reaches the slave, the preset node number upper limit termination algorithm.

Regarding the coverage K, a vector L = (L1, L2, ... LN) may be preset, where N is the number of grid points n2. The element Lj represents the coverage obtained by the grid point j during the running of the program, and the vector set L represents the set of coverage. Since all the grid points (Grid) are not covered by the node (Sink) at the beginning of the program, L = (0, 0, ..., 0) can be used to initialize the coverage set with the zero vector. L.

The key step of the two-dimensional greedy heuristic algorithm is iteration, that is, by deploying the nodes in the two-dimensional grid, the missed detection probability mij of all the grid points of the node is obtained, and the sum of the missed detection probabilities mij is minimized by each iteration, and the update coverage is The (i, j) position in the degree vector set L updates the missed detection probability matrix M, and when the coverage of a certain grid point satisfies the coverage degree K, the row corresponding to the grid point is deleted in the matrix M Columns to reduce the dimension of the matrix until the dimension of the matrix is zero. The overall calculation steps are as follows:

(1) Dividing the two-dimensional to-be-monitored plane area by n × n mesh, and the adjacent grid point distance can be determined according to the precision required by the sensor node configuration;

(2) Given the monitoring accuracy t: the input scalar value indicates that all nodes are the same; otherwise, the corresponding vector should be input. Given coverage k: the input scalar value indicates that all nodes are the same; otherwise, the corresponding vector should be input.

(3) Initialization:$N = size(D, 1); D = N * N$ The number of grid points is n = 0.

(4) Cycle start: deploy sensor nodes at current grid point k, request grid points $\sum k$ The smallest.

(5) If P(j) > T, then j = 1, ..., N, corresponding to Lj plus 1. Update vector L = (L$_1$, L$_2$, ... L$_N$)

(6) Sensor node plus 1;

(7) If the current Lj reaches a given coverage, the M matrix is deleted.

(8) Priority is given to coverage, where Lj > C, C represents the current sensor node coverage required for each grid.

(9) The algorithm terminates, and the process is as follows (Fig. 3):

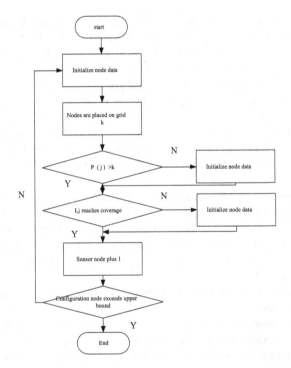

Fig. 3. Probabilistic two-dimensional greedy heuristic algorithm flow

2.3   Three-Dimensional Greedy Optimal Control Algorithm

In order to achieve optimal control, on the basis of the above, the design extends the greedy heuristic algorithm in two-dimensional space to three-dimensional space. Compared with two-dimensional space, three-dimensional space has more complexity. For example, in the two-dimensional space, the inequality of the geographical environment directly affects the spatial sensing node probability monitoring. This is especially true in the space environment. In addition, the three-dimensional space is more complex than the network topology of the two-dimensional space, simplifying the "obstacle" in the space as much as possible, and simplifying all the grid points into the ideal space. As with the two-dimensional space, we first need to give the premise and basic requirements of the algorithm. What is different from the two-dimensional space is that our pre-defined grid points are divided into three-dimensional to-be-monitored areas by n × n × n grids.

For the three-dimensional space, the design idea of the algorithm is to satisfy: to the three-dimensional to-be-monitored area divided by the mesh, to find the minimum number of sensor nodes, and determine the placement position in the grid, so that each grid point is the probability T implements K-coverage, that is, each target grid point of the monitoring area is monitored by at least K sensors, where K > 1, p(.I) > T, P is the maximum number of sensor nodes provided, K is called Coverage, T is called monitoring accuracy, is the three basic conditions of the algorithm (Fig. 4).

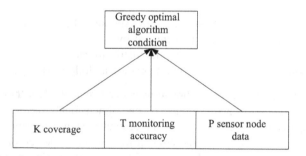

Fig. 4. Greedy optimal algorithm conditions

First, divide the three-dimensional space to be monitored with n × n mesh (the ideal spatial region is considered here, and other regions can be transformed into similar regions). The number of mesh points in space is N = n^3, from the upper part. It can be confirmed that the probability pij of the grid point J detected by the sensing node i in the area to be tested is called D = p_{ij}. The monitoring matrix has a total of n^6 elements. According to the probability of missed detection and the probability matrix D of monitoring, we can define $M = [m_{i,j}]_{N,N}$. For the current missed probability matrix.

In the three-dimensional greedy heuristic algorithm, the m value is initialized, where the elements of the matrix are all 1 means that all the grid points are not detected. The sensor nodes in the grid are determined by the iterative greedy heuristic algorithm.

Each step of the iteration must deploy the location of a sensor node to refresh the coverage probability $P(j)'$, whose expression is:

$$P(j)' = D[m_{i,j}]_{N,N}, j = 1, 2 \cdots, N \tag{4}$$

When the required grid points are covered by the degree of coverage k, the probability t reaches the requirement, or the total number of sensing nodes reaches ms, that is, the preset number of nodes is limited to the upper limit termination algorithm. Considering that there are special requirements in the three-dimensional space, for some grid points, there are priority coverage requirements (coverage priority and monitoring accuracy priority), etc., the following two-dimensional algorithm has the following modifications:

(1) Dividing the two-dimensional to-be-monitored planar area by a grid of n × n n, and the distance between adjacent grid points can be determined according to the precision required by the configuration of the sensing node;

(2) Given the monitoring accuracy t: the input scalar value indicates that all nodes are the same; otherwise, the corresponding vector should be input. Given coverage k: The input scalar value indicates that all nodes are the same; otherwise, the corresponding vector should be input.

(3) Initialization: N = size(D, 1); //D = N * N number of grid points

(4) Start of the loop: deploy the sensor node (Sensor) at the grid point (Grid), request $\sum k$ most

(5) If $P(j) > T, j = 1, 2 \ldots, N$, there is a corresponding l plus 1. update vector l;

(6) The number of sensor nodes is increased by 1;

(7) If l reaches the specified coverage, the matrix m can be deleted;

(8) The algorithm loop terminates, the process is as follows (Fig. 5):

Considering that the current three-dimensional spatial algorithm value of the wireless sensor network node model is one-dimensional vector than the two-dimensional space, it is necessary to set the difficulty when performing optimal control $O(mN)$. Where m represents the amount of network node sensors for the entire control core and N represents a boundary of the current m. Under ideal conditions, when the sensor nodes are deployed in the wireless sensor network, the minimum coverage of each grid point needs to be K, and the minimum coverage probability is T (or $1 - N1$). – Express).

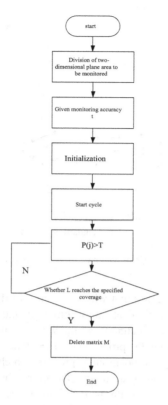

Fig. 5. Three dimensional greedy optimal control algorithm flow

3 Experimental Performance Analysis

3.1 Simulation

Consider a region of size 20 m × 20 m for simulation. Five beacon nodes are deployed in this region. The position of the beacon node is known. The coordinates are: (0, 0), (20, 0), (0, 20), (20, 20) and (10, 10), a target sensing network node is randomly deployed within the test area. S00 simulation experiments were performed on each algorithm on a desktop computer with Intel Corei5-4590, main frequency 3.3 GHz, memory 16 GB, 1600 MHz DDR3. For the existing MLE algorithm and LLS algorithm, it is assumed that the transmit power is known for each simulation. For the proposed optimal control model, the transmit power P_0 will be in the interval [5, 25] randomly selected between.

Figure 6 shows the sensor positioning error of different algorithms under different noise effects when the transmit power is unknown.

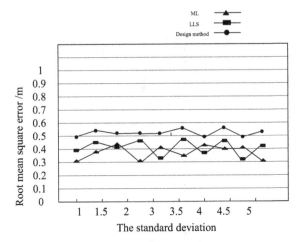

Fig. 6. Positioning error of different control methods

It can be seen from the figure that when the standard deviation of noise gradually increases, the positioning error of all algorithms will increase accordingly, and the MLE algorithm has the largest increase and the worst positioning performance. When the transmit power is unknown, the proposed control model treats the transmit power as an unknown variable during the positioning process, which reduces the impact of the transmit power on the positioning accuracy. Therefore, the proposed algorithm is far superior to the MLE and LLS algorithms. Positioning performance. When the noise is small, the proposed VTPNL algorithm and VTPLL algorithm have smaller sensing node positioning errors. Figure 7 depicts the positioning errors of different algorithms when the transmit power is different. In the simulation process, it is assumed that the

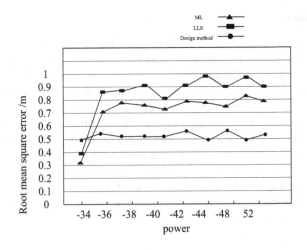

Fig. 7. Positioning error of different algorithms under unknown power

existing ML algorithm and the LLS algorithm both predict the true value of the power in advance, and the value of the value is set to −35 dBma, but the size of the po is actually unknown.

It can be seen from the figure that when the transmission power is unknown, when the difference between the real transmit power and the preset value is small, all the positioning algorithms can obtain better positioning performance. When the difference between the real transmit power and the preset value is gradually increased, the ML algorithm and the LLS algorithm may cause a large positioning error, and the accurate positioning of the target node cannot be achieved. Moreover, as the difference between the real value and the preset value becomes larger, the positioning errors of the two traditional wireless sensor network node control algorithms increase sharply. This is because as the difference between the true value and the preset value, the difference between the RSSI values becomes larger and larger, and the ranging is more and more sensitive. When the RSSI value exceeds 50 dBm, and smaller difference will bring larger ranging error, resulting in a sharp drop in positioning accuracy. The control model considers the unknown transmission power, no matter how much the difference between the real value and the preset value, the influence on the positioning perfor-mance is very small, and the corresponding positioning accuracy is better.

After two comparison experiments, the optimal control model of the wireless sensor network node can be determined. Through the greedy algorithm and the optimized probability algorithm, the current sensor node is directly regarded as an unknown variable to participate in the node location calculation. The optimization algorithm solves it to obtain the position information of the target node. In the calculation process, the transmission power is directly eliminated to reduce the influence of the transmission power on the positioning accuracy, and the nonlinear positioning problem is transformed into linear optimization by linear approximation method, which can effectively solve the traditional control model. Control problems and increase coverage.

In order to further verify the effectiveness of the method in this paper, the control accuracy of the control model in this paper and the control model in reference [5] are compared and analyzed, and the comparison results are shown in Fig. 8.

According to Fig. 8, the control accuracy of the wireless sensor network node of the control model in this paper can reach up to 90%, which is higher than that of the wireless sensor network node of the control model in reference [5], indicating that the control effect of the wireless sensor network node of the control model in this paper is better.

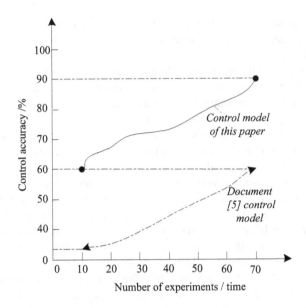

Fig. 8. Comparison of control accuracy of two systems

4 Conclusion

This paper introduces the origin of wireless sensor network and the research status at home and abroad. The related research on the coverage of wireless sensor networks at home and abroad is analyzed. The related technologies of wireless sensor networks are succinctly explained, and several characteristics of wireless sensor networks are summarized. The realization of the algorithm lays the foundation. This paper proposes a distributed algorithm based on probabilistic model. Under the condition that the coverage of each discrete grid point reaches k and the accuracy reaches T, the matlab simulation program is written. The simulation results of this algorithm are compared with those of average algorithm and random algorithm, the experimental results have significant advantages, and the algorithm is extended to the three-dimensional space. The experiment proves that the control method has practical performance.

References

1. Tang, B., Jiang, C.: A cost-sensitive method for wireless sensor network min beacon node set construction. Cluster Comput. **5**(1), 1–9 (2018)
2. Singh, P., Khosla, A., Kumar, A., et al.: Optimized localization of target nodes using single mobile anchor node in wireless sensor network. AEU – Int. J. Electron. Commun. **7**(91), 55–65 (2018)
3. Lee, W.K., Schubert, M.J.W., Ooi, B.Y., et al.: Multi-source energy harvesting and storage for floating wireless sensor network nodes with long range communication capability. IEEE Trans. Ind. Appl. **54**(3), 1 (2018)

4. Mansourkiaie, F., Ismail, L.S., Elfouly, T.M., et al.: Maximizing lifetime in wireless sensor network for structural health monitoring with and without energy harvesting. IEEE Access **5** (99), 2383–2395 (2017)
5. Gao, X.Y., He, Y.H.: Research on load balancing control of optical fiber network channel boundary node. Laser J. **6**, 184–187 (2019)
6. Wang, Y., Hang, J., Cheng, L., et al.: A hierarchical voting based mixed filter localization method for wireless sensor network in mixed LOS/NLOS environments. Sensors **18**(7), 2348 (2018)
7. Chen, Z., Teng, G., Zhou, X., et al.: Passive-event-assisted approach for the localizability of large-scale randomly deployed wireless sensor network. Tsinghua Sci. Technol. **24**(2), 14–26 (2019)
8. Tan, L., Tang, S.: Energy harvesting wireless sensor node with temporal death: novel models and analyses. IEEE/ACM Trans. Netw. **25**(2), 896–909 (2017)
9. Peng, Z., Wang, T., Wang, W., et al.: Survey of location-centric target tracking with mobile elements in wireless sensor networks. J. Central South Univ. **48**(3), 701–711 (2017)
10. Liu, S., Cheng, X., Fu, W., et al.: Numeric characteristics of generalized M-set with its asymptote. Appl. Math. Comput. **243**, 767–774 (2014)

A Decision Model for Substation Equipment Maintenance Based on Correlation Set Decomposition

Xue-sheng Li[✉]

Beifang Minzu University, Yinchuan 750021, China
lixuesheng651@163.com

Abstract. Substation equipment status maintenance background. Solve the problem of high cost in equipment maintenance. A decision model for substation equipment maintenance based on association set decomposition is proposed. Under the premise that the state of the device state is known. Analysis of the basic structure of substation equipment. And predict the operating status of the device. Starting from the functional association between devices, the association set is the basic unit. Realize the time-varying maintenance decision of the equipment. The conclusion is obtained by the model verification experiment: Compared with traditional equipment overhaul models. The method of substation equipment maintenance decision model based on association set decomposition can save 16.5% maintenance cost.

Keywords: Associative set decomposition · Transformer substation · Equipment overhaul · Maintenance decision

1 Introduction

As an important part of the power system, the substation refers to the place where the voltage is changed. The function of the equipment is to transmit the electric energy from the power plant to a distant place. The voltage must be raised into a high voltage, to the user's vicinity and then as needed to reduce the voltage, the work of this voltage rise and fall depends on the substation to complete. The main equipment in substations are switches and transformers. According to the size of the different, small called substation. Substations are larger than substations. A substation is an electric power facility in a power system that changes voltage, receives and distributes electric energy, controls the flow of electric power and adjusts voltage. It connects all levels of voltage to the grid through its transformers [1]. Substation has ac-dc-ac conversion process in specific environment. For example, undersea power transmission cables and long-distance transmission, some of which are in the form of high-voltage dc transmission and transformation, Dc transmission overcomes the capacitive and reactance loss of ac transmission and has the effect of energy saving. The main equipment and connection of substations vary according to their functions. Generally, the life cycle of a substation is as long as 30–50 years. Therefore, in the total life cycle cost of a substation, Operation cost, maintenance cost and failure cost often account for a large proportion.

© ICST Institute for Computer Sciences, Social Informatics and Telecommunications Engineering 2020
Published by Springer Nature Switzerland AG 2020. All Rights Reserved
Y.-D. Zhang et al. (Eds.): ICMTEL 2020, LNICST 327, pp. 228–240, 2020.
https://doi.org/10.1007/978-3-030-51103-6_20

It can be seen that good maintenance of various equipment in the substation is not only the basis for normal operation of the substation, but also can reduce the safety risks of the substation. An important way to achieve a balance between utility and cost. Under the guidance of the concept of network state maintenance, for the equipment maintenance decision problem topologically only associated with the substation, if the incoming and outgoing line state of the substation is assumed to be certain, the decision model of substation state maintenance is established. According to the concept of power grid state maintenance, the equipment under maintenance is all related to the substation. In terms of topological association, a type of equipment is associated with more than two substations, such as transmission lines. The other type is associated with only one substation, such as transformers, circuit breakers, disconnecting switches, etc. For substation, when the incoming and outgoing lines are given and the equipment only associated with the substation is in state maintenance, it is in line with the concept of network state maintenance. The timing of substation equipment maintenance is full of correlation and contradiction, so it is necessary to compromise between maintenance risk and fault risk. The research on the decision model of substation equipment maintenance is one of the important links in the process of adopting decomposition and coordination idea in the implementation of complex network state maintenance.

In reference [2], an early warning decision model of equipment preventive maintenance based on mean control chart is designed. Based on the economic design model of mean control chart, the economic model of mean control chart considering single system factor is constructed by combining the equipment preventive maintenance with mean control chart. Two thirds of the control boundary is set as warning area, Carry out preventive maintenance activities when the sample falls in the warning area. Through the parameter analysis of the model, it is found that the expected cost per unit time of preventive maintenance is lower than that of preventive maintenance when the alarm in the early warning area of control chart is given, which shows that the early warning decision model of equipment preventive maintenance based on mean control chart is of great value to reduce the cost and improve the quality control level of products. But the maintenance cost of the model is high and the maintenance time is long. Therefore, this paper proposes a decision-making model of substation equipment maintenance based on Association set decomposition.

2   Design of Substation Equipment Maintenance Decision Model

In order to minimize the maintenance cost of substation equipment, the maintenance strategy optimization of electrical equipment is realized by means of associative set decomposition in order to determine the optimal maintenance cycle and the upper limit of the optimal failure rate under the condition of maintenance.

2.1   Analyze the Basic Structure of Substation Equipment

Figure 1 shows the basic structure of substation equipment.

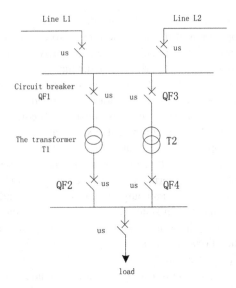

Fig. 1. Basic structure of substation equipment

In the basic structure {$QF1$, $T1$, $QF2$}, QF1 represents circuit breaker and T1 represents transformer. When transformer T1 is out of service due to maintenance or fault, all the air pressure associated with the equipment will also be out of service [3]. In this way, if the maintenance of other equipment is carried out after the shutdown of equipment T1, the maintenance cost and system outage loss caused by repeated shutdown of the system can be reduced.

2.2 Predict the Running State of the Equipment

The prediction of the running state of substation equipment is to judge whether the current running state of substation is a fault state. Generally speaking, the running state of substation equipment can be divided into normal running state, abnormal state, state transition state, fault state and failure-related transition state. The prediction of equipment running state is divided into two steps. Firstly, the state transition process of substation equipment is analyzed, and the probability of equipment in different states is calculated respectively, so as to get the prediction result of equipment running state.

(1) Device state transfer process

Transformers, circuits, circuit breakers and other equipment will be aging in use, maintenance is an important means to delay equipment aging and improve its reliability [4]. The aging process of the equipment is relatively long, and the state transition process of the equipment in a short time can be described by a two-state model. Figure 2 shows the state transition process of substation equipment.

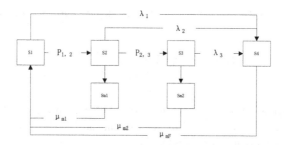

Fig. 2. Device state transition process

It can be seen from the figure that the substation equipment can be divided into six states, including S1 for good state, S2 indicates mild deterioration, S3 indicates severe deterioration status, S4 indicates functional failure status, S_{m1} indicates inspection m1 status, and S_{m2} indicates maintenance m2 status. When the equipment is in state S1, maintenance is not required. When the equipment is in state S2, S3, S4, maintenance can be carried out, expressed as m1 m2 and mF respectively, and all three maintenance methods make the equipment repaired to state S1 [5]. In addition, with the improvement of substation automation, the duration of switching state caused by active failure of equipment is short, which is negligible compared with the duration of equipment maintenance.

(2) Solve the equipment state probability

In the research cycle, the starting time of the next maintenance plan is set as M, and the substation equipment may be repaired or opportunity repaired after the failure occurs before the start of the maintenance plan, so as to change the failure rate of the equipment, thus affecting the subsequent risk level and decision-making level. The device state transition matrix is expressed as:

$$P_T = \begin{pmatrix} 1 - P_{1,2} - \lambda_1 & P_{1,2} & 0 & \lambda_1 & 0 & 0 \\ 0 & 1 - P_{2,3} - \lambda_2 & P_{2,3} & \lambda_2 & 0 & 0 \\ \mu_{mF} & 0 & 0 & 1 - \mu_{mF} & 0 & 0 \\ \mu_{m_1} & 0 & 0 & 0 & 1 - \mu_{m_1} & 0 \\ \mu_{m_2} & 0 & 0 & 0 & 0 & 1 - \mu_{m_2} \end{pmatrix} \qquad (1)$$

Where, parameter μ_{mF} represents the transfer rate of equipment failure state S4 to state S1; similarly, parameters μ_{m_1} and μ_{m_2} respectively represent the transfer rate of maintenance state Sm1 Sm2 to state S1 [6]. According to the nature of preventive maintenance, the equipment can be transferred to S1 state 100% after completion of m1 m2 maintenance. So the probability of state of device I in time period t can be calculated by formula 2.

$$\begin{cases} p_{i,S}(t) = p_{i,S}(0)P_{i,T}, \, t = 1 \\ p_{i,S}(t) = p_{i,S}(t-1)P_{i,T}, \, 1 < t \leq N_T \end{cases} \tag{2}$$

Where $p_{i,S}(t_0)$ is the state probability vector of equipment I in the current period, $P_{i,T}$ and $P_{i,M}$ are the state transition matrix and preventive maintenance transition matrix of substation equipment I, respectively.

2.3 Diagnose Substation Equipment Faults

Based on the prediction results of the running state of the equipment, the aging faults and sudden faults of the equipment are considered to diagnose whether the substation equipment has a fault state.

(1) Collect the operation signal of substation equipment

The signal sensor is installed on the equipment to be tested in the substation, and the collected signal is mainly the frequency signal generated during the operation of the equipment, so as to prove the failure of the equipment. The continuous working time of the sensor of substation equipment is set as 24 h, and the collection interval of sensor signals is set as 0.5 s [7]. In order to ensure the accuracy of the signal collection results, the selected sensor was modified. On the basis of the traditional sensor, a filter circuit was added to the sensor circuit to realize the signal filtering processing. As the equipment in the substation is interrelated, when a fault is found in the substation equipment, the equipment is the center to continue to detect the surrounding.

(2) Build the associated set risk assessment unit

The correlation between devices can be divided into three categories: functional correlation, economic correlation and random correlation. Function correlation means that when a device is out of service due to failure or maintenance, the related equipment will also be out of service. Economic correlation means that the maintenance cost will be reduced or increased if the related equipment is not in service at the same time. Random association refers to the random occurrence of external factors may make a certain device correlation. After the switching operation after the failure, the equipment that has a total outage belongs to an association set. Through analysis, it can be seen that the equipment set between two topological nodes can form an association set, which generally includes circuit or winding circuit breaker isolator, grounding switch arrester and high reactance. For substation lines, the constructed association set risk assessment unit, or association set, is shown in Fig. 3.

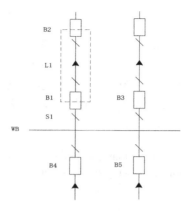

Fig. 3. Schematic diagram of association set

The association set contains three devices: circuit breaker B1 and B2 circuit L1. Under the premise of proper operation of the protection device, if there is a fault in line L1, the circuit breaker B1 and B2 will automatically jump to isolate the fault line, and the three devices will stop running at the same time until the fault is removed. If circuit breaker B1 fails, the circuit breaker B2 on the opposite side and the circuit breaker B3 B4 B5 on the bus WB will skip and all circuits will be out of service. Then, the disconnecting switch S1 will be disconnected to isolate the fault equipment and close the circuit breaker B3 B4 B5. Finally = only circuit breaker B1 and B2 line L1 will be out of service [8]. It can be seen that devices in the same association set are out of service. After switching operation, the association set will be out of service as a whole.

(3) The associative set can be solved by degree

The change of equipment performance is described as a time-varying shutdown model. Under the influence of correlation set, the state transition of substation equipment in the correlation set is shown in Fig. 4.

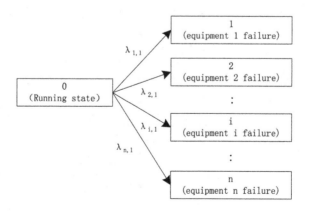

Fig. 4. Correlation set state transition process

Define the correlation set state transition rate as the matrix in formula 3.

$$v = \begin{bmatrix} -\mu_m & \mu_m & 0 & 0 & 0 \\ 0 & -\lambda_{12} & \lambda_{12} & 0 & 0 \\ 0 & 0 & -\lambda_{23} & \lambda_{23} & 0 \\ 0 & 0 & 0 & -\lambda_{34} & \lambda_{34} \\ 0 & \mu_f & 0 & 0 & -\mu_f \end{bmatrix} \qquad (3)$$

If p represents the probability that the association set is in state I, then the instantaneous state probability of solving the association set can be expressed as:

$$U_A = \frac{dp(t)}{dt} \qquad (4)$$

If the initial state of the association set is a, the instantaneous availability at time t can be expressed as:

$$A_a(t) = p_1(t) + p_2(t) + p_3(t) \qquad (5)$$

According to the setting of initial state and different state transition rate, the instantaneous solution of state probability distribution in different state transition process can be solved by this block diagram.

(4) Correlation set decomposition analysis

When the association set is determined to be available, the substation equipment in the association set is decomposed, and the specific processing process is shown in Fig. 5.

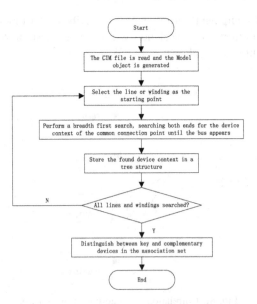

Fig. 5. Association set decomposition flow chart

The connection set and the branches in the network topology are one to one correspondence, and there is only one set of components in the circuit or winding. Taking the connection set as the basic unit can improve the efficiency of risk assessment.

(5) Calculate the associated set outage probability

Using the method of equipment risk assessment to calculate the outage probability of the correlation set completed by decomposition processing, and judging whether the substation equipment in the correlation set needs maintenance treatment based on the calculation result of probability [9]. The specific associated set device risk assessment technical architecture is shown in Fig. 6.

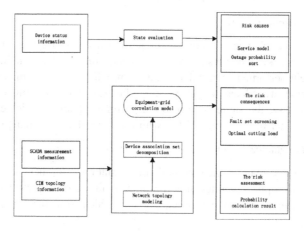

Fig. 6. Technical architecture diagram of equipment risk assessment

This technical architecture can realize the complete process from data acquisition to risk analysis to application decision. By collecting and analyzing the information of the data layer, the basic layer obtains the evaluation result of equipment state and the equipment-power network association model with topology modeling as the core and association set as the basic unit. The core layer obtains the real-time fault probability of the equipment by establishing the equipment outage model, and selects the correlation set with high outage probability to conduct the power grid operation risk assessment under the predicted fault set. Assuming that the correlation set is in the running state $P'(0)$ at the initial moment and its value is 1, the probability distribution of the state before the first failure in the corresponding period can be obtained through the decomposition of the correlation set, and the probability expression of the working state of the correlation set under different conditions is analyzed. If no equipment in the associated set fails before the maintenance plan, that is:

$$P_{case,0} = P_0(M) \qquad (6)$$

When this situation occurs, the equipment in the associated set can run reliably until the time of maintenance. If the solution result of $P_{case,0}$ does not meet the conditions in formula 6, it can be determined that the substation equipment in the correlation set needs maintenance treatment.

2.4   Realize the Maintenance Decision of Equipment in Time - Varying Shutdown

According to the fault diagnosis results of the equipment in the associated centralized substation, the maintenance treatment of the equipment with faults is carried out. Before the maintenance work, the maintenance decision scheme is formulated for the purpose of reducing the maintenance cost.

(1) State dependent decision

For unexpected changes in equipment performance, when the equipment state transition process is known, the state correlation decision is the specific implementation of the equipment performance correlation decision. All the states of the equipment that change with time are related to the possible changes of the scheduled maintenance decision. Obviously, the same state at different times should be viewed as different states, with a cluster of states associated with the same decision change.

(2) Select maintenance mode

In general, basic maintenance methods of substation equipment include complete maintenance, incomplete maintenance, minimal maintenance, post-maintenance and deterministic preventive maintenance, etc., among which some maintenance principles are shown in Fig. 7.

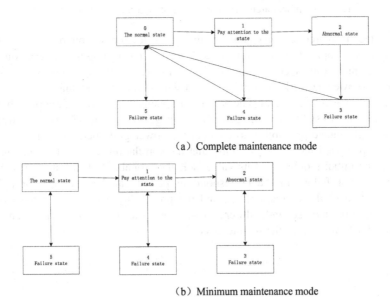

(a) Complete maintenance mode

(b) Minimum maintenance mode

Fig. 7. Maintenance schematic diagram

Take the minimal maintenance method as an example, the maintenance method is mainly to clear the equipment fault and make it in the running state after the equipment is repaired, but it will not improve the performance of the equipment or eliminate any influence of the aging of the equipment. In the working principle diagram of minimum maintenance, each failure state is transferred to the operating state before the equipment failure after the post-maintenance. Therefore, corresponding maintenance decisions can be selected according to the status diagnosis results of equipment and associated sets [10]. In addition to this state maintenance decision strategy, in order to ensure the safety of the use of substation equipment, it is also necessary to set a regular maintenance decision, and the corresponding maintenance decision expression is:

$$\begin{cases} \min C = f(T) \\ s.t. \quad 10\,years \leq T_1 \leq T_{st} \\ \quad\quad 5\,years \leq T_2 \leq 10\,years \\ \quad\quad T_1 - T_2 \geq 5\,years \end{cases} \tag{7}$$

Where T is T_1 two-dimensional decision vector, a represents the major repair period, and T_2 represents the minor repair period. T_{st} represents the upper bound of the overhaul cycle, and the value of this parameter is determined by the service life of the transformer equipment.

(3) Maintenance decision objective function and constraint condition

In order to minimize the total risk of the system, the functional expression of the state maintenance decision model of time-varying performance of related sets of substation equipment is obtained:

$$g = \min \left[\sum_{k=1}^{N} R_{1,k} + R_S \right] \tag{8}$$

In formula 8, is the operation risk of the k th device of $R_{1,k}$. The constraint conditions are set from three aspects of maintenance time simultaneous maintenance constraint and substation system safety constraint. The maintenance time constraint condition is:

$$\beta_i(t) = \begin{cases} 0, t < b_i \, or \, t > e_i \\ 0 \, or \, 1, \, b_i \leq t \leq e_i \end{cases} \tag{9}$$

Where, $\beta_i(t)$ is the state variable of equipment i in time period t. If the value is 1, it means the equipment is being repaired; if it is 0, it means the equipment is not being repaired. In addition, parameters b_i and e_i respectively represent the earliest and the latest time periods for the maintenance of equipment i. For the equipment in the same correlation set, it should be arranged together for maintenance to avoid repeated power failure of the system, that is:

$$x_i(t) = x_j(t) \tag{10}$$

The two variables in the above equation respectively represent the substation equipment in the same correlation set. In addition, the purpose of safety constraint is to control the maintenance power and control the actual power of power transmission and transformation equipment in the maintenance process within the maximum allowable transmission limit power.

3 Model Validation Analysis

In order to verify the feasibility and effectiveness of the proposed transformer substation equipment maintenance decision model based on association-set decomposition, the paper compares the different maintenance decision strategies of different equipment in the substation environment as shown in Fig. 8, and compares the maintenance costs consumed by different decision-making methods.

（a）Actual scene of substation environment

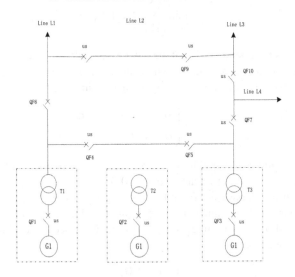

（b）Main wiring of substation equipment

Fig. 8. Substation environment

The study period was 60 days, divided into 60 periods in days. According to the monitoring results of equipment status, the relevant data of substation equipment scheduled for maintenance and repair in the cycle are shown in Table 1.

Table 1. Device-associated set initial state

Device name		Breaker kV211 220-3	Ming west line 211 switch	Ming west line	220 kV. #2 main transformer
Device type		Isolating switch	The circuit breaker	Line	Winding
The association set		Ming west	Ming west	The main direction of	The main direction of
The initial state	S1	0.90	0.10	0.0	0
	S2	0.10	0.90	0.0	0
	S3	0.00	0.80	0.2	0
	S4	0.00	0.70	0.3	0

Three different substations were selected as a, b and c respectively, and T1 and B2 were set as the equipment to be repaired in this experiment. Set and record the initial state of the equipment as shown in Table 1, and adjust the failover speed and state of the substation equipment through manual control. In order to highlight the decision performance of the proposed transformer substation equipment maintenance decision model based on correlation set decomposition, the literature [2] model was set as the control group in this verification experiment. The two decision models obtained two different maintenance methods, respectively recording the maintenance time and total risk cost of substation equipment, and the verification results obtained through statistics and calculation were shown in Table 2.

Table 2. Verify experimental data results

Maintenance decision method	Substation	Start time		Maintenance decision risk cost/ten thousand yuan
		T1	B2	
Literature [2] model	a	45	34	1554.9
	b			356.8
	c			165.7
A decision model for substation equipment maintenance based on correlation set decomposition	a	33	20	1303.4
	b			287.5
	c			142.6

After calculation, the average maintenance cost of the literature [2] model is 6.924 million yuan, and the average maintenance cost of the designed maintenance decision model is 5.778 million yuan, saving 1.46 million yuan. By comparing the two maintenance strategies, the corresponding results of substation a, b and c can be seen: the maintenance costs of the corresponding maintenance strategies of the substation equipment maintenance decision model based on the correlation set decomposition are the minimum, which can effectively make maintenance decisions on the substation equipment under the condition monitoring background.

4 Conclusion

On the basis of the existing research results, aiming at the functional correlation between equipment, the mathematical model of substation equipment state maintenance is constructed based on the idea of associative set decomposition, and the significance of model maintenance decision to improve the reliability and economy of maintenance cost is verified through experiments. Limited by time and space, it is necessary to conduct in-depth research on equipment condition maintenance of complex systems in the future research work around the correlation between multiple devices and the correlation with the system.

References

1. Latrache, H., Ouarzeddine, M., Souissi, B.: Adaptive model based on polarimetric decomposition using correlation coefficient in horizontal–vertical and circular basis. J. Appl. Remote Sens. **11**(3), 036006 (2017)
2. Zhao, Y.Q.: Research on early warning decision model of equipment preventive maintenance based on mean control chart. Comb. Mach. Tool Autom. Mach. Technol. **11**, 157–160 (2018)
3. Shuang, Z., Xiangchuan, Y.U., Lu, W.: Modified version of three-component model-based decomposition for polarimetric SAR data. J. Syst. Eng. Electron. **30**(02), 52–59 (2019)
4. Rong, K., Jiao, L., Shuang, W., et al.: Pansharpening based on low-rank and sparse decomposition. IEEE J. Sel. Top. Appl. Earth Obs. Remote Sens. **7**(12), 4793–4805 (2017)
5. Afzali, P., Keynia, F.: Lifetime efficiency index model for optimal maintenance of power substation equipment based on cuckoo optimisation algorithm. IET Gener. Transm. Distrib. **11**(11), 2787–2795 (2017)
6. Babu, A.S.: A model to derive software maintenance policy decision. Int. J. Product. Qual. Manag. **11**(3), 247–268 (2017)
7. Wang, L., Xu, N., et al.: Research on investment decision of substation project based on life cycle cost. In: IOP Conference Series Earth and Environmental Science, vol. 242, no. 2, p. 022016 (2019)
8. Kun, W.: Research on application of FMECA in missile equipment maintenance decision. MS&E **25**(2), 340 (2018)
9. Letot, C., Dehombreux, P., Fleurquin, G., et al.: An adaptive degradation-based maintenance model taking into account both imperfect adjustments and AGAN replacements. Qual. Reliab. Eng. Int. **33**(5), 1–15 (2017)
10. Liu, S., Cheng, X., Fu, W., et al.: Numeric characteristics of generalized M-set with its asymptote. Appl. Math. Comput. **243**, 767–774 (2014)

Research on Information Security Monitoring and Early Warning Mechanism of Internet Application Network Based on Particle Swarm Optimization

Feng Chen[1(✉)], Hong Zou[1], and Xue-sheng Li[2]

[1] Digital Grid Research Institute, CSG, Guangzhou 510507, China
cf9198@126.com
[2] Beifang Minzu University, Yinchuan 750021, China

Abstract. Due to the frequent occurrence of network security incidents, causing unnecessary losses to people, frequent network security incidents are worrying. For the problems of Internet application network information security, attackers use attacks to continuously threaten them. This paper studies the method of information security monitoring and early warning mechanism for Internet application network based on particle swarm optimization. Based on the support vector regression machine, a network security prediction model with multi-group chaotic particle optimization is established. The prediction results are obtained through the network information security monitoring and early warning mechanism, and the prediction results are analyzed and summarized. The results show that the Internet application network information security prediction model based on particle swarm optimization algorithm can provide guidance for the development of network security solutions and strategies, enhance the initiative of network security defense, reduce the losses caused by network attacks, and have better practicality Sex.

Keywords: Particle swarm optimization · Network information security · Monitoring and early warning

1 Introduction

The particle swarm optimization algorithm, also known as the particle swarm optimization algorithm or the flock foraging algorithm, abbreviated as PSO, is a new evolutionary algorithm developed by J. Kennedy and RC Eberhart in recent years. The PSO algorithm is a kind of evolutionary algorithm. It is similar to the simulated annealing algorithm. Starting from the random solution, it finds the optimal solution by iteration, and evaluates the quality of the solution according to the fitness. However, it is simpler than the genetic algorithm rule. It has no genetic algorithm. The "cross" and "mutation" operations seek global optimality by following the current searched optimal values. This kind of algorithm has attracted the attention of the academic community because of its high precision and fast convergence, and it has shown its superiority in

© ICST Institute for Computer Sciences, Social Informatics and Telecommunications Engineering 2020
Published by Springer Nature Switzerland AG 2020. All Rights Reserved
Y.-D. Zhang et al. (Eds.): ICMTEL 2020, LNICST 327, pp. 241–253, 2020.
https://doi.org/10.1007/978-3-030-51103-6_21

solving practical problems. Particle swarm optimization is a parallel algorithm. The information security situation value of Internet application network based on particle swarm optimization is an important indicator to measure network security. The value of its value directly affects the quality of network security [1]. Establishing an effective network information security monitoring and early warning mechanism model can prevent network security incidents and play an important role in network security protection.

Aiming at the shortcomings of particle swarm optimization algorithm, this paper improves it and proposes a multi-group chaotic particle optimization algorithm. The algorithm uses the randomness of chaotic principle, initializes the population particles in the initial stage, and divides the particles into three populations. Different populations adopt different updating strategies, and the convergence speed of the algorithm is accelerated by the synergy and information sharing among the three populations. The variance of the population fitness is used to judge whether the particles fall into local convergence, and chaotic processing is performed on the particles that are partially converged. It escapes from the local convergence point and avoids the phenomenon of "premature maturity" of the population with a certain probability, thus improving the optimization performance of the algorithm. The improved algorithm is tested by four standard test functions and compared with pso and ldw-pso to verify the algorithm has better performance.

2 Improved Particle Swarm Optimization for Network Information Security Prediction

Through the understanding and analysis of the particle swarm optimization algorithm, it can be seen that the exploration ability and development ability of the particle in the algorithm are a contradiction in the optimization process. If the particle swarm algorithm optimizes the simple problem, the global exploration ability of the particle itself and the local development capability is not high, and generally the optimal solution for the optimization problem can be found. However, for complex optimization problems, if the particle happens to be near the global best of the population, the development ability of the particle is weak due to the strong exploration ability of the particle at this time. So that the particles are farther away from the neighborhood where the optimal solution is located; if the particles are in the neighborhood of the most favorable locality, due to the strong development ability of the particles at this time, the exploration ability is weak, and it is easy for the particles to fall into local convergence [2]. Therefore, when the population is initialized, the randomness of the chaotic principle is used to make the particles evenly distributed in the search range, and then the population is divided into multiple populations, so that different populations can exhibit different exploration and development capabilities. Some particle swarms have the ability to explore and optimize. Global search in the process; part of the particles have the ability to develop, local search in the optimization process. Through the information sharing between the particles of the population, mention Optimization of the performance of the algorithm.

2.1 Building a Support Vector Regression Machine

Support vector regression is based on some complex relationship between sample data, to determine the functional relationship f (x) between the independent and dependent variables in the sample data for regression, the functional relationship f (x) can be divided into linear and non-linear Linear.

(a) Linear regression

Let the linear regression data be:

$$K = \{(x_1, y_1), (x_2, y_2), \ldots, (x_n, y_n)\}, \, x \in Rn, \, y \in R \tag{1}$$

Where x is the input of the regression function, the dimension is n, and the general case is n > 1. y is the output of the regression function, and the dimension is 1.

Let svr's regression function be:

$$f(x) = (w \cdot x) + b, \, w \in Rn, \, b \in R \tag{2}$$

Where w and b are regression functions and the unknown parameters of f (x) are real numbers.$(w \cdot x)$ The dot product of the vector, and the input vector x.

The goal of the support vector regression algorithm is to determine the regression function f (x) so that all sample data are near the regression function f (x), allowing some sample data to deviate from f (x) by a certain distance e, which has a good fit [3]. In the two-dimensional space, the regression function of the sample data is shown in Fig. 1.

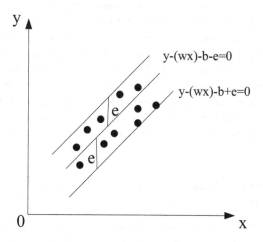

Fig. 1. Schematic diagram of the regression function of the two-dimensional space sample data

As you can see from Fig. 1, all the sample data is in
$Y - (w \cdot x) - be = 0$ and $y - (w \cdot x) - b + e = 0$ in the area of two parallel lines, that is, all sample data satisfy the formula (3).

$$-e < f(x_i) - (w \cdot x_i) - b < e, \, i = 1, 2, \ldots, n \qquad (3)$$

2.2   Network Information Security Monitoring and Early Warning Mechanism

The three parameters of the deviation parameter e, the penalty parameter C and the kernel function parameter g are determined to determine the network situation prediction model of the SVR. The deviation parameter e is the distance from which the sample data deviates from the regression line, and the number of sample data affecting the support vector. If the value of e is relatively large, the number of sample data of the support vector is relatively small, and the error of the regression estimation is increased, and the precision is lower. If the value of e is relatively small, the number of sample data of the support vector is relatively large, and the number of samples is reduced. The accuracy of the regression estimate is higher. Penalizing the size of the parameter C not only affects the training error of the sample, but also affects the complexity of the regression model [4]. If the value of C is relatively large, the phenomenon of "over-learning" will occur. The larger C value imposes a heavier penalty on the sample data outside the e-band, increasing the number of support vectors, thereby reducing the training error of the sample and increasing The generalization ability of the model; if the value of C is relatively small, it will produce the phenomenon of "under-learning". The smaller C value will lighten the penalty of the sample data outside the e-band, reducing the number of support vectors and increasing the training. Error, reducing the generalization ability of the model. The kernel function parameter g is the width of the RBF, and its magnitude affects the corresponding width of the inner product kernel function on the input variable. If the value of g is too large, the regression function or the discriminant function is too gentle; if the value of g is too small, the training data may have a certain memory ability or an "over-fitting" phenomenon.

In summary, setting reasonable parameters plays an important role in the generalization ability of the support vector regression model. By judging the error rate of the model algorithm, it can be judged whether the value of the model parameters is optimal [5].

2.3   Network Security Prediction Model for Multi-group Chaotic Particle Optimization

Set the number of hosts infected with network viruses to x_i. Number of websites that have been tampered with x_{min}. Total number of sites that were implanted in the back door x_{max}, the number of new information security vulnerabilities y_i As the independent variable of the network situation value, the network situation value is taken as the dependent variable. In order to reduce the different dimensions of the sample data and cause unnecessary errors in the results, the sample data needs to be normalized. Its formula is as shown in (4):

$$y_i = \frac{x_i - x_{min}}{x_{max} - x_{min}} \tag{4}$$

The network information security monitoring and early warning mechanism improves the shortcomings of particle swarm optimization algorithm. An MSCPO algorithm is proposed. The performance of the algorithm is verified by the test function, and the algorithm has better performance. Therefore, the parameters of the support vector regression machine model are optimized by MSCPO algorithm, and the model is trained for the sample data set used in this paper. Three parameters of deviating parameter e, penalty parameter C and kernel function parameter g are selected to establish MSCPO-SVR. The prediction model is expected to produce better prediction results [6].

The fitness function of the mscpo algorithm is the mean square error of the support vector regression model:

$$MSE = \frac{1}{n}\sum_{i=1}^{n} (x_i - x_i')^2 \tag{5}$$

among them, x_i For the true value, x_i' For the predicted value, n is the number of samples.

The establishment process of the network information security monitoring and early warning mechanism:

(1) Observe the original value of the input data and analyze it.
(2) Construct effective predictors, select independent and dependent variables.
(3) Preprocessing of the sample data, that is, normalization.
(4) The first 180 data of the sample data are used as the training set, and the last 10 data are used as the test set.
(5) Initialize the mscpo algorithm.
(6) Optimize the parameters of the support vector regression machine model according to the mscpo algorithm, and train the training set.

Practice, determine the MSCPO-SVR prediction model.

(7) Test the test sample data and output the test results of the model.

The flow chart for establishing the MSCPO-SVR prediction model is shown in Fig. 2.

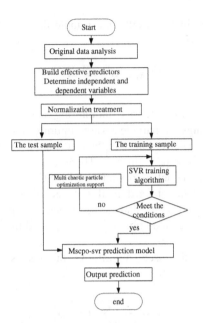

Fig. 2. MSCPO-SVR network situation prediction model flow

2.4   Prediction Results

Since the MSCPO-SVR prediction model has a small complexity and does not require prior knowledge, it can perform better generalization performance on randomly distributed sample data, and thus is applicable to any distributed sample data, so the MSCPO-SVR network situation prediction. The prediction results of the model [7]. MSCPO-SVR prediction model are shown in Table 1.

Table 1. Prediction results of the MSCPO-SVR prediction model

Serial number	Actual value	Predictive value	Absolute error
1	4	3.9150	0.0850
2	4	3.9442	0.1858
3	3	3.1303	0.2303
4	3	4.1372	0.1037
5	3	3.8613	0.2597
6	4	4.2018	0.2307
7	4	2.9263	0.1387
8	4	3.1597	0.3081
9	4	3.8867	0.2133
10	4	4.0932	0.1932

The absolute error of the prediction result of the MSCPO-SVR prediction model is shown in Fig. 3.

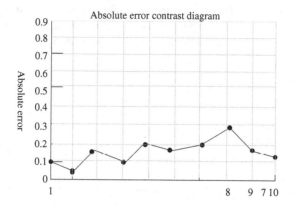

Fig. 3. Absolute error of the prediction result of the MSCPO-SVR prediction model

In order to visually represent the predictive performance of the model, compare its predicted value with the true value, such as

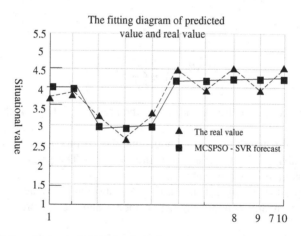

Fig. 4. Fitting the prediction results of the MSCPO-SVR prediction model with the real values

It can be seen from Fig. 3 and Fig. 4 that the absolute error of the prediction result obtained by the MSCPO-SVR prediction model proposed in this paper is relatively small, and the predicted value is close to the real value, and the fitting is better.

2.5   Comparative Analysis of Prediction Results

In order to verify the prediction performance of the MSCPO-SVR prediction model, it is compared with the prediction results of the bp neural network prediction model and the svr prediction model. This experiment uses the following commonly used evaluation indicators to evaluate the experimental prediction results, if the evaluation index is smaller, It indicates that the prediction result is more fitting to the true value.

(1) Absolute error e:

$$e = \left| x_i - x_i' \right| \tag{6}$$

(2) Average absolute error mae:

$$MAE = \frac{1}{n} \sum_{i=1}^{n} \left| x_i - x_i' \right| \tag{7}$$

(3) Average relative error mape:

$$MAPE = \frac{1}{n} \sum_{i=1}^{n} \left| x_i - x_i' \right| \frac{\left| x_i - x_i' \right|}{x_i} \tag{8}$$

(4) Average square root error rmse:

$$RMSE = \sqrt{\frac{1}{n} \sum_{i=1}^{n} (x_i - x_i')^2} \tag{9}$$

among them, x_i For the true value, x_i' For the predicted value, n is the total number of samples.

The bp neural network prediction model and the svr prediction model sample data are used for training, and the test results are obtained. The prediction results of the prediction model are shown in Table 2.

According to formulas (7), (8), (9), the evaluation index values of the prediction results of the bp neural network prediction model and the svr prediction model are shown in Table 3.

The comparison index between the bp neural network prediction model and the svr prediction model is shown in Fig. 5.

Table 2. Comparison prediction results of prediction models

Serial number	Actual value	MSCPO-SVR		BP	
		Predictive value	Absolute error	Predictive value	Absolute error
1	4	3.9150	0.0850	3.2150	0.7850
2	4	3.9442	0.1858	4.7442	0.7442
3	3	3.1303	0.2303	2.2303	0.7697
4	3	4.1372	0.1037	3.7459	0.7459
5	3	3.8613	0.2597	2.3524	0.6476
6	4	4.2018	0.2307	4.8372	0.8372
7	4	2.9263	0.1387	3.3613	0.6387
8	4	3.1597	0.3081	4.2965	0.2965
9	4	3.8867	0.2133	3.4867	0.5133
10	4	4.0932	0.1932	3.6932	0.3068

Table 3. Evaluation index value of the prediction model

Prediction model	MSCPO-SVR	BP
MAE	0.1189	0.6285
MAPE	0.0327	0.1751
RMSE	0.1167	0.6552

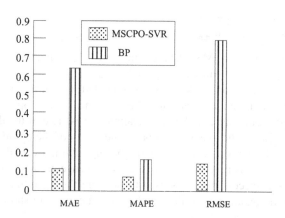

Fig. 5. Comparison chart of predictive model evaluation indicators

In summary, the predicted value of the MSCPO-SVR prediction model proposed in this paper is more fitting to the true value. Determine the three parameters of deviation parameter e, penalty parameter c and kernel function parameter g, establish the MSCPO-SVR prediction model, and test the 10 data after the sample data, and obtain the fitting comparison between the prediction result and the real value Strong.

Compared with BP neural network and MSCPO-SVR prediction model, the evaluation indexes of MSCPO-SVR prediction model are relatively small, which is better than the other two prediction models [8]. Through the BP neural network and MSCPO-SVR prediction model, the predicted value and the true value fit map show that the MSCPO-SVR prediction model has strong fitting and can maximize the optimization of network information security prediction [9, 10].

3   Simulation Test Experiment

3.1   Experimental Software and Hardware Environment

The experimental hardware configuration is computer configuration processor: Intel i5-3210 M CPU@2.SOGH; memory 10.0 GB; hard disk 1 TB; operating system 64-bit Windows 10 Professional.

The experimental software configuration is MATLAB environment: MATLABR 2016aVersion9.0.0.34136064-bit.

3.2   Comparison of Experimental Results

In order to verify the optimization performance of the MSCPO-SVR network information security monitoring and early warning algorithm proposed in this section, the algorithm of this section is used to test its algorithm and compare it with the pso and ldw-pso algorithms. The maximum number of iterations of the three optimization algorithms. Both are 1000, and the initialization parameters are set as follows.

PSO: w = 0.75, $c_1 = c_2 = 2$;
LDW-PSO: $w_{max} = 0.9$, $w_{min} = 0.4$, $c_1 = c_2 = 2$;

The first population in the MSCPO algorithm is w = 0.9, and the second population is w = 0.4, in the third population. $w_{max} = 0.9$, $w_{min} = 0.4$.

The three algorithms are independently operated 100 times in the 3 and 10 dimensions of the four standard test functions, and the average optimal solution, the average number of iterations, and the optimization success rate of the three algorithms are recorded. By finding the average number of iterations of the optimal solution, it is used to measure the convergence speed of the algorithm; by finding the average optimal solution size, it is used to measure the convergence accuracy of the algorithm; by the optimization success rate, it is used to measure the algorithm. Overall performance. The specific parameters of the standard test function are shown in Table 4.

Table 4. Standard test function parameter setting table

Test function name	Population size	Ranges	Convergence accuracy
Sphere	100	[−100, 100]	1E–10
Ackley	100	[−10, 10]	1E–6
Rastrigin	100	[−10, 10]	1E–6
Griewank	100	[−10, 10]	1E–6

The experimental results of the standard test functions of the three algorithms pso, ldw-pso and mscpo show that the mscpo algorithm finds the optimal solution with fewer iterations, and the optimization performance is better than the other two algorithms. For the peak function, the pso and ldw-pso algorithms also have better performance, and the average number of iterations is not much different. Because the unimodal function itself does not have a local optimal solution, the algorithm does not fall into local convergence. However, for the 3-D multi-peak test function, when the pso and ldw-pso algorithms are initialized, some particles may be in the vicinity of the optimal solution, so the optimal solution can be found with a certain probability, and the number of iterations is relatively small, but the population is found. The success rate of the optimal solution is relatively low. The average iteration number of the mscpo algorithm is lower than the other two algorithms, and the optimization success rate is significantly higher than the other two algorithms. Because the mscpo algorithm divides the particles into three populations during initialization. In the iterative process, the three populations cooperate and share the optimal value of each population. The chaotic processing of the particles that are trapped in the local optimum helps them escape the local best advantage, so that the algorithm finds the optimal solution with fewer iterations.

This paper is devoted to optimizing the shortcomings of particle swarm optimization algorithm. In order to better study the information security monitoring and early warning mechanism of Internet application network and improve its algorithm, a multi-group chaotic particle optimization algorithm is proposed. The 10D dimension of the MQC algorithm is tested and compared with the pso and ldw-pso algorithms. The experimental results show that:

(1) For the 3-dimensional test function, the pso and ldw-pso algorithms perform better for the single-peak test function because there is no local best advantage in the 3-dimensional single-peak test function. But for the 3-dimensional multi-peak test function Some particles may be in the vicinity of the optimal solution when they are initialized, so the optimal solution can be found, and the success rate of finding the optimal solution is relatively low. However, the optimization performance of the mscpo algorithm proposed in this paper is better than other in the single-peak test function. The two algorithms are not very obvious, but the optimization performance of the multi-peak test function is better than the pso and ldw-pso algorithms, showing good optimization performance.

(2) For the 10-dimensional test function, the pso and ldw-pso algorithms are only valid for the single-peak test function. For the multi-peak test function, the search success rate of the algorithm is very low, almost zero, and it is impossible to avoid the "precocity" of the population. Phenomenon. However, the mscpo algorithm proposed in this paper is better than the pso and ldw-pso algorithms for both single-peak and multi-peak functions. The population is initialized by chaotic mutation, and the population particles are divided into three different populations. Excellent particles adopt chaotic processing, which has certain efficiency to avoid the phenomenon of "early maturity" in the population. However, compared with the 3-dimensional multi-peak test function, although the average number of iterations has increased, the success rate of optimization has decreased, but still the optimal solution can be found with a certain probability.

The improved algorithm is used to optimize the parameters of the support vector regression machine model, the sample data set collected in this paper is trained, the MSCPO-SVR prediction model is established, and the sample data is predicted. The error between the predicted value and the real value is smaller. Compared with the bp neural network and the svr prediction model, the evaluation index of the model is lower than the other two prediction models. It can be seen that the prediction effect of the model is better, and the predicted value is the true value is more fitting.

In summary, the MSCPO-SVR prediction model can provide guidance for the development of network security solutions and strategies, enhance the initiative of network security defense, reduce the losses caused by network attacks, and have better practicability and enhance network security. The initiative of defense.

4 Conclusion

Through the analysis and summary of the research status of network information security monitoring and early warning model technology and particle swarm optimization algorithm, the regression principle of support vector machine is briefly explained. Analyze the situation report published on the website of the National Internet Emergency Center and establish a sample data set. The parameters of the MSCPO optimized support vector regression model are proposed. The first 100 data of the sample data are used as the training set to determine the three parameters of the deviation parameter e, the penalty parameter C and the kernel function parameter g. The MSCPO-SVR prediction model was established, and the 10 data after the sample data were tested. The prediction results obtained were more fitting with the real values. By comparing with BP neural network and MSCPO-SVR prediction model, it is verified that MSCPO-SVR prediction model has strong fitting and practicability for network information security monitoring.

Acknowledgment. China Southern Power Grid technology project "Information Operation Security System V1.0 product development" (2018030102dx00697).

References

1. Qi, Y., Tan, R.: Application of improved particle swarm optimization algorithm-based BP neural network to dam deformation analysis. Water Resour. Hydropower Eng. **48**(2), 118–124 (2017)
2. He, R., Luo, D.: Application of improved particle-swarm-optimization neural network in coalmine safety evaluation. Ind. Saf. Environ. Prot. **44**(11), 33–35 (2018)
3. Luo, S., Liu, C.: A detection method based on particle swarm optimization algorithm and SVM dealing with network intrusion. Mod. Electron. Tech. **40**(10), 31–34 (2017)
4. Xiao, Z.: Mobile Internet application platform of information security research of situation assessment. Comput. Simul. **34**(3), 423–426 (2017)
5. Baoren, Ch., Gu, W., Han Kuai, K., et al.: Research on ad hoc network optimization based on chaotic particle swarm optimization. Trans. Beijing Inst. Technol. **37**(4), 381–385 (2017)

6. Ren, P.-F., Gu, L.-K.: WSN node localization algorithm for power transmission networks based on particle swarm optimization. J. Shenyang Univ. Technol. **40**(5), 63–68 (2018)
7. Tao, C.P.Y.J.Y.J.Y.: Respiratory signals prediction based on particle swarm optimization and back propagation neural networks. Chin. J. Biomed. Eng. **37**(6), 714–719 (2018)
8. Wu, C., Liu, J.-M., Guo, Z.-D.: Use of hybrid fuzzy c-means and probabilistic neural network based on improved particle swarm optimization in the prediction of financial distress. Oper. Res. Manag. Sci. **27**(2), 106–114 (2018)
9. Yan, B., Wang, C.: Application of RBF neural network based on particle swarm optimization algorithm in crack width prediction of sluice pier. Water Power **44**(3), 33–36 (2018)
10. Liu, S., Cheng, X., Fu, W., et al.: Numeric characteristics of generalized M-set with its asymptote. Appl. Math. Comput. **243**, 767–774 (2014)

A Multiple Sclerosis Recognition via Hu Moment Invariant and Artificial Neural Network Trained by Particle Swarm Optimization

Ji Han[1] and Shou-Ming Hou[1,2(✉)]

[1] Henan Polytechnic University, Jiaozuo 454000, Henan, China
HanJi@home.hpu.edu.cn, housm@163.com
[2] Hebi Automotive Engineering Professional College,
Hebi 454030, Henan, China

Abstract. Multiple sclerosis can damage the central nervous system, and current drugs are difficult to completely cure symptoms. The aim of this paper was to use deep learning methods to increase the detection rate of multiple sclerosis, thereby increasing the patient's chance of treatment. We presented a new method based on hu moment invariant and artificial neural network trained by particle swarm optimization. Our method was carried out over ten runs of ten-fold cross validation. The experimental results show that the optimization ability of particle swarm optimization algorithm is superior to the genetic algorithm, simulated annealing algorithm and immune genetic algorithm. At the same time, compared with the HWT+PCA+LR method and the WE-FNN-AGA method, our method performs better in the performance of the detection.

Keywords: Multiple sclerosis · Hu moment invariant · Feedforward neural network · Particle swarm optimization

1 Introduction

Multiple sclerosis (MS) is an immune-mediated disease characterized by inflammatory demyelinating lesions of the white matter of the central nervous system [1]. Multiple sclerosis involves the effects of genetic, environmental and viral factors [2]. Multiple sclerosis is more likely to occur in young people, especially young women. Treating patients as early as possible can help slow the progression of the disease. Therefore, early detection of multiple sclerosis will benefit the treatment of patients.

In early stages of MS, there are neurodegenerative behaviors such as axons and loss of neurons. These behaviors can be used to determine permanent limb accumulation and cognitive impairment [3]. The treatment of MS mainly includes acute phase treatment and remission treatment. At present, the treatment plan for the acute phase is treated with high-dose glucocorticoid shock therapy, and the treatment for remission is recommended to use disease modification therapy.

As the advantages of deep learning methods in medical detection become more prominent, more researchers tend to use deep learning methods to perform multiple

Y.-D. Zhang et al. (Eds.): ICMTEL 2020, LNICST 327, pp. 254–264, 2020.
https://doi.org/10.1007/978-3-030-51103-6_22

sclerosis detection tasks. Alshayeji, Al-Rousan (2018) [4] proposed a high-efficiency detection system for multiple sclerosis based on neural network. Lopez (2017) [5] proposed a hybrid method combining principal component (PCA), Haar wavelet transform (HWT), and logistic regression (LR). Han and Hou (2019) [6] proposed a wavelet entropy (WE) with feedforward neural network (FNN), whose weights and biases were optimized through a new swarm intelligence method—adaptive genetic algorithm (AGA).

In the study, mathematical and morphological operations were used for feature extraction, and multi-layer feedforward neural networks were used to identify multiple sclerosis tissues. Currently, deep learning and transfer learning are widely used in medical image analysis [7–19]. Nevertheless, deep learning is not suitable for this study, because our dataset is too small.

In order to further improve the precision of detection, we use feedforward neural network in combination with hu moment invariant and particle swarm optimization algorithm to detect multiple sclerosis. The rest of the organization structure of this paper is as follows. Section 2 shows the source of the dataset used. Section 3 introduces the methods used to detect multiple sclerosis. Section 4 analyzes and discusses the experimental results. Section 5 summarizes the research contributions of the paper.

2 Dataset

Within our paper, the dataset we used were from Ref [17] and eHealth laboratory [20]. Table 1 shows the demographic characteristics of two datasets. 681 healthy slice brain images from 26 healthy controls and 676 MS slice brain images from 38 patients with multiple sclerosis were selected. All brain slice image data is guaranteed and confirmed by a professional brain surgeon. Therefore, the dataset used in the experiment is accurate and reliable.

Table 1. Demographic characteristics of two datasets.

Subject	Provider	NS1	NS2	Gender (m/f)	Age
Multiple sclerosis	eHealth laboratory	38	676	17/21	34.1 ± 10.5
Healthy control	Dr. Pan et al.	26	681	12/14	33.5 ± 8.3

(NS1 = number of subjects; NS2 = number of slice)

3 Methodology

The implementation of our detection method is based on Hu moment invariant (HMI), feedforward neural network, and particle swarm optimization (PSO). Among them, HMI is used to extract the enhanced dataset, feedforward neural network is used for feature learning and classification, and PSO is used to optimize the trained model because of its intelligent optimization ability.

3.1   HMI

In this study, image moment is used as the shape descriptor. For a 2D brain image $I(a, b)$, the raw moment M of order (p, q) is defined as

$$M_{pq} = \Sigma_a \Sigma_b a^p b^q I(a, b) \tag{1}$$

where u, v = 0, 1, 2,...

The central moment increases the translation invariance based on raw moment [21], and the central moment N is defined as

$$N_{pq} = \Sigma_a \Sigma_b [a - E(a)]^p [b - E(b)]^q I(a, b) \tag{2}$$

$$E(a) = \frac{M_{10}}{M_{00}} \tag{3}$$

$$E(b) = \frac{M_{01}}{M_{00}} \tag{4}$$

The normalized central moment increases the scale invariant based on the central moment [22], and the normalized central moment φ is defined as:

$$\varphi_{pq} = \frac{N_{pq}}{N_{00}\left(\frac{p+q}{2}+1\right)} \tag{5}$$

The hu moment increases rotation invariant based on normalized central moment [23]. Seven Hu moment invariants $[\omega_1, \omega_2, \omega_3, \omega_4, \omega_5, \omega_6, \omega_7]$ are defined as:

$$\omega_1 = \varphi_{02} + \varphi_{20} \tag{6}$$

$$\omega_2 = 4\varphi_{11}^2 + (\varphi_{20} - \varphi_{02})^2 \tag{7}$$

$$\omega_3 = (3\varphi_{21} - \varphi_{03})^2 + (\varphi_{30} - 3\varphi_{12})^2 \tag{8}$$

$$\omega_4 = (\varphi_{03} + \varphi_{21})^2 + (\varphi_{12} + \varphi_{30})^2 \tag{9}$$

$$\omega_5 = \left[3(\varphi_{12} + \varphi_{30})^2 - (\varphi_{03} + \varphi_{21})^2\right](\varphi_{03} + \varphi_{21})(3\varphi_{21} - \varphi_{03})$$
$$+ \left[(\varphi_{12} + \varphi_{30})^2 - 3(\varphi_{03} + \varphi_{21})^2\right](\varphi_{12} + \varphi_{30})(\varphi_{30} - 3\varphi_{12}) \tag{10}$$

$$\omega_6 = 4\varphi_{11}(\varphi_{03} + \varphi_{21})(\varphi_{12} + \varphi_{30})$$
$$+ \left[(\varphi_{12} + \varphi_{30})^2 - (\varphi_{03} + \varphi_{21})^2\right](\varphi_{20} - \varphi_{02}) \tag{11}$$

$$\omega_7 = \left[(\varphi_{12} + \varphi_{30})^2 - 3(\varphi_{03} + \varphi_{21})^2\right](\varphi_{12} + \varphi_{30})(3\varphi_{21} - \varphi_{03})$$
$$- \left[3(\varphi_{12} + \varphi_{30})^2 - (\varphi_{03} + \varphi_{21})^2\right](\varphi_{03} + \varphi_{21})(\varphi_{30} - 3\varphi_{12}) \tag{12}$$

Hu moment invariance (HMI) is successfully applied to object recognition, image detection, visual tracking, shape measurement, etc. due to its characteristics of translation, scale, and rotation invariants. [24]. In the future, we shall test other feature extractors [25–32].

3.2 Feedforward Neural Network

Feedforward neural networks (FNN) usually consist of an input layer, a hidden layer, and an output layer. The nodes of each adjacent layer are fully linked, and the size of the link weight reflects the closeness of the relationship between the two nodes. FNN can only propagate in the direction from input to output. The output of any neuron in the same layer will not affect other neurons, and there is a mapping relationship between the layers [33]. The structure of the general feedforward neural network is shown in Fig. 1.

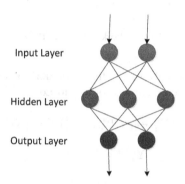

Fig. 1. Structure of feedforward neural network.

3.3 Particle Swarm Optimization

Particle Swarm Optimization (PSO) is an algorithm for intelligent optimization by simulating the predation of birds [34]. Particles represent the solution to the problem, and fitness functions are used to evaluate the performance of the solution. During each iteration, local and global optimal solutions are selected to update the position of the particles. After several iterations, the particle will search for the optimal solution. The workflow of the PSO is shown in Fig. 2.

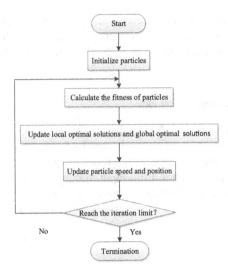

Fig. 2. Workflow of the PSO.

Assuming that there is only one optimal solution in region T, the position and velocity of each particle are updated according to Eqs. (13) and (14).

$$V_{it}^{n+1} = wV_{it}^{n} + a_1 r_1 (P_{best}^{n} - y_{it}^{n}) + a_2 r_2 (G_{best}^{n} - y_{it}^{n}) \tag{13}$$

$$Y_{it}^{n+1} = Y_{it}^{n} + V_{it}^{n+1} \tag{14}$$

where V_{it}^{n+1} represents the speed reached by the i-th particle in the $(n+1)$-th iteration, Y_{it}^{n} is the position of the i-th particle after nth iteration, w is inertia factor, a_1 and a_2 are acceleration constants, r_1 and r_2 are random numbers in the range [0, 1], t represents the position and speed of the t-th dimension. P_{best}, G_{best} represents the local optimal solution and the global optimal solution, respectively.

3.4   10-fold Cross Validation

The 10-fold cross validation in this paper is a method for assessing the accuracy of models built on datasets [35]. In 10-fold cross-validation, the data set is divided into 10 random subsets. In each iteration, one subset is selected as a test subset in turn, and the remaining nine subsets are used for training. After 10 iterations, the results of the verification are averaged to analyze the performance of the model [36]. This experiment will run 10 times 10-fold cross validation to ensure more accurate results.

4   Results and Discussions

4.1   Statistical Analysis

This proposed MS detection method "HMI-FNN-PSO" is implemented over 10 runs of 10-fold cross validation. Table 2 shows the results. Our method achieved satisfactory results on five indicators: sensitivity (Se), specificity (Sp), precision (Pr), accuracy (Ac), and F1 score (Fs). The test results of indicators Se, Sp, Pr, Ac, and Fs were $91.67 \pm 1.41\%$, $91.73 \pm 0.77\%$, $91.70 \pm 0.78\%$, $91.70 \pm 0.97\%$,   and $91.67 \pm 1.00\%$.

Table 2. 10 runs of 10-fold cross validation of our method.

Run	Se	Sp	Pr	Ac	Fs
1	93.05	92.51	92.53	92.78	92.78
2	94.38	92.95	93.00	93.66	93.68
3	91.12	90.75	90.71	90.94	90.91
4	92.75	92.36	92.37	92.56	92.55
5	89.05	91.49	91.29	90.27	90.14
6	90.98	91.63	91.54	91.30	91.26
7	91.86	91.93	91.88	91.90	91.87
8	91.86	90.31	90.39	91.08	91.12
9	91.27	91.34	91.27	91.30	91.27
10	90.38	92.08	92.03	91.23	91.16
Average	91.67 ± 1.41	91.73 ± 0.77	91.70 ± 0.78	91.70 ± 0.97	91.67 ± 1.00

4.2   Training Algorithm Comparison

We compared PSO with three training algorithms: genetic algorithm (GA), simulated annealing (SA), and immune genetic algorithm (IGA). The results of 10 runs of 10-fold cross validation using GA, SA, and IGA are shown in Table 3, Table 4, and Table 5, respectively. Table 6 shows the results of the comparison and Fig. 3 shows the results in the form of a histogram. The experimental results confirm that the performance of our training algorithm is superior to the other three training algorithms, and our test standards are more than 90%.

Table 3. 10 runs of 10-fold cross validation using GA.

Run	Se	Sp	Pr	Ac	Fs
1	85.21	85.76	85.59	85.48	85.38
2	84.62	81.35	81.83	82.98	83.20
3	84.02	78.71	79.73	81.36	81.78
4	84.32	84.43	84.31	84.38	84.30
5	83.88	85.16	85.00	84.52	84.38
6	85.21	86.34	86.09	85.78	85.64
7	81.36	81.64	81.52	81.50	81.42
8	84.91	86.93	86.60	85.92	85.74
9	82.54	85.47	85.01	84.01	83.74
10	84.91	85.91	85.71	85.41	85.30
Average	84.10 ± 1.19	84.17 ± 2.54	84.14 ± 2.18	84.13 ± 1.60	84.09 ± 1.47

Table 4. 10 runs of 10-fold cross validation using SA.

Run	Se	Sp	Pr	Ac	Fs
1	79.44	78.71	78.75	79.07	79.09
2	81.07	79.14	79.51	80.10	80.25
3	82.10	82.96	82.72	82.53	82.41
4	77.22	81.20	80.31	79.22	78.73
5	81.66	82.38	82.17	82.02	81.90
6	78.25	76.07	76.47	77.16	77.33
7	81.07	78.71	79.07	79.88	80.06
8	77.37	80.47	79.73	78.92	78.53
9	79.59	77.68	77.97	78.63	78.77
10	80.47	81.94	81.56	81.21	81.00
Average	79.82 ± 1.66	79.93 ± 2.12	79.83 ± 1.83	79.87 ± 1.56	79.81 ± 1.53

Table 5. 10 runs of 10-fold cross validation using IGA.

Run	Se	Sp	Pr	Ac	Fs
1	87.57	88.40	88.23	87.99	87.90
2	84.32	85.76	85.44	85.04	84.88
3	87.43	86.49	86.53	86.96	86.97
4	88.61	90.16	89.98	89.39	89.28
5	86.54	86.93	86.79	86.74	86.66
6	85.06	83.26	83.46	84.16	84.25
7	83.58	81.06	81.41	82.31	82.48
8	84.17	84.58	84.42	84.38	84.30
9	86.83	85.61	85.70	86.22	86.25
10	86.69	89.43	89.09	88.06	87.87
Average	86.08 ± 1.60	86.17 ± 2.64	86.11 ± 2.48	86.12 ± 2.04	86.08 ± 1.97

Table 6. Comparison of different training algorithms.

Method	Se	Sp	Pr	Ac	Fs
GA	84.10 ± 1.19	84.17 ± 2.54	84.14 ± 2.18	84.13 ± 1.60	84.09 ± 1.47
SA	79.82 ± 1.66	79.93 ± 2.12	79.83 ± 1.83	79.87 ± 1.56	79.81 ± 1.53
IGA	86.08 ± 1.60	86.17 ± 2.64	86.11 ± 2.48	86.12 ± 2.04	86.08 ± 1.97
PSO (ours)	91.67 ± 1.41	91.73 ± 0.77	91.70 ± 0.78	91.70 ± 0.97	91.67 ± 1.00

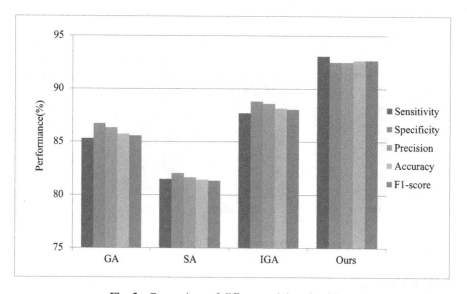

Fig. 3. Comparison of different training algorithms

4.3   Comparison with State-of-the-Art Algorithms

We compared our method with two state of the art approaches: one based on Haar
wavelet transform (HWT), principal component analysis (PCA), and logistic regression
(LR). Another based on wavelet entropy (WE), feedforward neural network (FNN), and
adaptive genetic algorithm (AGA). Table 7 shows the results of the comparison. In
Fig. 4, the comparison results are more intuitive in the form of a histogram. From an
precision point of view, the experimental results of HWT+PCA+LR [5] and WE-FNN-
AGA [6] were 87.51% and 86.32 ± 0.86%, respectively. However, the precision of
our method is 91.70 ± 0.78%, so it is significantly better than the two state-of-the-art
approaches.

Table 7. Comparison with state-of-the-art approaches.

Method	Se	Sp	Pr	Ac	Fs
HWT+PCA+LR [5]	89.04	90.24	87.51	89.72 ± 1.18	88.27
WE-FNN-AGA [6]	86.32 ± 1.09	86.36 ± 1.06	86.32 ± 0.86	86.34 ± 1.12	n/a
HMI-FNN-PSO (ours)	91.67 ± 1.41	91.73 ± 0.77	91.70 ± 0.78	91.70 ± 0.97	91.67 ± 1.00

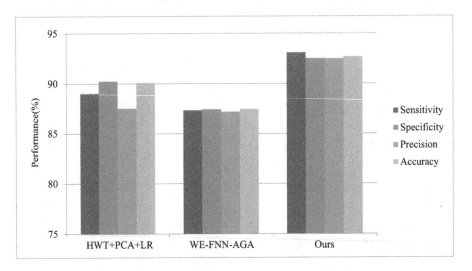

Fig. 4. Comparison to state-of-the-art approaches.

5   Conclusions

In this paper, we conduct research based on the detection of brain images of multiple sclerosis. Combine hu moment invariant and particle swarm optimization algorithms with feedforward neural networks. Our main research contributions are shown below:

(1) For the first time, we applied Hu Moment invariant and feedforward neural networks to the detection of multiple sclerosis, which improved the detection rate and helped patients with early treatment.
(2) In order to improve the optimization ability, a particle swarm algorithm is proposed and the performance is superior.
(3) Compared with other state-of-the-art approaches, the proposed method achieves the best performance in terms of index Se, Sp, Pr and Ac. Test results show that our detection performance is significantly better than the HWT+PCA+LR and WE-FNN-AGA approach.

In future research, we will further innovate in algorithms to improve detection accuracy and make more contributions to the early detection of multiple sclerosis.

Acknowledgement. This paper was supported by Key research and development and technology promotion projects in Henan Province, China (No. 172102210273, 182102210086, 182102310629), Youth backbone training program for colleges and universities in Henan, China (No. 2018GGJS298).

References

1. Barro, C., et al.: Serum neurofilament as a predictor of disease worsening and brain and spinal cord atrophy in multiple sclerosis. Brain **141**(8), 2382–2391 (2018)
2. Tintore, M., et al.: Treatment of multiple sclerosis—success from bench to bedside. Nat. Rev. Neurol. **15**(1), 53–58 (2019)
3. Pietroboni, A.M., et al.: The loss of macular ganglion cells begins from the early stages of disease and correlates with brain atrophy in multiple sclerosis patients. Mult. Scler. J. **25**(1), 31–38 (2019)
4. Alshayeji, M.H., et al.: An efficient multiple sclerosis segmentation and detection system using neural networks. Comput. Electr. Eng. **71**, 191–205 (2018)
5. Lopez, M.: Multiple sclerosis slice identification by haar wavelet transform and logistic regression. Adv. Eng. Res. **114**, 50–55 (2017)
6. Han, J., Hou, S.-M.: Multiple sclerosis detection via wavelet entropy and feedforward neural network trained by adaptive genetic algorithm. In: Rojas, I., Joya, G., Catala, A. (eds.) IWANN 2019. LNCS, vol. 11507, pp. 87–97. Springer, Cham (2019). https://doi.org/10. 1007/978-3-030-20518-8_8
7. Govindaraj, V.V.: High performance multiple sclerosis classification by data augmentation and AlexNet transfer learning model. J. Med. Imaging Health Inform. **9**(9), 2012–2021 (2019)
8. Jiang, X.: Chinese sign language fingerspelling recognition via six-layer convolutional neural network with leaky rectified linear units for therapy and rehabilitation. J. Med. Imaging Health Inform. **9**(9), 2031–2038 (2019)
9. Yu, X., Zeng, N., Liu, S., Zhang, Y.-D.: Utilization of DenseNet201 for diagnosis of breast abnormality. Mach. Vis. Appl. **30**(7–8), 1135–1144 (2019)
10. Li, Z.: Teeth category classification via seven-layer deep convolutional neural network with max pooling and global average pooling. Int. J. Imaging Syst. Technol. **29**, 577–583 (2019). https://doi.org/10.1002/ima.22337
11. Hong, J.: Detecting cerebral microbleeds with transfer learning. Mach. Vis. Appl. **30**(7–8), 1123–1133 (2019)
12. Tang, C.: Cerebral micro-bleeding detection based on densely connected neural network. Front. Neurosci. **13**, 422 (2019)
13. Xie, S.: Alcoholism identification based on an AlexNet transfer learning model. Front. Psychiatr. **10**, 205 (2019)
14. Jia, W.: Five-category classification of pathological brain images based on deep stacked sparse autoencoder. Multimed. Tools Appl. **78**(4), 4045–4064 (2017)
15. Muhammad, K.: Image based fruit category classification by 13-layer deep convolutional neural network and data augmentation. Multimed. Tools Appl. **78**(3), 3613–3632 (2019)
16. Sangaiah, A.K.: Alcoholism identification via convolutional neural network based on parametric ReLU, dropout, and batch normalization. Neural Comput. Appl. **32**(3), 665–680 (2019). https://doi.org/10.1007/s00521-018-3924-0
17. Pan, C.: Multiple sclerosis identification by convolutional neural network with dropout and parametric ReLU. J. Comput. Sci. **28**, 1–10 (2018)

18. Hou, X.-X.: Seven-layer deep neural network based on sparse autoencoder for voxelwise detection of cerebral microbleed. Multimed. Tools Appl. **77**(9), 10521–10538 (2018)

19. Pan, C.: Abnormal breast identification by nine-layer convolutional neural network with parametric rectified linear unit and rank-based stochastic pooling. J. Comput. Sci. **27**, 57–68 (2018)

20. MRI Lesion Segmentation in Multiple Sclerosis Database, in eHealth Laboratory, University of Cyprus. http://www.medinfo.cs.ucy.ac.cy/index.php/downloads/datasets

21. Wang, Y.Q., et al.: The optimal fractional S transform of seismic signal based on the normalized second-order central moment. J. Appl. Geophys. **129**, 8–16 (2016)

22. Yang, J.: Pathological brain detection in MRI scanning via Hu moment invariants and machine learning. J. Exp. Theor. Artif. Intell. **29**(2), 299–312 (2017)

23. Hu, M.-K.: Visual pattern recognition by moment invariants. IRE Trans. Inf. Theory **8**(2), 179–187 (1962)

24. Hou, X.-X.: Alcoholism detection by medical robots based on Hu moment invariants and predator-prey adaptive-inertia chaotic particle swarm optimization. Comput. Electr. Eng. **63**, 126–138 (2017)

25. Ramirez, J.: Unilateral sensorineural hearing loss identification based on double-density dual-tree complex wavelet transform and multinomial logistic regression. Integr. Comput.-Aided Eng. **26**, 411–426 (2019). https://doi.org/10.3233/ICA-190605

26. Yang, J.: An adaptive encoding learning for artificial bee colony algorithms. J. Comput. Sci. **30**, 11–27 (2019)

27. Gorriz, J.M.: Multivariate approach for Alzheimer's disease detection using stationary wavelet entropy and predator-prey particle swarm optimization. J. Alzheimer's Dis. **65**(3), 855–869 (2018)

28. Sun, J.: Preliminary study on angiosperm genus classification by weight decay and combination of most abundant color index with fractional Fourier entropy. Multimed. Tools Appl. **77**(17), 22671–22688 (2018)

29. Zhao, G.: Smart pathological brain detection by synthetic minority oversampling technique, extreme learning machine, and Jaya algorithm. Multimed. Tools Appl. **77**(17), 22629–22648 (2018)

30. Lu, S.: Pathological brain detection in magnetic resonance imaging using combined features and improved extreme learning machines. J. Med. Imaging Health Informat. **8**, 1486–1490 (2018)

31. Li, Y.-J.: Single slice based detection for Alzheimer's disease via wavelet entropy and multilayer perceptron trained by biogeography-based optimization. Multimed. Tools Appl. **77**(9), 10393–10417 (2018)

32. Kong, F.Q.: Ridge-based curvilinear structure detection for identifying road in remote sensing image and backbone in neuron dendrite image. Multimed. Tools Appl. **77**(17), 22857–22873 (2018)

33. Chowdhury, N.: A comparative analysis of feed-forward neural network & recurrent neural network to detect intrusion. In: 2008 International Conference on Electrical and Computer Engineering, pp. 488–492. IEEE (2008)

34. Eberhart, R., et al.: A new optimizer using particle swarm theory. In: Proceedings of the Sixth International Symposium on Micro Machine and Human Science, MHS 1995, pp. 39–43 (1995)

35. Kohavi, R.: A study of cross-validation and bootstrap for accuracy estimation and model selection. In: IJCAI, Montreal, Canada, pp. 1137–1145 (1995)

36. Sharma, M., et al.: A new approach to characterize epileptic seizures using analytic time-frequency flexible wavelet transform and fractal dimension. Pattern Recogn. Lett. **94**, 172–179 (2017)

Digital Image Processing, Analysis and Application Based on Machine Learning

Teeth Category Classification by Fractional Fourier Entropy and Improved Hybrid Genetic Algorithm

Siyuan Lu[1,2(✉)] and Liam O'Donnell[3]

[1] Guangxi Key Laboratory of Manufacturing System, Advanced Manufacturing Technology, Guilin 541004, Guangxi, China
siyuan_lu@foxmail.com
[2] School of Informatics, University of Leicester, Leicester LE1 7RH, UK
[3] School of Engineering, University of Limerick, Limerick, Ireland

Abstract. It is significant to classify teeth categories in dental treatment. A novel teeth classification method was proposed in this paper, which combined fractional Fourier entropy and feedforward neural network. Firstly, fractional Fourier transform was performed on the teeth CT images and the obtained spectrums were used to extract entropies as the features. Then, a feedforward neural network was employed for automatic classification. To train the parameters in the network, improved hybrid genetic algorithm was leveraged. Experiment results suggested that our method achieved state-of-the-art performance.

Keywords: Teeth classification · Fractional fourier transform · Entropy · Neural network · Genetic algorithm

1 Introduction

Teeth are important for human beings. They help us to chew. Teeth can grind food into small pieces which is the first step of digestion [1]. This decomposition can stimulate the further digestion in stomach. Our communication also depends on teeth. Teeth, tongue and lip must work harmoniously for pronunciation of human language. Meanwhile, teeth are playing a significant role for facial expression. Teeth can serve as weapon in emergent situations, they are the gifts from nature. Generally, there are four kinds of teeth in human, they are incisor, canine, premolar and molar.

Human beings have paid a lot of attention to protecting their teeth, but many still suffer from dental illness [2]. The recognition of teeth category is the first step for diagnosis of dental diseases, which is vital for clinical treatment. Manual classification is weak in reproducibility, so researchers try to develop automatic and accurate teeth classification systems using artificial intelligence.

Mahoor and Abdel-Mottaleb (2005) [3] developed their teeth classification system based on bitewing images. They firstly performed Fourier transform on the images for feature extraction. Then, the obtained feature vectors were fed into a Bayesian classifier for training and testing. The labels of the images were binary, which include molar and premolar. To boost the detection precision of their system, the spatial information was

Y.-D. Zhang et al. (Eds.): ICMTEL 2020, LNICST 327, pp. 267–276, 2020.
https://doi.org/10.1007/978-3-030-51103-6_23

used as well. There are totally 50 images in their dataset. Their system achieved promising accuracy on experiment. Lu (2018) [4] used principal component analysis (PCA) and extreme learning machine (ELM) to classify four tooth categories. Payne (2018) [5] combined Hu moment invariant (HMI) with extreme learning machine (ELM).

The above teeth classification methods achieved good performance, but the accuracy can be improved. In this paper, we proposed a novel teeth classification method based on fractional Fourier entropy (FRFE) feedforward neural network, and improved hybrid generic algorithm (IHGA). FRFE was employed for feature extraction and feedforward neural network was the classification algorithm. The parameters in the network were optimized by IHGA. Our approach yielded better classification performance than several state-of-the-art methods.

2 Dataset

To evaluate our approach, computed tomography teeth image was used in experiment. There are four categories of samples in the dataset: incisor, canine, premolar and molar, and each category contain 50 samples. Some samples are given in Fig. 1.

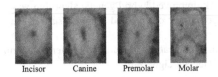

Incisor   Canine   Premolar   Molar

Fig. 1. Teeth images

3 Methodology

Image based machine learning systems usually solve two major challenges: feature extraction and classifier training. Feature extraction is to generate some representations from images because images contain too much information. Classifier training is to determine the parameters in classification models to achieve the best accuracy. So, in our scheme, FRFE was leveraged for feature extraction and feedforward neural network was employed for classification. We proposed a novel IHGA for optimization of the parameters in neural network for better generalization ability.

3.1 FRFE

Fractional Fourier entropy (FRFE) [6–15] is a combination of fractional Fourier transform (FRFT) and Shannon entropy. It has been successfully applied into many fields. Given a function $f(x)$, its α-angle FRFT can be expressed as [16]:

$$r_\alpha(v) = \int_{-\infty}^{+\infty} G_\alpha(x, v) f(x) dx \tag{1}$$

where x represents the time and v the frequency. The definition of transform kernel G is:

$$G_\alpha(x, v) = \sqrt{1 - i \cot \alpha} \times \exp\left[i\pi\left(v^2 \cot \alpha - 2vx \cot \alpha + x^2 \cot \alpha\right)\right] \tag{2}$$

where i is the imaginary unit. As teeth images are two dimension signals, the FRFT needs to be extended with two angles: α and β. The FRFE denoted by F can be obtained easily after FRFT by entropy operator E:

$$F = E \cdot r \tag{3}$$

3.2 Feedforward Neural Network

Feedforward neural network is a basic structure with all the nodes in adjacent layers linked to each other [17]. Feedforward neural network is widely applied in practical problems because it does not require prior distribution information of the simples [18–22]. The architecture of feedforward neural network is shown in Fig. 2, where N_I, N_H and N_O denote the numbers of nodes in input, hidden and output layers, respectively.

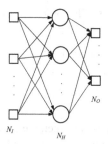

Fig. 2. Feedforward neural network structure

The training of the network is to determine the parameters. Although back propagation is a good choice for training algorithm, it cannot guarantee to converge at the global best solution, because back propagation belongs to a greedy algorithm which depends on gradient descent. Therefore, we try to solve the training problem using optimization methods [23–28].

3.3 Standard Genetic Algorithm

Standard genetic algorithm (GA) belongs to an optimization method, which was inspired by the Darwin's famous evolution theory [29]. GA mimics the behavior of reproduction, crossover and mutation to get the fittest solution.

The solutions are randomly initialized and encoded into strings of '0' and '1' stored in chromosomes. In each generation, the fitness of all the potential solutions will be calculated and sorted. Elitism is used to preserve the high quality solutions, in which several top ranking solutions shall be copied in to the next generation directly. Some solutions will be moved into the next generation by crossover and mutation operations. The stopping criterion for GA can be a pre-defined fitness value or the maximum generation. The top ranking solution in the last generation can be the global best one [30, 31].

However, GA is weak in local searching, and it sometimes cannot converge because the fitness of descendants becomes worse with the generations.

3.4 Improved Hybrid Genetic Algorithm

IHGA is proposed by Ahmad, Isa (2013) [32], which not only trains the weights in network but also finds the best number of hidden nodes and feature subset. The chromosome in IHGA is divided into three gene segments (GS) for three purposes, shown in Fig. 3.

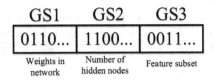

Fig. 3. Chromosome in IHGA

The fitness of the solutions is defined as:

$$fit = w \times acc + (1 - r) \times \frac{1}{cp} \tag{4}$$

where w is the weight to compromise between the testing accuracy acc and the network complexity inverse $1/cp$. The definitions of acc and cp are given below:

$$acc = 100 \times \left(\frac{N_C}{N_T}\right) \tag{5}$$

$$cp = (f \times h) + (h \times o) + h + o \tag{6}$$

Where N_C and N_T denote the number of correctly classified samples and total sample number, respectively. The f, h and o represent the number of selected features, the hidden nodes number and the output nodes number, respectively.

IHGA employs segmented multi-chromosome crossover (SMCC) to generate offspring chromosome that gets gene from more than one couple parent chromosomes. The diagram of IHGA is summarized in Fig. 4.

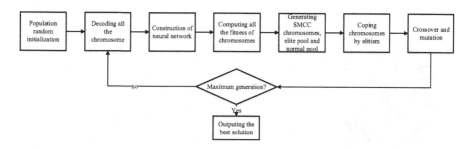

Fig. 4. Steps of IHGA

3.5   10-Fold Cross Validation

Cross validation is a method for training. It is often used to prevent overfitting when the dataset is small. The k-fold cross validation divides the dataset into k groups of the same size [33–35]. Then, every time, one group is chosen for testing and the other k-1 groups serve for training [36, 37]. The training will be run for k times and each group will be used for testing. Then, the average performance can be obtained. The 10-fold cross validation is illustrated in Fig. 5.

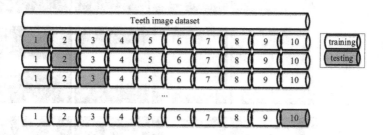

Fig. 5. 10-fold cross validation

4   Experiment Results and Discussions

4.1   FRFE Results

Here is an example of FRFT. The input was a teeth CT age and 25 pairs of angles were used, shown in Fig. 6. The result of FRFT was given in Fig. 7.

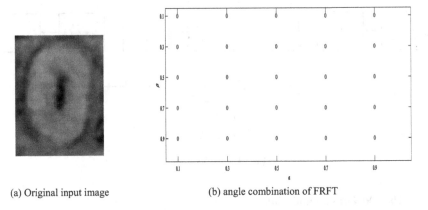

(a) Original input image   (b) angle combination of FRFT

Fig. 6. Input and Parameter of FRFT

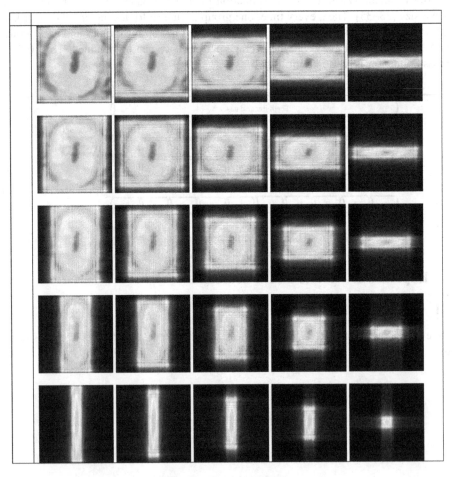

Fig. 7. Results of FRFT

Then, the entropies of the above 25 FRFT spectrums were calculated, shown in Table 1. These 25 FRFEs were fed into the feedforward neural network for training and testing.

Table 1. Entropy values of FRFT

6.3632	6.8078	7.2109	7.0521	5.3368
6.8068	7.0813	7.3951	7.1713	5.5611
7.2050	7.3947	7.4329	6.7633	5.6494
6.9844	7.1126	6.7732	5.7245	5.2942
5.2210	5.4708	5.5977	5.3015	4.6309

4.2 GA Versus IHGA

The confusion matrixes of our FRFE-IHGA and FRFE-GA were given in Fig. 8 and Fig. 9, respectively. FRFE-IHGA achieved overall accuracy of 80.0% while the accuracy of FRFE-GA was 74.0%. Meanwhile, for sensitivity and specificity of all four classes, FRFE-IHGA outperformed FRFE-GA. So, IHGA was a better option for network parameter optimization than GA in this study.

	c_1	c_2	c_3	c_4	Sen
c_1	41	4	5	0	82.0%
c_2	4	40	4	2	80.0%
c_3	3	2	40	5	80.0%
c_4	3	4	4	39	78.0%
Prc	80.4%	80.0%	75.5%	84.8%	Acc 80.0%

Fig. 8. Confusion matrix of our FRFE-IHGA

	c_1	c_2	c_3	c_4	Sen
c_1	37	6	4	3	74.0%
c_2	4	38	4	4	76.0%
c_3	3	4	37	6	74.0%
c_4	5	4	5	36	72.0%
Prc	75.5%	73.1%	74.0%	73.5%	Acc 74.0%

Fig. 9. Confusion matrix if replace IHGA with ordinary GA

4.3   Comparison to State-of-the-Art Approaches

We compared our FRFE-IHGA with two state-of-the-art methods: PCA-ELM [4] and HMI-ELM [5]. As shown in Table 2 and Fig. 10, FRFE-IHGA was better than the other two methods in detecting incisor teeth and premolar teeth. For canine, both FRFE-IHGA and PCA-ELM reached the best sensitivity. Though sensitivity for molar teeth of our method was lower than PCA-ELM, FRFE-IHGA achieved the best overall accuracy of the three approaches. Therefore, our FRFE-IHGA is a good solution for automatic teeth classification.

Table 2. Comparison with state-of-the-art approaches

Approach	Incisor	Canine	Premolar	Molar	Total
PCA-ELM [4]	80.00	**80.00**	79.60	**79.40**	79.75
HMI-ELM [5]	78.25	78.00	79.25	78.75	78.56
FRFE-IHGA (Ours)	**82.0**	**80.0**	**80.0**	78.0	**80.0**

(bold means the best)

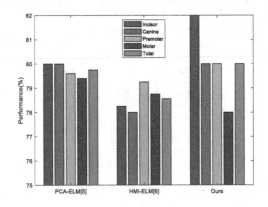

Fig. 10. Comparison against PCA-ELM and HMI-ELM

5   Conclusions

In this paper, a new teeth classification method was proposed based on teeth CT images. Fractional Fourier entropy was employed to generate features from images, and a feedforward neural network trained by improved hybrid genetic algorithm served as the classifier. 10-fold cross validation was employed to avoid overfitting. Our method yielded overall accuracy of 80.0%, which was comparable to state-of-the-art methods.

For future research, we shall collect more teeth images and build bigger dataset to re-test our method. We will try to introduce deep learning to improve the classification accuracy.

Acknowledgement. This paper is supported by Open Fund of Guangxi Key Laboratory of Manufacturing System & Advanced Manufacturing Technology (17-259-05-011K).

References

1. Dodds, M.W.J., et al.: An induced extrinsic tooth stain prevention model to investigate whitening potential of sugar-free chewing gums. Am. J. Dentistry **30**(6), 309–315 (2017)
2. Vasconcelos, A.G., et al.: Mammal tooth traces in a ferruginous cave in southeastern Brazil and their relevance to cave legal protection. Anais Da Academia Brasileira De Ciencias, **91**, 11 (2019). Article ID. UNSP e20180861
3. Mahoor, M.H., et al.: Classification and numbering of teeth in dental bitewing images. Pattern Recogn. **38**(4), 577–586 (2005)
4. Lu, S.: Teeth classification based on extreme learning machine. In: World Conference on Smart Trends in Systems, Security & Sustainability (WS4), pp. 198–202. IEEE, London (2018)
5. Payne, R.: Teeth category classification via Hu moment invariant and extreme learning machine. Adv. Intell. Syst. Res. **151**, 220–223 (2018)
6. Zhang, Y.-D., Sun, J.: Preliminary study on angiosperm genus classification by weight decay and combination of most abundant color index with fractional Fourier entropy. Multimedia Tools Appl. **77**(17), 22671–22688 (2017). https://doi.org/10.1007/s11042-017-5146-3
7. Cheng, H.: Multiple sclerosis identification based on fractional Fourier entropy and a modified Jaya algorithm. Entropy, **20**(4) (2018). Article ID. 254
8. Li, J.: Texture analysis method based on fractional fourier entropy and fitness-scaling adaptive genetic algorithm for detecting left-sided and right-sided sensorineural hearing loss. Fundamenta Informaticae **151**(1–4), 505–521 (2017)
9. Yang, J.F.: Comprehensive survey on fractional fourier transform. Fundamenta Informaticae **151**(1–4), 1–48 (2017)
10. Sun, Y.: A Multilayer perceptron based smart pathological brain detection system by fractional fourier entropy. J. Med. Syst. **40**(7) (2016). Article ID. 173
11. Li, J., Detection of Left-Sided and Right-Sided Hearing Loss via Fractional Fourier Transform. Entropy, 2016. **18**(5): Article ID. 194
12. Cattani, C., et al.: Tea category identification using a novel fractional fourier entropy and jaya algorithm. Entropy, **18**(3) (2016). Article ID. 77
13. Liu, G.: Computer-aided diagnosis of abnormal breasts in mammogram images by weighted-type fractional Fourier transform. Adv. Mech. Eng. **8**(2) (2016). Article ID. 11
14. Yang, X., et al.: Pathological brain detection by a novel image feature—fractional fourier entropy. Entropy **17**(12), 8278–8296 (2015)
15. Chen, S., et al.: Magnetic resonance brain image classification based on weighted-type fractional Fourier transform and nonparallel support vector machine. Int. J. Imag. Syst. Technol. **25**(4), 317–327 (2015)
16. Ben Farah, M.A., et al.: A novel chaos based optical image encryption using fractional Fourier transform and DNA sequence operation. Optics Laser Technol. **121**, 8 (2020). Article ID. 105777
17. Yavari, M., et al.: Fractional infinite-horizon optimal control problems with a feed forward neural network scheme. Network-Comput. Neural Syst. **30**, 23 (2019)
18. Chen, Y.: Cerebral micro-bleeding identification based on a nine-layer convolutional neural network with stochastic pooling. Concurr. Comput. Pract. Experience **31**(1), e5130 (2020)
19. Govindaraj, V.V.: High performance multiple sclerosis classification by data augmentation and AlexNet transfer learning model. J. Med. Imag. Health Inform. **9**(9), 2012–2021 (2019)
20. Jiang, X.: Chinese sign language fingerspelling recognition via six-layer convolutional neural network with leaky rectified linear units for therapy and rehabilitation. J. Med. Imag. Health Inform. **9**(9), 2031–2038 (2019)

21. Ramirez, J.: Unilateral sensorineural hearing loss identification based on double-density dual-tree complex wavelet transform and multinomial logistic regression. Integr. Comput. Aided Eng. (2019). https://doi.org/10.3233/ICA-190605

22. Li, Z.: Teeth category classification via seven-layer deep convolutional neural network with max pooling and global average pooling. Int. J. Imag. Syst. Technol. (2019). https://doi.org/10.1002/ima.22337

23. Hong, J., Cheng, H., Zhang, Y.-D., Liu, J.: Detecting cerebral microbleeds with transfer learning. Mach. Vis. Appl. **30**(17), 1123–1133 (2019). https://doi.org/10.1007/s00138-019-01029-5

24. Xie, S.: Alcoholism identification based on an AlexNet transfer learning model. Front. Psychiatry, **10** (2019). Article ID. 205

25. Tang, C.: Cerebral micro-bleeding detection based on densely connected neural network. Front. Neurosci. **13** (2019). Article ID. 422

26. Jia, W., Muhammad, K., Wang, S.-H., Zhang, Y.-D.: Five-category classification of pathological brain images based on deep stacked sparse autoencoder. Multimedia Tools Appl. **78**(4), 4045–4064 (2017). https://doi.org/10.1007/s11042-017-5174-z

27. Zhang, Y.-D., et al.: Image based fruit category classification by 13-layer deep convolutional neural network and data augmentation. Multimedia Tools Appl. **78**(3), 3613–3632 (2017). https://doi.org/10.1007/s11042-017-5243-3

28. Wang, S.-H., Muhammad, K., Hong, J., Sangaiah, A.K., Zhang, Y.-D.: Alcoholism identification via convolutional neural network based on parametric ReLU, dropout, and batch normalization. Neural Comput. Appl. **32**(3), 665–680 (2018). https://doi.org/10.1007/s00521-018-3924-0

29. Cappelletti, M.A., et al.: Extraction of the minority carrier transport properties of solar cells using the Hovel model and genetic algorithms. Measure. Sci. Technol. **31**(2), 9 (2020). Article ID. 025601

30. Ji, G.L.: A rule-based model for bankruptcy prediction based on an improved genetic ant colony algorithm. Math. Prob. Eng. (2013). Article ID. 753251

31. Ji, G.: Genetic pattern search and its application to brain image classification. Math. Prob. Eng. (2013). Article ID. 580876

32. Ahmad, F., et al.: Intelligent medical disease diagnosis using improved hybrid genetic algorithm–multilayer perceptron network. J. Med. Syst. **37**(2), 9934 (2013)

33. Stappenbeck, L.E., et al.: Pretesting of the German Vocal Fatigue Index (VFI-D)-transcultural translation and cross validation. Hno **68**, 7 (2020)

34. Jiang, X., et al.: Classification of Alzheimer's Disease via eight-layer convolutional neural network with batch normalization and dropout techniques. J. Med. Imag. Health Inform. **10**(5), 1040–1048 (2020)

35. Kang, C., et al.: A heuristic neural network structure relying on fuzzy logic for images scoring. IEEE Trans. Fuzzy Syst. **1**, 1 (2020)

36. Wang, S.-H., et al.: Cerebral micro-bleeding identification based on a nine-layer convolutional neural network with stochastic pooling. Concurr. Comput. Pract. Experience **32**(1), e5130 (2020)

37. Mao, X., et al.: Knowledge-Aided 2-D autofocus for spotlight SAR filtered backprojection imagery. IEEE Trans. Geosci. Remote Sens. **57**(11), 9041–9058 (2019)

Hearing Loss Identification via Fractional Fourier Entropy and Direct Acyclic Graph Support Vector Machine

Liying Wang[1,2] and Zhiqiang Xu[3(✉)]

[1] Department of Educational Technology,
Nanjing Normal University, Nanjing 210097, China
wangliying@njnu.edu.cn
[2] Department of Informatics, University of Leicester, Leicester LE1 7RH, UK
[3] Unit of Urology, Tongliao Hospital of Inner Mongolia,
Tongliao 028000, China
403970025@qq.com

Abstract. With the risk of hearing loss being higher than before since the digital device is more popular, it becomes more urgent to identify the sensorineural hearing loss from the view of changes in internal brain structure. Based on 180 brain MRI of three categories of hearing loss balanced dataset, one schema with fractional Fourier transform entropy and direct acyclic graph support vector machine is proposed and applied to identify the features and predict the categories of hearing loss. The experiments prove this schema rather promising when the dataset is not large since the overall accuracy is up to 94.06 ± 1.08% which is higher than those of some previous methods in scope of traditional machine learning.

Keywords: Hearing loss identification · Fractional fourier transform entropy · Direct acyclic graph support vector machine

1 Introduction

Hearing Loss (HL) is a general term for decreased auditory sensitivity, increased hearing threshold, hearing impairment and even hearing deafness. It is estimated by the World Health Organization (WHO) ahead of World Hearing Day (3 March), the risk of hearing loss is higher and higher up to nearly 50% of people aged 12–35 years or 1.1 billion young people due to prolonged and excessive exposure to loud sounds, including music they listen to through personal audio devices. It is estimated that by 2050 over 0.9 billion people or 10% people will have disabling hearing loss.

Moreover, permanent hearing loss can lead to changes in brain structure and function, such as brain signal deterioration, auditory cortex degeneration, loss of neurons and branches of neurons, and reduction of overall brain volume. These structural and functional changes may affect the brain's ability to process and perceive sound, and may lead to cognitive decline. These changes can be captured in Magnetic Resonance Imaging (MRI). But it is still challenging for human to recognize these slight changes without computer aided classification.

© ICST Institute for Computer Sciences, Social Informatics and Telecommunications Engineering 2020
Published by Springer Nature Switzerland AG 2020. All Rights Reserved
Y.-D. Zhang et al. (Eds.): ICMTEL 2020, LNICST 327, pp. 277–289, 2020.
https://doi.org/10.1007/978-3-030-51103-6_24

Therefore, this paper focuses on the identification of Sensorineural Hearing Loss (SHL) whose lesions occur in the cochlea, auditory nerve or auditory center. Since MRI is the key medical modality to check and analyze the brain structure, researchers proposed different machine learning algorithms to identify SHL from the data source of MRI in recent years.

Because the medical dataset is relatively small, it is not suitable to directly apply the currently hot method of deep learning neural networks, but to depend on traditional machine learning methods. These works mainly solved the three sub-problems of one machine learning algorithm, which are feature extraction, learning model construction and optimization solution.

For example, Ref. [1–3] extracted images features respectively by applying Fractional Fourier Transform (FRFT) as a 25-dimension vector, Wavelet Entropy (WE) as a 10-dimension vector, FRFT as a 12-dimension vector. They supposed the learning model as a Single Hidden Layer-Feedforward Neutral Network (SHL-FNN), statistical model called Support Vector Machine (SVM). To search the optimal parameters of the learning models, they not only use the traditional back-propagate method, but also with the Levenberg-Marquardt algorithm or fitness-scaling adaptive Genetic Algorithm (GA). These methods demonstrated average overall accuracy as 95%, 95.1% and 95.51%. The strength of these methods is they applied the medium-dimension vectors to express features and obtained the acceptable good classification results.

Later, Wang et al. used the discrete wavelet transform [4] and dual-tree wavelet transform [5] to extract entropy feature. Then the overall accuracy to classify reached to 95.31% and 96.17 ± 2.49% respectively. Ref. [6] use the Principal Component Analysis (PCA) and SVM to classify the hearing loss whose overall accuracy reached to 95.71%. In addition, Nayak [7] applied stationary wavelet entropy to input one SHL-FNN classifier to detect the unilateral hearing loss which is better than the biorthogonal wavelet transform and get the accuracies of HC, LHL, and RHL are 96.94%, 97.14%, and 97.35%, respectively. Bao, Nakamura [8] combined Wavelet Entropy (WE) and particle swarm optimization (PSO) approach. Tang, Lee [9] proposed a novel method that combines Tabu search (TS) and PSO method. In addition, Nayeem [10] used wavelet entropy and genetic algorithm to detect hearing loss with the sensitivity for HC of 81.25 ± 4.91%, for left-sided hearing loss of 80.42 ± 5.57%, for right-sided hearing loss of 81.67 ± 6.86%, and an overall accuracy of 81.11 ± 1.34%. Gao, Liu [11] tried to use wavelet entropy and Cat Swarm Optimization (CSO) to identify the hearing loss, the overall accuracy achieved 84.50 ± 0.81%.

From the review of articles, we found the works with higher accuracy need higher-dimension feature expression. Meanwhile, optimization algorithms need to be designed carefully to avoid the local minimum. In addition, most of the above studies are based on the same brain MRI database with the number of 49 images. Since the scale of training dataset is small the overfitting phenomenon may be inevitable.

In order to enlarge the dataset, Jia [12] augmented dataset number from total 49 images to each category 420 images, Nevertheless, they tried to design a deep Stacked Sparse Autoencoder to identify the image of unilateral hearing loss. Although the time consuming of this method is longer than others, the result of the overall accuracy reached to 99.5%. It revealed the trend that deep learning method is rather exciting and promising for future medical classification though it is less interpretable or uninterpretable.

In order to explore and improve the identification performance of hearing loss from a clinical MRI dataset, we attempt to increase the dimension of the features extracted and rearrange the learning model for the classification of unilateral hearing loss. Firstly, the Fractional Fourier Entropy is applied to extract the time-frequency domain features that can differ the hearing loss images from healthy hearing images. After that the 36-dimension feature vector is input to the Direct Acyclic Graph Support Vector Machine (DAG-SVM) to predict the category label of one MRI whether it is unilateral hearing loss.

The contribution of this paper is (i) We applied FRFT to extract features from brain images; (ii) We used an advanced SVM to create the classifier; (iii) Our system shows the superiority of our method to state-of-the-art approaches.

The rest of this paper is arranged as followed. Section 2 introduces the methodology and related dataset the experiment involved. Section 3 shows the experiments and results of our work in detail. Section 4 discusses the performance, finally gives the conclusion.

2 Material and Methodology

2.1 Dataset

We use the dataset which includes 180 images in format of MRI, which come from 60 subjects of each of three categories respectively with left-sided hearing loss (LHL), right-sided hearing loss (RHL) and age-, sex- and education-matched healthy controls (HC). The small size of our dataset cannot provide sufficient data for deep learning [13–20].

This dataset contains same number of three categories because balanced dataset to favor to learning model. This study on the dataset got all the subjects formal written consent approved by the Ethics Committee of Zhongda Hospital which is associated with Southeast University. In order to label the category of hearing loss, the subject was diagnosed by a pure tone audiometry with six different octave frequencies (2.5, 5.0, 10.0, 20.0, 40.0 and 80.0 kHz) to evaluate the Pure Tone Average (PTA). No subjects used any hearing aid over the impaired ear during the pure tone audiometry. The difference between PTA and the average of ISO standard under the corresponding frequency reflects hearing performance. If the difference is more than 20db, it indicates the hearing losing happened. The images from MRI are processed through the pipeline below. The distinct slice images from the three categories are shown in Fig. 2.

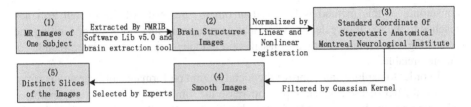

Fig. 1. MRI dataset process pipeline

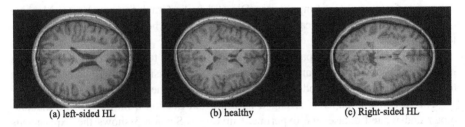

(a) left-sided HL   (b) healthy   (c) Right-sided HL

Fig. 2. Distinct slices of MR images from three categories

2.2 FRFE

There are many feature descriptors in the field of signal processing, such as image statistical measures, moments [21, 22], color features [23], wavelet features [24–28], texture features [29], etc. Fractional Fourier entropy (FRFE) is an effective feature descriptor and can achieve excellent performances.

As is known, Fractional Fourier Transform (FRFT) is the general form of traditional Fourier Transform (FT) defined as formula (1) with the parameter a, the range of which is [0, 1]. When $a = 1$, the FRFT is just the traditional FT on the original signal x. If the parameter a is equivalently transformed to $2\varphi/\pi$, a new parameter φ representing rotation angle is introduced, the range of which is [0, $\pi/2$]. In order to compute the FRFT, one kernel function $K(t, u, \varphi)$ with three parameters of time t, frequency u and rotation φ, is defined as below. FRFT has the advantage to transform a signal to any intermediate one between time domain and frequency domain which is called the unified time-frequency domain [30]. It has been applied in many application field since FRFT was proposed in 1980, such as signal or image processing, detection, recognition and classification [31].

$$F^a(x) = F(x, a) = F^{2\varphi/\pi}(x) = F(t, u, \varphi) = \int_{-\infty}^{\infty} K(t, u, \varphi) x(t) dt, \quad (1)$$

where

$$K(t, u, \varphi) = \begin{cases} the\ diract\ delta\ function\ at\ (t - u) : \delta(t - u) & \varphi = 2k\pi, k \in Z \\ \delta(t + u) & \varphi = (2k + 1)\pi \\ \sqrt{1 - jcot\varphi}\exp\left(j\pi\left[(t^2 + u^2)cot\varphi - 2utcsc\ \varphi\right]\right) & other \end{cases}$$

To extend the above one-dimensional FRFT to two-dimensional FRFT, two parameters a_x and a_y should be introduced to represent the transformation of each dimension. Therefore, 2D FRFT is denoted as $F\left(x, y, u, a_x, a_y\right) = F\left(x, y, u, \varphi_x, \varphi_y\right)$, in which the parameters φ_x, φ_y are used to express he rotation angles respectively along x and y axes. In our work 2D FRFT is calculated in the image space. We choose 6 angles for φ_x and φ_y respectively in range of 0 to $\pi/2$ with a step of $\pi/10$, which are equivalent to set the parameters of a_x and a_y respectively from 0 to 1 with the step of 0.2. Therefore, there are 36 FRFT images obtained after 2D FRFT for each MRI. The pixel value of one FRFT image is the modulus of the complex of FRFT. For each of the 36 FRFT images, the FRFT Entropy (FRFE) feature is calculated to represent the spatial frequency spectrum energy of each FRFT image, which is denoted as formula (2).

$$\text{FRFE}\left(u, \varphi_x, \varphi_y\right) = -\sum_{i} p(u_i) \log\left(p(u_i)\right) \tag{2}$$

In this study, we use the FRFT programs provided by computer science of Ku Leuven which can be downloaded from the website [32]. The 2D FRFT images of the slice MRI of one subject are shown in Fig. 3. The corresponding FRFE feature values are shown in Table 1.

Table 1. FRFE results for one subject's MRI

a_y	a_x					
	0	0.2	0.4	0.6	0.8	1.0
0	5.5342	6.0946	6.2403	6.2445	5.9114	5.1656
0.2	6.6437	6.8245	6.8389	6.6928	6.2125	5.3588
0.4	6.9273	6.9591	6.8893	6.6878	6.1044	5.3193
0.6	6.8326	6.7984	6.7080	6.4431	5.8177	5.2331
0.8	6.4155	6.3415	6.2174	5.8789	5.4163	5.1364
1.0	5.6751	5.6341	5.5927	5.4615	5.2522	5.0410

Next, we organize the 36 FRFE into a vector to represent the key features of the brain MRI of one hearing loss subject. This vector and its category label will be input into the Direct Acyclic Graph support vector machine (DAG-SVM) as the training dataset.

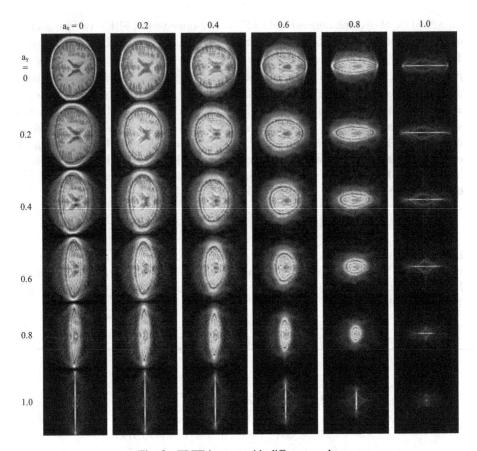

Fig. 3. FRFT images with different angles

2.3   DAG-SVM

Support Vector Machine is one binary classification model combined empirical loss and structural risk minimization together. It has been successfully used in many tasks such as text classification, signal prediction [33]. Besides, SVMs are proven to be able to achieve better performances than other traditional classifiers [34, 35]. For N-classes multiple classification problem, there are two methods to make use of the binary classification model. The first one of them constructs N binary classification to identify one from the rest. The second one constructs $N(N\text{-}1)/2$ binary classification to identify one from the other.

Comparatively the former would encounter errors like label overlap and unknown, while the latter would give definite label. In our work, we will use the latter to construct a three-class classification model using the Direct Acyclic Graph shown as Fig. 4 to organize three binary SVM classifications with the soft margin. The soft-margin binary SVM is described as the formula (3). In that formula, the input data include the FRFE features of MRI and the hearing loss labels (HC, LHL or RHL) of the subject from the dataset to train three binary SVM models.

$$\min_{w,b,\varepsilon} \frac{\|w\|^2}{2} + C\sum_{i=1}^{N} \varepsilon_i$$
$$\text{st. } y_i(wx_i + b) > 1 - \varepsilon_i$$
$$\varepsilon_i > 0, i = 1, 2, \ldots, N$$

(3)

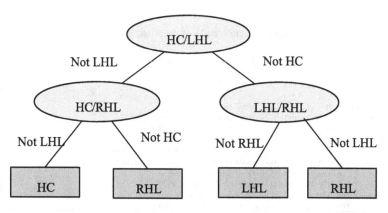

Fig. 4. DAG-SVM

2.4 K-Fold Cross Validation

K-fold cross validation is the most common method to evaluate the classification model performance. Especially for the relatively small dataset it is helpful to overcome overfitting through random equally splitting the dataset into K sets and in order choose one as test set and others as train sets. In addition, the accuracy can be estimated. We run 10 times K-fold cross-validation repetitively to alleviate the random effects, but more repetitions will burden the computation needed. Because there are 180 images in our dataset, if we set the default 10-fold, then each fold only contains 18 images, which is too small. Figure 5 shows the index of 6-fold cross validation, which will repeat 10 times with different initialization.

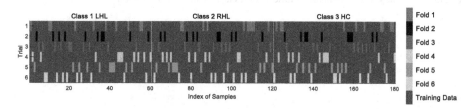

Fig. 5. Index of 6-fold cross validation

In order to measure the identification performance of one learning model, the Sensitivity of one category and the Overall Accuracy are used. The Sensitivity of one category is defined by the percentage of the correct prediction within one category on the basis of testing dataset. After one K-fold cross-validation run, we can get the prediction category of each testing data. We calculate the sensitivity of categories of LHL, HC, and RHL. The Overall Accuracy is defined by the percentage of the correct prediction of all categories of the testing dataset. After runs 10 times, the average sensitivity and overall accuracy can be calculated. Table 2 showed the pseudo code of proposed algorithm.

Table 2. Pseudo code of proposed algorithm

```
Input: dataset={(xi,yi), i=1,2, …,180},yi is the category label of MRI xi.
1. Do FRFT for each image in dataset:
       Fj=FRFT(xi,ax,ay),  ax,ay=0,0.2,0.4,0.6,0.8,1
                       j=1,2, …,36
2. Calculate the entropy Ej =FRFE(Fj)
3. Organize the entropy Ej as the image features Ei = (Ej).
       Get  the feature dataset={(Ei,yi), i=1,2, …,180}.
4. Run 10 times
       Run 6-fold cross validation of DAG-SVM on feature dataset.
       Calculate the confusion matrix of prediction on validation dataset.
       Calculate the sensitivity of one category.
       Calculate the overall accuracy of the DAG-SVM.
Output: the average sensitivity of one category.
        the average overall accuracy of the DAG-SVM.
```

3   Experiments and Results

We carry out the 10 times 6-fold cross-validation on the 180 MRI dataset to identify three categories of hearing loss. The confusion matrix of the three categories are obtained to calculate the sensitivity of each category and overall accuracy. Next we show the statistical results on this small data, but also give the comparison with the related other traditional machine learning methods.

3.1   Statistical Analysis

As shown in Table 3, the mean and standard deviation of the sensitivity reach to 93.83 ± 1.77, 94.17 ± 2.12 and 94.17 ± 2.75 for HC, LHL and RHL respectively. From the overall accuracy the mean and standard deviation is 94.06 ± 1.08.

Table 3. 10 times 6-fold sensitivity and overall accuracy results

	Sensitivity-LHL	Sensitivity-HC	Sensitivity-RHL	Overall accuracy
R1	96.67	95.00	96.67	96.11
R2	91.67	96.67	91.67	93.33
R3	95.00	91.67	96.67	94.44
R4	93.33	91.67	95.00	93.33
R5	95.00	93.33	90.00	92.78
R6	93.33	96.67	96.67	95.56
R7	95.00	96.67	90.00	93.89
R8	95.00	95.00	93.33	94.44
R9	91.67	91.67	96.67	93.33
R10	91.67	93.33	95.00	93.33
Mean $\pm$ SD	93.83 $\pm$ 1.77	94.17 $\pm$ 2.12	94.17 $\pm$ 2.75	94.06 $\pm$ 1.08

3.2 Comparison to State-of-the-Art Approaches

To validate the effectiveness of our method, we do experiments about the above-mentioned methods of WE+PSO [8], TS-PSO [9] and CSO [11] to compare with ours. From Table 4 and Fig. 6, it is evident that the overall accuracy of our method is up to 8% better than those of these previous methods. That proved that the stability of the proposed method is better than others. Meanwhile it shows the FRFE can capture more difference of the texture features of the brain structure in MRI of sensorineural hearing loss. The traditional machine learning is enough to train an acceptable and interpretable learning model on a small dataset.

Table 4. Comparison of overall accuracy with our method

Approach	Sensitivity-LHL	Sensitivity-HC	Sensitivity-RHL	Overall accuracy
WE+PSO [8]	85.20 $\pm$ 4.64	85.20 $\pm$ 3.79	86.40 $\pm$ 5.06	85.60 $\pm$ 0.84
TS-PSO [9]	86.50 $\pm$ 2.77	86.50 $\pm$ 3.09	85.50 $\pm$ 2.23	86.17 $\pm$ 0.41
CSO [11]	84.50 $\pm$ 4.97	85.50 $\pm$ 6.85	83.50 $\pm$ 5.80	84.50 $\pm$ 0.81
Ours: FRFE+DAG-SVM	93.83 $\pm$ 1.77	94.17 $\pm$ 2.12	94.17 $\pm$ 2.75	94.06 $\pm$ 1.08

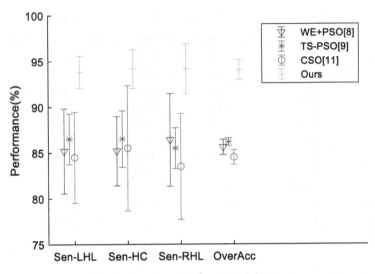

Fig. 6. Error bar of approach comparison

4 Conclusion

This paper proposes one schema with the fractional Fourier transform entropy and direct acyclic graph support vector machine to identify the features of hearing loss on brain MRI. Based on 180 brain MRI of three categories of hearing loss balanced dataset, the experiments prove the schema, though belonging to the traditional machine learning, still a rather promising direction when the dataset is not large, since the overall accuracy is up to $94.06 \pm 1.08\%$ which is higher than those of some previous methods in scope of traditional machine learning. In the future, the deep learning method, such as transfer learning, auto encoder would be explored to find more efficient solution with the time and labor consuming trade-off.

Acknowledgement. The paper is supported by the educational science plan foundation "in 12th Five-Year" of Jiangsu province (B-a/2015/01/010), Henan Key Research and Development Project (182102310629), National key research and development plan (2017YFB1103202), Open Fund of Guangxi Key Laboratory of Manufacturing System & Advanced Manufacturing Technology (17-259-05-011 K), Guangxi Key Laboratory of Trusted Software (kx201901).

References

1. Li, J.: Detection of left-sided and right-sided hearing loss via fractional fourier transform. Entropy, **18**(5) (2016), Article ID: 194. https://doi.org/10.3390/e18050194
2. Gorriz, J.M., Ramírez, J.: Wavelet entropy and directed acyclic graph support vector machine for detection of patients with unilateral hearing loss in MRI scanning. Front. Comput. Neurosci., **10** (2016). Article ID: 160. https://doi.org/10.3389/fncom.2016.00106

3. Li, J.: Texture analysis method based on fractional fourier entropy and fitness-scaling adaptive genetic algorithm for detecting left-sided and right-sided sensorineural hearing loss. Fundamenta Informaticae 151(1–4), 505–521 (2017)

4. Wang, S., Du, S., Li, Y., Lu, H., Yang, M., Liu, B., Zhang, Yudong: Hearing loss detection in medical multimedia data by discrete wavelet packet entropy and single-hidden layer neural network trained by adaptive learning-rate back propagation. In: Cong, F., Leung, A., Wei, Q. (eds.) ISNN 2017. LNCS, vol. 10262, pp. 541–549. Springer, Cham (2017). https://doi.org/10.1007/978-3-319-59081-3_63

5. Wang, S., Zhang, Y., Yang, M., Liu, B., Ramirez, J., Gorriz, J.M.: Preliminary study on unilateral sensorineural hearing loss identification via dual-tree complex wavelet transform and multinomial logistic regression. In: Ferrández Vicente, J.M., Álvarez-Sánchez, J.R., de la Paz López, F., Toledo Moreo, J., Adeli, H. (eds.) IWINAC 2017. LNCS, vol. 10337, pp. 289–297. Springer, Cham (2017). https://doi.org/10.1007/978-3-319-59740-9_28

6. Chen, Y., Yang, M., Chen, X., Liu, B., Wang, H., Wang, Shuihua: Sensorineural hearing loss detection via discrete wavelet transform and principal component analysis combined with generalized eigenvalue proximal support vector machine and Tikhonov regularization. Multimedia Tools Appl. 77(3), 3775–3793 (2016). https://doi.org/10.1007/s11042-016-4087-6

7. Nayak, D.R.: Detection of unilateral hearing loss by Stationary Wavelet Entropy. CNS Neurol. Disorders - Drug Targets 16(2), 15–24 (2017). https://doi.org/10.2174/1871527315666161026115046

8. Bao, F., Nakamura, K.: Hearing Loss via wavelet entropy and particle swarm optimized trained support vector machine. In: 3rd International Conference on Electrical, Control and Automation Engineering (ECAE), pp. 158–165. DEStech Publication Inc, Wuhan, China (2018)

9. Tang, C., Lee, E.: Hearing loss identification via wavelet entropy and combination of Tabu search and particle swarm optimization. 23rd International Conference on Digital Signal Processing (DSP), pp. 1–5. IEEE, Shanghai (2018)

10. Nayeem, A.: Hearing loss detection based on wavelet entropy and genetic algorithm. Adv. Intell. Syst. Res. 153, 49–53 (2017)

11. Gao, R., Liu, J.: Hearing loss identification by wavelet entropy and cat swarm optimization. AIP Conference Proceedings, 2073 (2019). Article ID: 020082

12. Jia, W.: Three-category classification of magnetic resonance hearing loss images based on deep autoencoder. J. Med. Syst. 41 (2017). Article ID: 165

13. Xie, S.: Alcoholism identification based on an AlexNet transfer learning model. Front. Psychiatry, 10 (2019), Article ID: 205. https://doi.org/10.3389/fpsyt.2019.00205

14. Tang, C.: Cerebral micro-bleeding detection based on densely connected neural network. Front. Neurosci. 13 (2019), Article ID: 422. https://doi.org/10.3389/fnins.2019.00422

15. Zhang, Y.-D., Dong, Z., Chen, X., Jia, W., Du, S., Muhammad, K., Wang, S.-H.: Image based fruit category classification by 13-layer deep convolutional neural network and data augmentation. Multimedia Tools Appl. 78(3), 3613–3632 (2017). https://doi.org/10.1007/s11042-017-5243-3

16. Wang, S.-H., Sun, J., Phillips, P., Zhao, G., Zhang, Y.-D.: Polarimetric synthetic aperture radar image segmentation by convolutional neural network using graphical processing units. J. Real-Time Image Process. 15(3), 631–642 (2017). https://doi.org/10.1007/s11554-017-0717-0

17. Pan, C.: Multiple sclerosis identification by convolutional neural network with dropout and parametric ReLU. J. Comput. Sci. 28, 1–10 (2018). https://doi.org/10.1016/j.jocs.2018.07.003

18. Jiang, X., Chang, L.: Classification of Alzheimer's Disease via Eight-layer convolutional neural network with batch normalization and dropout techniques. J. Med. Imag. Health Inform. **10**(5), 1040–1048 (2020)

19. Wang, S.-H., Muhammad, K., Hong, J., Sangaiah, A.K., Zhang, Yu-Dong: Alcoholism identification via convolutional neural network based on parametric ReLU, dropout, and batch normalization. Neural Comput. Appl. **32**(3), 665–680 (2018). https://doi.org/10.1007/s00521-018-3924-0

20. Chen, Y.: Cerebral micro-bleeding identification based on a nine-layer convolutional neural network with stochastic pooling. Concurr. Comput. Pract. Experience **31**(1), e5130 (2020). https://doi.org/10.1002/cpe.5130

21. Hou, X.-X.: Alcoholism detection by medical robots based on Hu moment invariants and predator-prey adaptive-inertia chaotic particle swarm optimization. Comput. Electric. Eng. **63**, 126–138 (2017)

22. Yang, J.: Pathological brain detection in MRI scanning via Hu moment invariants and machine learning. J. Exper. Theor. Artif. Intell. **29**(2), 299–312 (2017). https://doi.org/10.1080/0952813X.2015.1132274

23. Pan, H., Zhang, C., Tian, Y.: RGB-D image-based detection of stairs, pedestrian crosswalks and traffic signs. J. Vis. Commun. Image Representation **25**(2), 263–272 (2014). https://doi.org/10.1016/j.jvcir.2013.11.005

24. Zhou, X.-X., Sheng, H.: Combination of stationary wavelet transform and kernel support vector machines for pathological brain detection. Simulation **92**(9), 827–837 (2016). https://doi.org/10.1177/0037549716629227

25. Wang, S., Du, S., Atangana, A., Liu, A., Lu, Z.: Application of stationary wavelet entropy in pathological brain detection. Multimedia Tools Appl. **77**(3), 3701–3714 (2016). https://doi.org/10.1007/s11042-016-3401-7

26. Wu, X., Yang, J., Wang, S.: Tea category identification based on optimal wavelet entropy and weighted k-Nearest Neighbors algorithm. Multimedia Tools Appl. **77**(3), 3745–3759 (2016). https://doi.org/10.1007/s11042-016-3931-z

27. Zhan, T.M., Chen, Y.: Multiple sclerosis detection based on biorthogonal wavelet transform, RBF kernel principal component analysis, and logistic regression. IEEE Access **4**, 7567–7576 (2016). https://doi.org/10.1109/ACCESS.2016.2620996

28. Chen, Y., Zhang, Yin, Lu, H.-M., Chen, X.-Q., Li, J.-W., Wang, S.-H.: Wavelet energy entropy and linear regression classifier for detecting abnormal breasts. Multimedia Tools Appl. **77**(3), 3813–3832 (2016). https://doi.org/10.1007/s11042-016-4161-0

29. Lu, S., Lu, Z., Yang, J., Yang, M., Wang, Shuihua: A pathological brain detection system based on kernel based ELM. Multimedia Tools Appl. **77**(3), 3715–3728 (2016). https://doi.org/10.1007/s11042-016-3559-z

30. Kang, C., Yu, X., Wang, S., Guttery, D., Pandey, H., Tian, Y., Zhang, Y.: A heuristic neural network structure relying on fuzzy logic for images scoring. IEEE Trans. Fuzzy Syst. **1**, 1 (2020). https://doi.org/10.1109/TFUZZ.2020.2966163

31. Yang, J.F.: Comprehensive survey on fractional fourier transform. Fundamenta Informaticae **151**(1–4), 1–48 (2017). https://doi.org/10.3233/fi-2017-1477

32. Calculation of the Fractional Fourier Transform. http://nalag.cs.kuleuven.be/research/software/FRFT/. Accessed 18 May 2019

33. Sharifzadeh, M., Sikinioti-Lock, A., Shah, N.: Machine-learning methods for integrated renewable power generation: a comparative study of artificial neural networks, support vector regression, and Gaussian Process Regression. Renew. Sustain. Energy Rev. **108**, 513–538 (2019). https://doi.org/10.1016/j.rser.2019.03.040

34. Chen, Y.: A Feature-free 30-disease pathological brain detection system by linear regression classifier. CNS Neurol. Disorders - Drug Targets **16**(1), 5–10 (2017)

35. Li, W.: A gingivitis identification method based on contrast-limited adaptive histogram equalization, gray-level co-occurrence matrix, and extreme learning machine. Int. J. Imag. Syst. Technol. **29**(1), 77–82 (2019). https://doi.org/10.1002/ima.22298

Gingivitis Classification via Wavelet Entropy and Support Vector Machine

Cui Li$^{(\boxtimes)}$ and ZhiHai Lu

School of Education Science, Nanjing Normal University,
Nanjing 210024, Jiangsu, China
{997268314,1030609013}@qq.com

Abstract. Gingivitis is usually detected by a series of oral examinations. In this process, the dental record plays a very important role. However, it often takes a lot of physical and mental effort to accurately detect gingivitis in a large number of dental records. Therefore, it is of great significance to study the classification technology of gingivitis. In this study, a new gingivitis classification method based on wavelet entropy and support vector machine is proposed to help diagnose gingivitis. The feature of the image is extracted by wavelet entropy, and then the image is classified by support vector machine. The experimental results show that the average sensitivity, specificity, precision and accuracy of this method are 75.17%, 75.29%, 75.35% and 75.24% respectively, which are superior to the other three methods This method is proved to be effective in the classification of gingivitis.

Keywords: Wavelet · Entropy · Support vector machine · Gingivitis · Classification

1 Introduction

Teeth are one of the most important parts of human mouth. There are many diseases of teeth. Gingivitis is a common disease of teeth. Gingivitis mainly refers to the acute or chronic inflammation on the gum. The common symptoms of gingivitis mainly include swelling, bleeding and pain. Some patients may also have local itching and bad breath. Plaque accumulation, lack of nutrients or improper brushing may cause gingivitis.

In recent years, many scholars have carried out research on gingivitis and achieved many new results. Li (2019) [1] identified gingivitis through the method based on gray level co-occurrence matrix (GLCM) and extremum learning machine (ELM). In the study, this method was used to identify 52 teeth images, and different information about teeth was obtained through segmentation and classification from teeth images. It is found that this method is superior to other techniques in average sensitivity, specificity, precision and accuracy. Li (2019) [1] studied a method to identify gingivitis based on contrast-limited adaptive histogram equalization (CLAHE), gray level co-occurrence matrix (GLCM) and extremum learning machine (ELM). By comparing 58 pictures of gingivitis with 35 pictures of healthy teeth, we found that compared with the most advanced method, this method has higher classification accuracy and more sensitive results. Supranoto, Slot (2015) [2] studied the effect of chlorhexidine denifrice or

© ICST Institute for Computer Sciences, Social Informatics and Telecommunications Engineering 2020
Published by Springer Nature Switzerland AG 2020. All Rights Reserved
Y.-D. Zhang et al. (Eds.): ICMTEL 2020, LNICST 327, pp. 290–300, 2020.
https://doi.org/10.1007/978-3-030-51103-6_25

gel versus chlorhexidine moushwash on gingivitis by retrieving databases such as PudMed-MEDLINE. Through screening 2256 samples, 5 publications meeting the standards were obtained. The research found that: by using chlorhexidine denifrice or gel versus chlorhexidine moushwash, we can effectively inhibit gingivitis. A randomized clinical trial was conducted by Sangeetha (2017) [3] to investigate the effect of triclosan containing tooth paste and conventional fluoride tooth paste on gingivitis. In the study, 56 children were randomly divided into two groups, experimental group used the triclosan containing tooth paste and control group used conventional fluoride tooth paste. The results showed that the experimental group was better than the control group in reducing the incidence of gingivitis. Triclosan containing tooth paste can inhibit gingivitis more effectively. Feng, Zhang (2015) [4] used wavelet energy (WavEnr) to identify brain images. Brown (2018) [5] used extreme learning machine (ELM) to identify gingivitis lesions of teeth images.

Through the above research results, we can find that the research of gingivitis mainly includes exploring the recognition methods of gingivitis, studying the methods of inhibiting gingivitis.

The main contribution of this study is to combine wavelet entropy (WE) and support vector machine (SVM) to propose a new method for gingivitis classification. wavelet entropy (WE) can not only get the image features, but also reduce the dimension of the features. As a powerful classifier, support vector machine (SVM) can achieve the image classification. We have achieved good results for combining WE with SVM.

The other parts of this paper are as follows: the second section describes the dataset, the third section briefly introduces the research methods, the fourth section is the experimental results and discussion, the fifth section is the conclusion of this study.

2 Dataset

In this study, we selected 5 patients with gingivitis from Nanjing Stomatological hospital to observe their gingivitis [1]. In this study, two digital single lens reflex (DSLR), i.e. A and B, were used to randomly select different teeth of each patient for image collection. A total of 170 teeth pictures were obtained, including 85 gingivitis pictures and 85 healthy teeth pictures. In the image, we mark three regions: near, middle and far. The field of view is 51–200 mm in diameter, and the voxel resolution is 0.1–0.39 mm.

In the study, we will manually adjust the length and width of the area of interest to make its appearance similar to the simulated image, in which we can clearly see the tooth area. The average length of the 12 bit image is 456732 and the average width is 567833. Figure 1 shows two sample images in the dataset. Figure 1(a) shows a gingivitis image, and Fig. 1(b) shows a healthy teeth image.

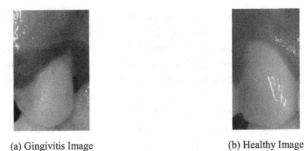

(a) Gingivitis Image   (b) Healthy Image

Fig. 1. Samples of our dataset

3   Methodology

3.1   Wavelet

Wavelet transform [6–10] is an improved frequency transform method based on Fourier Transform (FFT). Wavelet transform overcomes the defect that the unstable signal can't be processed in Fourier transform [11–15]. The infinite trigonometric function base is replaced by the finite attenuation wavelet base, which can not only capture the frequency of the unstable signal, but also locate the time.

The image is a two-dimensional matrix. We get four components (LL1, LH1, HL1, HH1) after wavelet transform. One component (LL1) represents the blurred image, and the other three components (LH1, HL1, HH1) represent the detailed image. Then we get another four components (LL2, LH2, HL2, HH2) of LL1 by double wavelet transform (2D-DWT). After the double wavelet transform (2D-DWT), we get an entropy value. The formula of wavelet is:

$$WT(\alpha, \tau) = \frac{1}{\sqrt{\alpha}} \int_{-\infty}^{\infty} f(t) * \varphi\left(\frac{t - \tau}{\alpha}\right) dt \tag{1}$$

Wavelet transform can effectively decompose images of different pixels and retain image information. But after decomposition, it contains too many image features, which not only takes up a lot of storage space, but also increases the calculation time [16–20]. So we need to reduce the dimension of the feature by introducing entropy.

3.2   Entropy

Entropy is the uncertainty degree of information that Shannon quoted from thermodynamics. Entropy is a measure of disorder, which is used to represent the average value of information of probability distribution [21–25].

Suppose a random variable X, the value of X in dataset D is $X = \{x_1, x_2, \cdots, x_n\}$, $P(R)$ is the probability function, then the entropy is:

$$H(X) = E[I(X)] = E[-ln(K(X))] \tag{2}$$

Where E is the expected value. If D is an infinite set, then the entropy of the random variable X is:

$$H(X) = -\sum_i P(x_i) \log_b P(x_i) \tag{3}$$

As shown in Fig. 2, after wavelet transform, the image is decomposed into seven components, and then the entropy of these seven components is calculated to get the eigenvector. Through the method of wavelet entropy, we can not only get the image features, but also effectively reduce the dimension of image features.

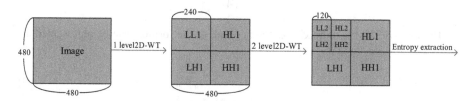

Fig. 2. Diagram of 2D-WT

3.3 Support Vector Machine

Like classified learning, the most basic idea of image classification is to find a partition hyperplane to separate different images. Support Vector Machine (SVM) is the latest classification method based on machine learning theory [26–29]. It is a two-classification model to find a hyperplane for image segmentation. The function of SVM is to help to build a hyperplane with a maximum interval. In support vector machine, a training sample set D is given.

$$D = \{(x_1, y_1), (x_2, y_2), \cdots, (x_m, y_m), y_i \varepsilon \{-1, +1\}\}. \tag{4}$$

Based on the sample set D, a partition hyperplane is found to separate the categories of different samples. The linear equation of partition hyperplane is expressed as:

$$w^T x + b = 0 \tag{5}$$

Where w is the normal vector, which determines the direction of the hyperplane, and b is the displacement, which determines the distance between the hyperplane and the origin.

Supposing the hyperplane can classify the training samples correctly, for the training samples (x_i, y_i), the following formula is satisfied:

$$\begin{cases} W^T x_i + b \geq +1 y_i = +1 \\ W^T x_i + b \leq -1 y_i = -1 \end{cases} \tag{6}$$

The formula is equivalent to:

$$y_i \left(W^T x_i + b \right) \geq +1 \tag{7}$$

The sample points which are closest to the hyperplane and meet the fomula of $y_i(W^T x_i + b) \geq +1$ are support vector.

From the above formula, we can get the interval:

$$\gamma = \frac{2}{\|W\|} \tag{8}$$

The idea of SVM is to maximize the interval, so the formula can be represented as:

$$max \frac{2}{\|W\|}$$
$$s.t. y_i \left(W^T x_i + b \right) \geq +1 \tag{9}$$

From the above formula, we can know that maximizing $\frac{2}{\|W\|}$ is equivalent to minimizing $\|W\|$, so the basic type of SVM can be expressed as:

$$\min \frac{1}{2} \|W\|^2$$
$$s.t. y_i (w^T x_i + b) \geq 1, i = 1, 2, \ldots, n \tag{10}$$

The Lagrange multiplier method is used to solve the basic dual problem:

$$L(w, b, \alpha) = \min \frac{1}{2} \|W\|^2 - \sum_{i=1}^{n} \alpha^{(i)} \left[y^{(i)} \left(w^T \phi \left(x^{(i)} \right) + b \right) - 1 \right]$$
$$s.t. y_i \left(w^T \phi \left(x^{(i)} \right) + b \right) \geq 1, i = 1, 2, \ldots, n \tag{11}$$

After deriving the w and b of the above formula, make equal to 0, and bring them into the Lagrange multiplier method, we can get:

$$L(w, b, \alpha) = \sum_{i=1}^{n} \alpha_i - \frac{1}{2} \sum_{i=1}^{n} \sum_{j=1}^{n} \alpha_i \alpha_j y_i y_j x_i x_j$$
$$s.t. \sum_{i=1}^{n} \alpha_i y_i = 0, \alpha_i \geq 0, i = 1, 2, \ldots, n \tag{12}$$

After solving the above problems, we can get the optimal classification function as follows:

$$f(x) = w^T x + b = \sum_{i=1}^{n} \alpha_i y_i x_i^T x + b \tag{13}$$

This WE and SVM combination belongs to traditional feature extraction + classifier combination. We do not use deep learning methods [30–38] because the small-size dataset.

3.4 10-Fold Cross-Validation

10-fold cross validation is a commonly used data accuracy test method in statistics. This method is to divide the data set into 10 groups as shown in Fig. 3, one group as the test set in turn, and the other nine groups as the training set for experiments. Each experiment will produce a result, and the average value of the 10 results is the accuracy value of the algorithm. In this study, 10-fold cross validation is used to verify the accuracy of image classification.

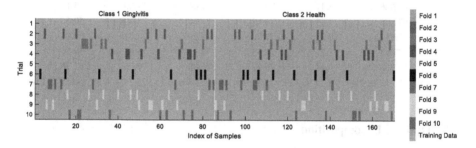

Fig. 3. Index of 10-fold cross validation

This 10-fold cross validation will be run 10 times, and we use those measures to performance the performance of proposed algorithm. There are six evaluation indexes in classification: sensitivity, specificity, accuracy, accuracy, F1 and MCC. Sensitivity is the proportion that the test is correctly recognized as positive; specificity is the proportion that the test is correctly recognized as negative. Precision is the ratio of the number of positive samples correctly predicted to the number of positive samples predicted. Accuracy represents the ratio of the number of correctly predicted samples to the total number of predicted samples. F1 is the harmonic average of precision and recall. MCC is essentially the phase between the observed value and the predicted value Relation number.

4   Experiment Results and Discussions

4.1   Statistical Results

In the first experiment, we set the decomposition level of WE as three, and the results are shown in Table 1. From Table 1, we can see that WE-SVM method has achieved good results as a whole. At the 6th run, the sensitivity, accuracy, F1 and MCC of the samples were all good, and the values were the highest; at the 8th run, the specificity and precision of the samples reached a peak with the values of 80.01% and 77.90% respectively, but the values of sensitivity, accuracy, F1 and MCC were relatively bad.

Table 1. Statistical analysis of our WE-SVM method.

Run	Sensitivity	Specificity	Precision	Accuracy	F1	MCC
1	72.98	78.82	77.49	75.90	75.14	51.91
2	77.63	72.95	74.14	75.29	75.85	50.64
3	72.90	76.47	75.55	74.71	74.17	49.44
4	76.47	71.76	73.03	74.11	74.71	48.28
5	77.63	70.60	72.51	74.12	74.98	48.35
6	77.66	76.47	76.74	77.06	77.20	54.13
7	72.95	77.63	76.55	75.29	74.70	50.65
8	70.57	80.01	77.90	75.29	74.05	50.80
9	77.63	71.73	73.38	74.71	75.42	49.50
10	75.30	76.44	76.23	75.88	75.76	51.76
Mean + SD	75.17 ± 2.49	75.29 ± 3.11	75.35 ± 1.85	75.24 ± 0.85	75.20 ± 0.88	50.55 ± 1.69

4.2   Optimal Decomposition

In this experiment, we compared the best decomposition levels. We suppose the decomposition level L vary from 1 to 4, and the corresponding results are shown below in Table 2.

Table 2. Results of different composition levels.

Decomposition level (L)	Sensitivity	Specificity	Precision	Accuracy	F1	MCC
1	71.41 ± 2.30	69.88 ± 2.53	70.39 ± 1.52	70.65 ± 1.16	70.85 ± 1.25	41.35 ± 2.32
2	73.52 ± 1.61	72.82 ± 1.93	73.04 ± 1.33	73.17 ± 1.02	73.26 ± 1.01	46.37 ± 2.06
3	**75.17 ± 2.49**	**75.29 ± 3.11**	**75.35 ± 1.85**	**75.24 ± 0.85**	**75.20 ± 0.88**	**50.55 ± 1.69**
4	73.52 ± 3.21	74.23 ± 4.28	74.22 ± 2.61	73.88 ± 1.40	73.77 ± 1.34	47.88 ± 2.78

(bold means the best)

In Table 2, we can see that when the value of L is 3, the values of sensitivity, specificity, precision, accuracy, F1 and MCC are the highest, so the effect of image at the third level is the best. we regard the third level as the best level of image decomposition.

4.3   Comparison with Other Classifiers

Decision tree (DT) and naive Bayesian classifier (NBC) are two classical classifiers. We compared using SVM with DT and NBC, at the condition of using 3-level decomposition. The results of DT and NBC are listed in Table 3. The different classifier comparisons are shown in Table 4.

From Table 3, it can be seen that DT and NBC are quite different in all aspects, of which the difference between the two methods in sensitivity and specificity is the most obvious. The highest value of DT is 78.82% and 76.47% respectively, NBC is 72.92% and 72.98%, DT is obviously superior to NBC. In Table 4, when DT, NBC and SVM are compared, SVM is superior to the other two methods in all aspects and the difference is large, which shows that SVM has obvious advantages as a classifier. So the SVM method used in this study is effective.

Table 3. Results using DT and NBC.

DT	Sensitivity	Specificity	Precision	Accuracy	F1	MCC
1	67.05	76.47	74.02	71.76	70.37	43.72
2	72.92	74.11	73.79	73.53	73.35	47.05
3	78.82	64.70	69.07	71.76	73.62	43.96
4	74.09	69.44	70.82	71.76	72.38	43.62
5	74.11	70.60	71.60	72.36	72.83	44.74
6	67.05	74.11	72.15	70.58	69.51	41.27
7	71.73	70.60	70.92	71.17	71.30	42.36
8	72.92	71.76	72.08	72.35	72.50	44.70
9	72.92	69.41	70.44	71.17	71.66	42.36
10	74.11	70.57	71.60	72.35	72.83	44.72
Mean + SD	72.57 ± 3.29	71.18 ± 3.08	71.65 ± 1.41	71.88 ± 0.78	72.04 ± 1.25	43.85 ± 1.54
NBC	Sensitivity	Specificity	Precision	Accuracy	F1	MCC
1	67.03	70.57	69.50	68.82	68.22	37.66
2	67.08	70.54	69.58	68.82	68.31	37.66
3	71.76	69.41	70.11	70.58	70.93	41.18
4	69.38	69.38	69.38	69.38	69.38	38.76
5	69.41	68.22	68.60	68.81	69.00	37.63
6	68.24	68.22	68.24	68.24	68.24	36.46
7	67.05	69.41	68.67	68.24	67.85	36.47
8	72.92	69.41	70.44	71.18	71.66	42.37
9	65.89	72.98	70.99	69.44	68.33	38.99
10	70.57	69.41	69.75	69.99	70.16	39.98
Mean + SD	68.93 ± 2.18	69.75 ± 1.31	69.53 ± 0.81	69.35 ± 0.93	69.21 ± 1.23	38.72 ± 1.86

Table 4. Comparison with different classifiers.

Classifier	Sensitivity	Specificity	Precision	Accuracy	F1	MCC
DT	72.57 ± 3.29	71.18 ± 3.08	71.65 ± 1.41	71.88 ± 0.78	72.04 ± 1.25	43.85 ± 1.54
NBC	68.93 ± 2.18	69.75 ± 1.31	69.53 ± 0.81	69.35 ± 0.93	69.21 ± 1.23	38.72 ± 1.86
SVM	**75.17 ± 2.49**	**75.29 ± 3.11**	**75.35 ± 1.85**	**75.24 ± 0.85**	**75.20 ± 0.88**	**50.55 ± 1.69**

4.4   Comparison to State-of-the-Art Approaches

We compared our WE-SVM method with three state-of-the-art approaches: WavEnr [4], ELM [5], GLCM [1]. The results are shown in Table 5. As shown in Table 5, when compared with the most advanced methods, we can find that the value of WE-SVM is higher than that of other methods in terms of sensitivity, specificity, precision and accuracy, which proves the effectiveness of WE-SVM and shows that WE-SVM is the optimal algorithm.

Table 5. Comparison with state-of-the-art approaches.

Approach	Sensitivity	Specificity	Precision	Accuracy
WavEnr [4]	62	68	65.96	65
ELM [5]	72	70	70.59	71
GLCM [1]	75	73	74	74
WE-SVM (Ours)	75.17 ± 2.49	75.29 ± 3.11	75.35 ± 1.85	75.24 ± 0.85

5   Conclusions

In this paper, a method of gingivitis classification based on wavelet entropy and support vector machine is proposed. By using this method, we can accurately classify the teeth pictures. Through the analysis of experimental data, WE-SVM can find that the sensitivity, specificity, accuracy and accuracy of the method are higher than 75%, which is more accurate and sensitive than the three most advanced methods.

However, due to the small number of samples in the database, it may cause over fitting phenomenon, which will be improved and avoided in future research, so that this research will be more helpful for dentists to carry out gingivitis testing.

References

1. Li, W.: A gingivitis identification method based on contrast-limited adaptive histogram equalization, gray-level co-occurrence matrix, and extreme learning machine. Int. J. Imag. Syst. Technol. **29**(1), 77–82 (2019)
2. Supranoto, S.C., et al.: The effect of chlorhexidine dentifrice or gel versus chlorhexidine mouthwash on plaque, gingivitis, bleeding and tooth discoloration: a systematic review. Int. J. Dental Hygiene **13**(2), 83–92 (2015)
3. Sangeetha, K.M.: Effect of triclosan containing tooth paste and conventional fluoride tooth paste on plaque and gingivitis: a randomized clinical trial. Int. J. Sci. Study **4**(10), 58–61 (2017)
4. Feng, C., et al.: Automated classification of brain MR images using wavelet-energy and support vector machines. In: Liu, C., Chang, G., Luo, Z. (eds.) International Conference on Mechatronics, Electronic, Industrial and Control Engineering, pp. 683–686. Atlantis Press, USA (2015)

5. Brown, M.: Gingivitis identification via grey-level cooccurrence matrix and extreme learning machine. Adv. Soc. Sci. Educ. Humanit. Res. **250**, 486–492 (2018)

6. Ramirez, J.: Unilateral sensorineural hearing loss identification based on double-density dual-tree complex wavelet transform and multinomial logistic regression. Integr. Comput. Aided Eng. (2019). https://doi.org/10.3233/ICA-190605

7. Gorriz, J.M.: Multivariate approach for Alzheimer's disease detection using stationary wavelet entropy and predator-prey particle swarm optimization. J. Alzheimer's Disease **65** (3), 855–869 (2018)

8. Li, Y.-J.: Single slice based detection for Alzheimer's disease via wavelet entropy and multilayer perceptron trained by biogeography-based optimization. Multimedia Tools Appl. **77**(9), 10393–10417 (2018)

9. Han, L., Identification of Alcoholism based on wavelet Renyi entropy and three-segment encoded Jaya algorithm. Complexity, **2018** (2018). Article ID. 3198184

10. Phillips, P.: Intelligent facial emotion recognition based on stationary wavelet entropy and Jaya algorithm. Neurocomputing **272**, 668–676 (2018)

11. Farsi, C., et al.: Spectral triples and wavelets for higher-rank graphs. J. Math. Anal. Appl. **482**(2), 39 (2020). Article ID. 123572

12. Li, P., et al.: Pathological brain detection via wavelet packet tsallis entropy and real-coded biogeography-based optimization. Fundamenta Informaticae **151**(1–4), 275–291 (2017)

13. Li, Y.: Detection of dendritic spines using wavelet packet entropy and fuzzy support vector machine. CNS Neurol. Disorders - Drug Targets **16**(2), 116–121 (2017)

14. Nayak, D.R.: Detection of unilateral hearing loss by Stationary Wavelet Entropy. CNS Neurol. Disorders - Drug Targets **16**(2), 15–24 (2017)

15. Lu, H.M.: Facial emotion recognition based on biorthogonal wavelet entropy, fuzzy support vector machine, and stratified cross validation. IEEE Access **4**, 8375–8385 (2016)

16. Bertoluzza, S., et al.: Wavelets and convolution quadrature for the efficient solution of a 2D space-time BIE for the wave equation. Appl. Math. Comput. **366**, 21 (2020). Article ID. Unsp 124726

17. Gorriz, J.M., et al.: Wavelet entropy and directed acyclic graph support vector machine for detection of patients with unilateral hearing loss in MRI scanning. Frontiers in Computational Neuroscience, **10** (2016). Article ID. 160

18. Zhou, X.-X.: Comparison of machine learning methods for stationary wavelet entropy-based multiple sclerosis detection: decision tree, k-nearest neighbors, and support vector machine. Simulation **92**(9), 861–871 (2016)

19. Yang, M.: Dual-tree complex wavelet transform and twin support vector machine for pathological brain detection. Appl. Sci. **6**(6) (2016). Article ID. 169

20. Sun, P.: Preliminary research on abnormal brain detection by wavelet-energy and quantum-behaved PSO. Technol. Health Care **24**(s2), S641–S649 (2016)

21. Zhou, X., Zhang, G., Dong, Z., Wang, S., Zhang, Y.: Tea category classification based on feed-forward neural network and two-dimensional wavelet entropy. In: Xie, J., Chen, Z., Douglas, C.C., Zhang, W., Chen, Y. (eds.) HPCA 2015. LNCS, vol. 9576, pp. 48–54. Springer, Cham (2016). https://doi.org/10.1007/978-3-319-32557-6_5

22. Mondal, H., et al.: Entropy generation of variable viscosity and thermal radiation on magneto nanofluid flow with dusty fluid. J. Appl. Comput. Mech. **6**(1), 171–182 (2020)

23. Zhang, Y.D., Sun, J.: Preliminary study on angiosperm genus classification by weight decay and combination of most abundant color index with fractional Fourier entropy. Multimedia Tools Appl. **77**(17), 22671–22688 (2017). https://doi.org/10.1007/s11042-017-5146-3

24. Cheng, H.: Multiple sclerosis identification based on fractional Fourier entropy and a modified Jaya algorithm. Entropy, **20**(4) (2018). Article ID. 254

25. Li, J.: Texture analysis method based on fractional fourier entropy and fitness-scaling adaptive genetic algorithm for detecting left-sided and right-sided sensorineural hearing loss. Fundamenta Informaticae 151(1–4), 505–521 (2017)
26. Arahmane, H., et al.: Neutron-gamma discrimination based on support vector machine combined to nonnegative matrix factorization and continuous wavelet transform. Measurement, 149, 9 (2020). Article ID. Unsp 106958
27. Chen, Y.: Pathological brain detection by wavelet-energy and fuzzy support vector machine. In: 8th International Symposium on Computational Intelligence and Design (ISCID), p. 409–412. IEEE, Hangzhou, China (2015)
28. Chen, M.: Morphological analysis of dendrites and spines by hybridization of ridge detection with twin support vector machine. PeerJ, 4 (2016). Article ID e2207
29. Liu, G.: Pathological brain detection in MRI scanning by wavelet packet Tsallis entropy and fuzzy support vector machine. SpringerPlus, 4(1) (2015). Article ID. 716
30. Jiang, X.: Chinese sign language fingerspelling recognition via six-layer convolutional neural network with leaky rectified linear units for therapy and rehabilitation. J. Med. Imag. Health Inform. 9(9), 2031–2038 (2019)
31. Govindaraj, V.V.: High performance multiple sclerosis classification by data augmentation and AlexNet transfer learning model. J. Med. Imag. Health Inform. 9(9), 2012–2021 (2019)
32. Li, Z.: Teeth category classification via seven-layer deep convolutional neural network with max pooling and global average pooling. Int. J. Imag. Syst. Technol. (2019). https://doi.org/10.1002/ima.22337
33. Zhang, Y.D., Dong, Z., Chen, X., Jia, W., Du, Sidan, Muhammad, K., Wang, S.H.: Image based fruit category classification by 13-layer deep convolutional neural network and data augmentation. Multimedia Tools Appl. 78(3), 3613–3632 (2017). https://doi.org/10.1007/s11042-017-5243-3
34. Wang, S.H., Sun, J., Phillips, P., Zhao, G., Zhang, Y.D.: Polarimetric synthetic aperture radar image segmentation by convolutional neural network using graphical processing units. J. Real-Time Image Process. 15(3), 631–642 (2017). https://doi.org/10.1007/s11554-017-0717-0
35. Huang, C.: Multiple sclerosis identification by 14-layer convolutional neural network with batch normalization, dropout, and stochastic pooling. Front. Neurosci. 12 (2018). Article ID. 818
36. Pan, C.: Multiple sclerosis identification by convolutional neural network with dropout and parametric ReLU. J. Comput. Sci. 28, 1–10 (2018)
37. Pan, C.: Abnormal breast identification by nine-layer convolutional neural network with parametric rectified linear unit and rank-based stochastic pooling. J. Comput. Sci. 27, 57–68 (2018)
38. Zhang, Y.-D., Khan, M., Tang, C.: Twelve-layer deep convolutional neural network with stochastic pooling for tea category classification on GPU platform. Multimedia Tools Appl. 77(17), 22821–22839 (2018). https://doi.org/10.1007/s11042-018-5765-3

Data Fusion Filter and Machine Learning for Statistical Signal Processing

LiDAR/DR-Integrated Mobile Robot Localization Employing IMM-EKF/PF Filtering

Ning Feng, Yong Zhang, Yuan Xu$^{(\boxtimes)}$ ⓘ, Shuhui Bi, and Tongqian Liu

School of Electrical Engineering, University of Jinan, Jinan 250022, Shandong, China
xy_abric@126.com

Abstract. In order to solve the problems that indoor mobile robots have parking during the traveling process and the Extended Kalman filter (EKF) receives too much influence on parameter selection, this paper proposes an Interacting Multiple Model (IMM)-EKF/Particle Filtering (PF) adaptive algorithm for the tightly inertial navigation system (INS)/Light Detection And Ranging (LiDAR) integrated navigation. The EKF and PF calculate the position of the robot respectively, then the smaller *Mahalanobis* distance-based filter's output is selected as the initial value of the next iteration, which improves the accuracy of the positioning for the robot. Based on that, the two motion equations of the static and normal motion models are dsigned at the same time. A *Markov* chain for converting the two state of the model, and the weighting filtering result of the filtered is used to provide distance estimates. The real experimental results show that the IMM-EKF/PF adaptive algorithm improves the positioning accuracy of mobile robots in the presence of parking.

Keywords: Indoor robot location · Kalman filter · Particle filtering · Interacting multiple model

1 Introduction

Today, mobile robots play an important role in medical care, home cleaning, indoor rescue, warehousing and transportation. Moreover, mobile robots are often used to serve people in complex environments, so the positioning accuracy of mobile robots is particularly important.

Usually robot positioning is achieved by a single sensor, such as Global Positioning System (GPS), Radio Frequency Identification (RFID), Inertial Navigation System (INS), and so on. But using only a single sensor can't get rid of its inherent drawbacks when used, for example, GPS is the most widely used

This work was supported by the Shandong Key Research and Development Program under Grants 2019GGXI04026 and 2019GNC106093.

positioning sensor Wu (2019). In the case of indoor positioning, it is difficult to complete positioning due to difficulty in receiving positioning signals Wheeler (2018).

RFID is a wireless communication technology that uses radio signals to identify specific targets and read and write related data. There is no need to identify mechanical or optical contact between the system and a particular target. The cost is relatively low, and the shortcoming anti-interference ability is poor Liu (2006). INS is an autonomous navigation system that does not rely on external information, but the positioning error of inertial navigation technology accumulates over time, and finally the positioning cannot be completed due to excessive error.

In order to overcome the shortcomings of single navigation technology, many universities and enterprises began to study integrated navigation technology. Therefore, the integrated navigation technology has been rapidly developed Fern (2018).

The integrated navigation system can not only make full use of the information of the subsystem, but also complement each other's navigation methods to improve the accuracy of navigation and positioning information Rafatnia (2019). It is also possible to arbitrarily switch the operating modes of each subsystem to make it flexible and enhance the reliability of the system.

For example, Liu (2006) and Xu (2012) used IEKF for the INS/Wireless Sensors Networks (WSNs)-integrated navigation for indoor robot, from the paper we can see that the EKF is fast for nonlinear systems, but its output is greatly affected by parameter selection.

Integrated navigation used IMM-PF algorithm in Ming (2012), Particle Filtering (PF) has obvious advantages for the processing of state estimation problems under nonlinear and non-Gaussian conditions Xu (2018), but because of its large amount of calculation, the real-time performance is poor Sovic (2011).

In this paper, we proposed a method to combine the advantages of EKF and PF. And then, the interactive multi-model method is used to improve the positioning accuracy of the robot in the stationary/moving situation.

The rest of this article is organized as follows. The second section introduces the INS/LiDAR tight combination model structure based on the EKF/PF adaptive algorithm. Besides section introduces the IMM algorithm under two different models of stationary motion model and moving motion model. A *Markov* chain for converting the two state of the model, and the weighting filtering result of the filtered is used to provide distance estimates. The third chapter mainly introduces the experimental environment and experimental results. Finally, fourth chapter gives the conclusion.

2   The Introduce of IMM-EKF/PF Adaptive Algorithm

In this section, we will introduce the INS/LiDAR tight combination model and the adaptive method of EKF/PF, and then, the model selection of IMM will be investigated. Figure 1 shows integrated navigation strategy combining INS and LiDAR measurement.

In this model, firstly, INS and LiDAR separately calculate the distance from the feature point to the robot, in this paper, the feature points use the corner points extracted by the radar from the environmental information, and $(d_m^{Lidar})^2$, $(d_m^{INS})^2$ means the LiDAR and INS calculated distance from the feature point to the robot. Then the optimal error estimates $\mathbf{P}^{INS}$ for INS and LiDAR are calculated by the IMM-EKF/PF adaptive filter. Finally, the error $\mathbf{P}^{INS}$ is used to compensate for the position where the INS is settled separately.

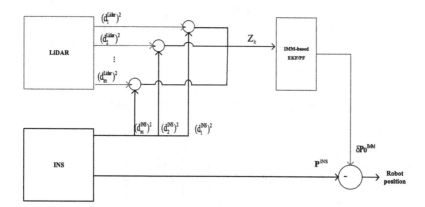

Fig. 1. The integrated navigation strategy of the robot localization system.

2.1 The State Equation and Observation Equation of the EKF/PF Adaptive Algorithm

In this section, we will introduce the state equation and observation equation of EKF/PF adaptive algorithm used in this article, and how to select EKF or PF filter will be designed.

Equation 1 is the state equation of the PF and EKF for the INS/LiDAR-based localization model.

$$\underbrace{\begin{bmatrix} \delta x_k \\ \delta y_k \\ \delta V_k \\ \delta \phi_k \end{bmatrix}}_{\mathbf{X}_{k|k-1}} = \underbrace{\begin{bmatrix} \delta x_{k-1} + T \cdot \delta V_{k-1} \sin(\phi_{k-1}) \\ \delta y_{k-1} + T \cdot \delta V_{k-1} \cos(\phi_{k-1}) \\ \delta V_{k-1} \\ \delta \phi_{k-1} \end{bmatrix}}_{f(\mathbf{X}_{k-1})} + \omega_k, \tag{1}$$

where the k means the time index, δx_k and δy_k mean the robot's position error in east and north directions, V means the robot velocity error, φ means the heading angle error, T represents the sample time, ω_k is the system noise. The observation equation is listed as Eq. 2.

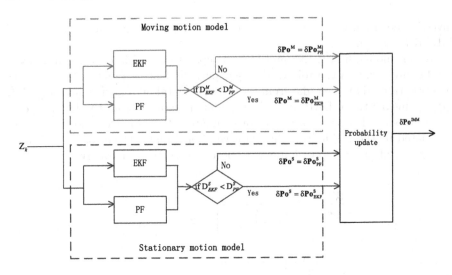

Fig. 2. The IMM-EKF/PF adaptive algorithm.

$$
\underbrace{\begin{bmatrix} (d_{1,k}^{INS})^2 - (d_{1,k}^{LiDAR})^2 \\ (d_{2,k}^{INS})^2 - (d_{2,k}^{LiDAR})^2 \\ \vdots \\ (d_{i,k}^{INS})^2 - (d_{m,k}^{LiDAR})^2 \end{bmatrix}}_{Z_k} = \underbrace{\begin{bmatrix} h_1(X_k) \\ h_2(X_k) \\ \vdots \\ h_m(X_k) \end{bmatrix}}_{h(X_k)} + v_k \,, \tag{2}
$$

where v_k is the observed noise matrix of the system, and its covariance matrix is $\mathbf{R}$.

2.2 IMM-EKF/PF Algorithm

In order to solve the situation that the mobile robot has a pause in the indoor motion state, we use the IMM algorithm to deal with it. Figure 2 shows the flow chart of IMM-EKF/PF Algorithm.

Firstly, the optimal error estimate $\delta\mathbf{Po}_{EKF}^M$, $\delta\mathbf{Po}_{PF}^M$ is calculated by EKF and PF under the moving motion model. The *Mahalanobis* distance of the two values D_{PF}^M, D_{EKF}^M is calculated, and the value with a smaller *Mahalanobis* distance is selected as the output of the filter. Meanwhile, the stationary motion model $\delta\mathbf{Po}^S$ can be calculated in a similar way.

Finally, the optimal error estimate $\delta\mathbf{Po}^S$, $\delta\mathbf{Po}^M$ are calculated in two states for output interaction, and the optimal error estimate $\delta\mathbf{Po}^{IMM}$ including the motion model and the stationary model is obtained. Through the above steps, we get the optimal position error estimate calculated by the EKF/PF adaptive filter under stationary motion model and moving stationary motion model states.

For each *ith*, the *Markov* transition probability matrix has been listed as follow.

$$\prod = \begin{bmatrix} \pi_{11} & \pi_{12} \\ \pi_{21} & \pi_{22} \end{bmatrix}, \tag{3}$$

where π is the conversion probability of the filter.

The weight matrix of each motion model is used as follows.

$$v = \begin{bmatrix} v_1 \\ v_2 \end{bmatrix}, \tag{4}$$

The moving motion model state equation is listed as follows.

$$\underbrace{\begin{bmatrix} \delta x_k \\ \delta y_k \\ \delta V_k \\ \delta \phi_k \end{bmatrix}}_{\mathbf{X}_{k|k-1}^M} = \underbrace{\begin{bmatrix} \delta x_{k-1} + T \cdot \delta V_{k-1} \sin(\phi_{k-1}) \\ \delta y_{k-1} + T \cdot \delta V_{k-1} \cos(\phi_{k-1}) \\ \delta V_{k-1} \\ \delta \phi_{k-1} \end{bmatrix}}_{f(\mathbf{X}_{k-1}^M)}, \tag{5}$$

The stationary motion model state equation is listed as follows.

$$\underbrace{\begin{bmatrix} \delta x_k \\ \delta y_k \end{bmatrix}}_{\mathbf{X}_k^S} = \underbrace{\begin{bmatrix} \delta x_{k-1} \\ \delta y_{k-1} \end{bmatrix}}_{f(\mathbf{X}_{k-1}^S)}, \tag{6}$$

The observation equations of the stationary motion model and the moving motion model are consistent with Eq. 2.

3 Experiment

To verify the validity of the IMM-EKF/PF algorithm, we did a real experiment in the 1st Teaching Building, University of Jinan. The real experimental environment is shown in Fig. 3.

This experiment mainly uses the following equipment, one LiDAR, one INS, one mobile robot and one computer. The LiDAR and INS are fixed on the mobile robot, the computer is used to collect the data returned by the sensor. In order to synchronize the sampling time of the LiDAR rotation, the INS sampling time T is set to 0.15 s, same with LiDAR. In addition, the IMM-EKF/PF algorithm initial parameters are as follows

$$\mathbf{\Pi} = \begin{bmatrix} 0.5 & 0.5 \\ 0.5 & 0.5 \end{bmatrix}, \tag{7}$$

and

Fig. 3. The real experimental environment

$$\mathbf{v} = \begin{bmatrix} 0.5 \\ 0.5 \end{bmatrix}, \tag{8}$$

The model of sensor and mobile robot combination is shown in Fig. 4. From Fig. 5 we can see four tracks, the black dotted line is the robot's preset reference track. The solid green line is the trajectory calculated using only the LiDAR to the corner point distance least squares method, the solid blue line is the robot position calculated by INS only, and the red solid line is the trajectory calculated by the tight combination EKF, the parameters $\mathbf{Q} = 10^{-2} \times \mathbf{I}$, $\mathbf{R} = 10^{-2} \times \mathbf{I}$. It can be seen from the figure that in the case where the position of the robot is estimated using only the INS, the error of the INS gradually diverges with time, and the positioning cannot be caculated. So we can know that INS is not suitable for positioning alone. Compared with INS, LiDAR uses the position of the corner points to calculate the position of the robot much more accurately but the accuracy of the individual points is poor. Tightly combined EKF and the least squares method calculate the position, the burr is less, the precision is higher.

Since the EKF is greatly affected by the parameters and the PF is less affected by the parameter selection, we combine the two filters, and select the output of the double filters with a smaller *Mahalanobis* distance as the initial value of the next iteration.

Under parameters $\mathbf{Q} = 10^0 \times \mathbf{I}$, $\mathbf{R} = 10^{-2} \times \mathbf{I}$, the sampling point of the PF is 6000, the EKF, and EKF/PF adaptive algorithm effects are shown in Fig. 6. In Fig. 6, we can see the EKF employs the parameters $\mathbf{Q} = 10^0 \times \mathbf{I}$, $\mathbf{R} = 10^{-2} \times \mathbf{I}$, performance is significantly worse than EKF under parameters

Fig. 4. The mobile robot and the sensors

$\mathbf{Q} = 10^0 \times \mathbf{I}$, $\mathbf{R} = 10^{-2} \times \mathbf{I}$. And the EKF/PF adaptive algorithm performance is much better than EKF. The error was shown in Table 1.

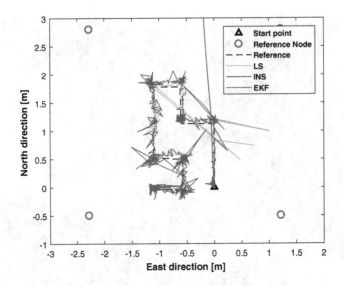

Fig. 5. INS trajectory and LiDAR trajectory (Color figure online)

Table 1. Localization error comparing.

Model	East Mean [m]	East Std. [m]	North Mean [m]	North Std. [m]
EKF	0.0459	0.0644	0.0708	0.0578
IMM-EKF/PF	**0.0294**	**0.0261**	**0.0587**	**0.0338**

Figure 7 shows the EKF/PF adaptive algorithm and the EKF algorithm with effect of the mobile robot in the presence of parking conditions. The red line is the robot trajectory calculated by the IMM-EKF/PF algorithm, and the blue line is the trajectory calculated by the EKF alone, two trajectories are closer. And Table 2 lists the localization error between EKF and IMMM-EKF/PF.

Table 2. Localization error comparing.

Model	East Mean [m]	East Std. [m]	North Mean [m]	North Std. [m]
EKF	0.0320	0.02235	0.0314	0.0232
IMM-EKF/PF	0.0318	0.02224	0.0309	0.0229

From the Fig. 7 and the Table 2, it can be seen that the performance of the proposed IMM method is smaller than the EKF tight combination method and the precision is higher.

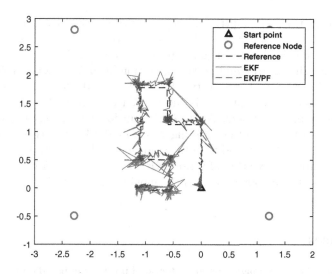

Fig. 6. EKF trajectory and EKF/PF trajectory

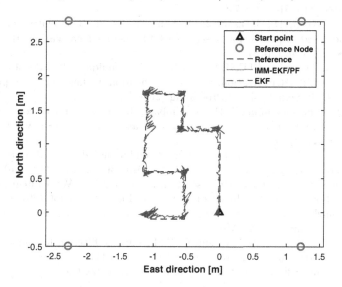

Fig. 7. EKF/PF trajectory and IMM-EKF/PF trajectory (Color figure online)

4   Conclusion

The algorithm in this paper is based on a tightly coupled integrated positioning strategy of INS/LiDAR mobile robot. The integrated navigation strategy effectively solves the problem of indoor robot accumulating errors over time using a single sensor INS. In this paper, an adaptive filter of EKF and PF is proposed to solve the problem that EKF is greatly affected by parameter selection.

Then based on this, an adaptive algorithm based on IMM-EKF/PF is proposed to solve the phenomenon that indoor robot have parking during the traveling process. The effectiveness of the adaptive algorithm of IMM-EKF/PF is verified by a real experiment to simulate the operation of indoor mobile robot. In the filter operation phase, there are two EKF/PF hybrid filters, representing two motion models of the robot, one is the normal motion model and the other is the stationary motion model. A *Markov* chain for converting the two state of the model, and the weighting filtering result of the filtered is used to provide distance estimates. Compared with the traditional EKF tight combination algorithm, the adaptive algorithm of IMM-EKF/PF obtains better robot positioning accuracy when the indoor robot has the moving motion and stationary motion double model.

References

Wu, Z., Wang, W.: INS/magnetometer integrated positioning based on neural network for bridging long-time GPS outages. GPS Solut. **23**(3), 1–11 (2019). https://doi.org/10.1007/s10291-019-0877-4

Wheeler, D., Koch, D., Jackson, J., Mclain, T., Beard, R.: Relative navigation: a keyframe-based approach for observable GPS-degraded navigation. IEEE Control Syst. **38**(4), 30–48 (2018)

Fernndez, A., et al.: ATENEA : Advanced techniques for deeply integrated GNSS/INS/LiDAR navigation. Satellite Navigation Technologies European Workshop on GNSS Signals Signal Processing (2018)

Rafatnia, S., Nourmohammadi, H., Keighobadi, J.: Fuzzy-adaptive constrained data fusion algorithm for indirect centralized integrated SINS/GNSS navigation system. GPS Solut. **23**(3), 62 (2019)

Liu, G., Gao, E., Fan., C.: Algorithm of imm combining kalman and particle filter for maneuvering target tracking. Int. J. Inf. Acquisit. **3**(04), 321–328 (2006)

Xu, Y., Chen, X.Y., Li, Q.H.: Unbiased tightly-coupled INS/WSN integrated navigation based on extended Kalman filter. J. Chin. Inertial Technol. (2012)

Ming, L., Binzhou, D.: Center, Binzhou College, Inertial/geomagnetic integrated navigation algorithm based IMM-PF

Sovic, V., Athalye, A., Bolic, M., Djuric, P.M.: Particle filtering for indoor RFID tag tracking. Statistical Signal Processing Workshop (2011)

Yan, Y.P., Wong, S.F.: Particle filtering for indoor RFID tag tracking. Cluster Comput. (2018)

Review on Flocking Control

Ku Ge and Jin Cheng$^{(\boxtimes)}$ (iD)

University of Jinan, Jinan 250022, People's Republic of China
cse_chengj@ujn.edu.cn

Abstract. Nowadays, significant changes have taken place in the field of information technology and industry and robot research is also deepening. The realization of multi-robot flocking control problem has far-reaching significance. This paper mainly introduces the development status of flocking control at home and abroad and summarizes several commonly used distributed flocking control strategies. In this paper, on the basis of summarizing the development of flocking research at home and abroad, forecasts its development prospect in the field of aviation and so on.

Keywords: Multi-robot · Flocking control · Distributed

1 Introduction

With the development of robot technology and the further improvement of social needs, people's requirements on robots are no longer limited to a single robot, but more and more interested in the system composed of multiple robots. This is not only because some tasks cannot be undertaken by a single robot, but more and more examples show that for some dynamic and complex tasks, the development of a single robot is far more complex and expensive than the development of multiple robot systems. At the same time, with the emergence of robot production line, the desire of autonomous operation of multiple robot systems becomes more and more strong. In the late 1970s, some robotics researchers applied the multi-agent theory of artificial intelligence to the research of multi-robot systems, thus starting the research of multi-robot technology in the field of robotics. Multi-robot system is not a simple stack of a single robot, but an organic combination of multiple robots. It effectively avoids the shortcomings of a single robot and gives full play to the advantages of group robots. With the continuous deepening of robot research and the continuous development of military, security, industrial production and other application fields, as one of the important bases and research directions of multi-robot cooperation and coordination, the realization of multi-robot crowd control problem plays an increasingly important role.

Multi-robot system research began in the 1970s, since then many foreign universities and research institutes extensively studied the multi-robot system, they

Supported by National Nature Science Foundation under Grant 61203335, and partly by National Natural Science Foundation of China (Nos. 61603150).

Y.-D. Zhang et al. (Eds.): ICMTEL 2020, LNICST 327, pp. 313–319, 2020.
https://doi.org/10.1007/978-3-030-51103-6_27

set up specifically for multi-robot system research project, MARTHA (Multiple Autonomous Robots for Transporting and Handling Application) in [12]. Under the study of foreign experts and scholars, this kind of flocking control has been widely applied in network topology, obstacle avoidance of autonomous mobile agents and other aspects. After nearly 40 years' research, flocking control has been well developed. In recent years, Chinese researchers have also begun to have a strong interest in studying robot swarm motion control. Flocking control has achieved remarkable results in military and other aspects.

2 Research and Analysis

Flocking control has been paid great attention by intelligent system theorists at home and abroad. Formation control refers to the control technology that multiple mobile robots maintain a certain formation while adapting to environmental constraints (such as the existence of obstacles or physical restrictions of space) when they reach the destination. Through research and development and practical application, the great application prospect of this technology in industrial and agricultural production, flexible manufacturing, unmanned exploration (ocean, space, nuclear environment), especially in the national defense industry is gradually reflected. These areas include moving large objects in fixed formations, encircling/capturing intruders in arcs, and completing space missions. Autonomous formation control is also applied to vehicles, mainly in formation movement according to the specified path, including collision avoidance between vehicles. This multi-robot formation is also used in exploration, rescue and environmental monitoring of unknown environments [17].

2.1 Flocking Control Based Leader-Follower

In the follower-leader formation, one robot plays the role of leader and has absolute control, coordinating with other robots, while other intelligent robots play the role of follower. The navigator is responsible for sending commands and navigation information to other robots, and then leads other robots to maintain a formation so that the robots in the formation will not collide with each other. At the same time, the artificial potential field method is used to maintain a certain distance and angle between the robots. The downside of this approach is that we don't get enough information, navigation information only, cannot obtain the following information such as the position and posture, and the lack of information communication between followers. It can cause the connectivity between the two is not good, when the navigator and followers have a communications problem between, both will lose connection, which can cause the failure of the formation, and are more prone to local minima. But the advantage is to the communication request is not so high, the real time is strong. A fuzzy controlled multi-agent distributed control is adapted in [1].

Due to the poor stability of the traditional formation method of leaders and followers, for example, when the communication between the leader and the following robot goes wrong, the formation of the whole robot will be wrong. In order to solve this problem, a method of re-electing leaders is proposed in [16]. The method of connection problems when the formation of the leader and other robots, the system will calculate the average energy of the whole robot team formation to level, and based on the energy level of each robot in the robot to elect a leader in the group, then discard leader of problems before, this will greatly improve the stability of the robot formation, significantly reduce formation error. The method in this paper is based on the leader follower control method proposed in [19].

A distributed control method for aggregation and formation of machine fish is proposed in [8]. Leaders in this approach have no external input. In this case, the fish inside need of attraction and repulsion method is introduced to avoid the fish within the collision, the number of leaders is far less than the follower, and the leader is not affected by followers, thus caused the unidirectional communication between leader and followers, in order to improve the school internal stability and avoid collisions, has also taken by reducing the speed and increase the intensity of potential function of individual fish.

The advantage of this method is that the behavior of the whole robot formation can be controlled only by the behavior and trajectory of the given navigator, and the control is simple. The disadvantage is that the navigator is not easy to get the tracking error feedback of the follower. If the leader moves too fast or the follower is blocked by the obstacle, the formation may be damaged, which will affect the quality of task completion in severe cases. In the case of leader failure, if the design is not fully considered and simply relies on it, the consequences will be very serious. Usually, this requires the adoption of strategies such as designated alternate leader and leader replacement according to the situation in the design process to improve the system capacity.

2.2   Behavior-Based Approach

Different from the traditional formation method, the behavior-based architecture does not need to be modeled, but directly outputs the expected speed and direction through sensors to feel the external stimuli, which has the advantages of fast response speed and strong adaptability to the environment. The behavior-based approach consists of three parts. When the robot movement through its positioning module for current position, and then through the communication module for data broadcasting, the robot for leadership position information of the whole system, finally, following its own position information and the position of the leader and by motion control module to realize their own update speed and angle. The triangle control method of multi-intelligent aircraft cluster is described in [11]. The whole aircraft group has the leader and the following aircraft. The leader is the center of the whole aircraft group, and the other following aircraft keep the relative position with the center aircraft. The standard linear quadratic method is adopted to design the final following aircraft of the aircraft group.

In [2], Olfati Saber algorithm is adopted, and the change of constant formula and gronwall inequality proves that the whole swarm system will increase the probability of collision within the swarm if the distribution of robots is too dense.

There is a self-triggering distributed control algorithm in [7]. Each agent can determine the adaptive sampling time, thus reducing communication time and controlling behavior time. Dynamic decoupling can also be carried out between multiple agents, so that each agent has a local controller, so as to alleviate the problem of irregular position distance and unstable speed between robots. This method makes the formation of the group more stable, and also alleviates the problem of collision between robots to some extent.

In [18], they focused on the cucker-smale multi-intelligence model prediction cluster control scheme, and mainly studied the cluster prediction problem of the model in the discrete time domain. The velocity can be changed by adjusting the control force, so that the velocity of the group can be unified and the movement of the group can be relatively stable The disadvantage of this method is that the group behavior is not clearly defined, it is difficult to carry out mathematical analysis and it is difficult to ensure the stability of formation.

2.3 Flocking Control Without Leader

A formation control of amigobot robot has been proposed in [10], which is a discontinuous cluster control for tracking γ-agent. The formation control of five robots was realized by tracking the trajectory of five γ-agents. The asymptotic stability of this approach has been demonstrated by treating a classical non-intact robot system as α-agent and then tracking γ-agent.

Haibo studied the flocking problem for flying robots in three-dimensional space in [9]. Each robot's attitude dynamics is considered. Motivated by consensus algorithms, they propose a distributed control law to achieve the flocking of flying robots. Each robot is regarded as an agent which make decisions just based on limited neighbours' information. Under the condition of undirected and connected communication topology, every robot moves in the same direction and same speed finally.

The advantage of this method is that it is easy to specify the behavior of the robot group (virtual structure behavior), and can make formation feedback. The disadvantage is that it requires the formation to be a virtual structure, which is relatively fixed and rigid, so it cannot consider the overall obstacle avoidance problem. Therefore, it is generally applied in the barrier-free environment.

2.4 Formation Control Algorithm

In view of amigobot, a nonlinear and incomplete system, Newton method is proposed for the cluster control and formation of robots. In [3], a method can help the artificial potential field method to jointly complete the adjacent robots to maintain a certain distance and avoid obstacles by acting together with the attractive and repulsive forces. It makes the swarm control and motion control of the robot more regular. This method makes the robots in the group react quickly and swarm well, but it is more affected by the environment.

Many previous clustering algorithms (including MPC clustering algorithm [21]) have involved the acquisition of robot position and speed state, but not all robots are equipped with speed sensors. Some studies have explored algorithms that only need location information. MPC clustering algorithm based on global measured location information is proposed in [22]. This not only saves the high cost of communication, but also improves the performance of the cluster.

In [6], it presents a distributed model predictive algorithm for a multi-agent system with communication delays between agents. Considering a multi-agent system in the presence of communication delays, we have proposed a novel DMPC in [5] based flocking algorithm by introducing the waiting mechanism to cope with communication delays. In the proposed flocking algorithm, the communication cost is greatly reduced because all the agents do not have to exchange information at each discrete sampling step. However, the proposed algorithm does not involve the external disturbances, which is another issue urgent to be solved.

Most of the methods to achieve multi-agent clustering control require the location and speed information of the robot, but Lu proposed a method of speed estimation and driving based on local location measurement in [13], which is conducive to cluster control and solving the problems of cluster speed estimation.

Previous cluster control did not do very well in terms of connectivity between robots. MAO proposed a distributed cluster control strategy with self-maintenance function for connectivity in [14]. Different from the previous method of fixing one edge, this method can allow any edge to be destroyed. It can be seen that this method improves the stability and flexibility of the whole system and makes the connectivity between robots better.

2.5 Formation Control Method Based on Virtual Structure

In [15], a network pinning strategy was proposed by analyzing the topological structure of the flocking agents and based on the knowledge of graph theory and control, which could greatly improve the clustering control of the flocking agents. This method applies the information transfer matrix to the multi-agent network topology. By using PBH criterion to set up specific nodes, a cluster control strategy is designed to improve the control of robot crowd.

In [4], it addresses the bipartite flock control problem with or without a virtual leader. In such a collective motion, the whole group separates into two clusters, in each of which all individuals move with the same direction. Meanwhile, every pair of agents in different clusters moves with opposite directions. Moreover, all agents in the two separated clusters approach a common velocity magnitude, and collision avoidance among each cluster is ensured as well.

A distributed event-triggered hybrid control is proposed in [20] to investigate flocking problem with a virtual leader in multi-agent systems. In the proposed control algorithm, the continuous relative position information are used while the relative velocity information are sampled at the instant determined by event-triggered mechanism. A distributed event-triggered hybrid control algorithm is proposed to investigate the flocking problem, in which continuous position and sampled-velocity information are utilized. The main contribution of this paper is that neighbors' information are transmitted only at discrete event-triggered instants for each agent. This method can ensure the connectivity of the system, make the speed of the robot in the system reach the same, and improve the stability of the system.

3   Discussion

Multi-robot distributed flocking control has been a hot topic in the field of robot research and it has been paid more and more attention by academic circles. Formation of robot swarm movement is still the focus and difficulty in the field of robot research, and it is necessary to further improve the autonomy and adaptability of robots, as well as enhance the communication and collaboration ability between multiple robots. Flocking control has been used in many fields: unmanned aerial vehicles, aerospace, logistics and transportation. Multi-robot path planning is an important research direction of navigation and control at any time. Autonomous formation control is also applied to vehicles, mainly in formation movement according to the specified path, including collision avoidance between vehicles. This multi-robot formation is also used in exploration, rescue and environmental monitoring of unknown environments. According to group of coalescence behavior rules, the robot has developed some control algorithm, the network topology, flocking in future studies should play a bigger role, especially in unmanned combat aircraft flight control algorithms, and it will greatly improves the operational capacity. Someday in the future, it can coordinate to complete the mission of the fleet and reduce the casualty rate and time. It is very useful.

References

1. Bhowmick, C., Behera, L., Shukla, A., Karki, H.: Flocking control of multi-agent system with leader-follower architecture using consensus based estimated flocking center. In: Conference of the IEEE Industrial Electronics Society (2016)
2. Bian, W., Zhou, J., Qian, H., Lu, X.: Further properties of second-order multi-agent flocking under olfati-saber's algorithms. In: Control Conference (2016)
3. Cheng, J., Wang, B., Xu, Y.: Flocking control of amigobots with newton's method. In: 2017 IEEE International Conference on Robotics and Biomimetics (ROBIO), pp. 2372–2376, December 2017. https://doi.org/10.1109/ROBIO.2017.8324774
4. Fan, M.C., Zhang, H.T.: Bipartite flock control of multi-agent systems. In: Control Conference (2013)
5. Gurtovenko, A.A., Patra, M.M., Vattulainen, I.: Cationic DMPC/DMTAP lipid bilayers: molecular dynamics study. Biophys. J. **86**(6), 3461–3472 (2004)

6. Hu, Y., Zhan, J., Yuan, Q., Li, X.: A multi-agent flocking system with communication delays via distributed model predictive control. In: 2017 36th Chinese Control Conference (CCC), pp. 8449–8454, July 2017. https://doi.org/10.23919/ChiCC.2017.8028696

7. Hu, Y., Zhan, J., Li, X.: Self-triggered distributed model predictive control for flocking of multi-agent systems. IET Control Theory Appl. **12**(18), 2441–2448 (2018)

8. Jia, Y., Long, W.: Leader-follower flocking of multiple robotic fish. IEEE/ASME Trans. Mechatron. **20**(3), 1372–1383 (2015)

9. Jian, D., Haibo, J., Kun, L., Kaihong, Y., Wang, Y.: Flocking control of flying robots considering model's dynamics processes. In: 2017 36th Chinese Control Conference (CCC), pp. 817–821, July 2017. https://doi.org/10.23919/ChiCC.2017.8027445

10. Jin, C., Yong, Z., Hui, Q.: A tracking control method for flocking of amigobots. In: Control Conference (2015)

11. Joelianto, E., Sagala, A.: Swarm tracking control for flocking of a multi-agent system. In: Control, Systems & Industrial Informatics (2012)

12. Kvarchelia, L., Gaina, A.: Questions to sergej kovalev. Questions to Sergej Kovalev (2008)

13. Lu, X., Jin, Z., Zhou, J., Qin, B., Qian, H.: Flocking control of multi-agents based on self-adaptively weighting observers driven only by local position measurements. In: Control & Decision Conference (2017)

14. Mao, Y., Dou, L., Fang, H., et al.: Distributed flocking of Lagrangian systems with global connectivity maintenance. In: 2013 IEEE 3rd Annual International Conference on Cyber Technology in Automation, Control and Intelligent Systems (CYBER). IEEE (2013)

15. Mei, Y., Chen, S.: Flocking algorithm for directed multi-agent networks via pinning control. In: Chinese Automation Congress (2016)

16. Prasad, B.K.S., Manjunath, A.G., Ramasangu, H.: Flocking trajectory control under faulty leader: Energy-level based election of leader. In: IEEE International Conference on Power Electronics (2017)

17. Reyes, L.A.V., Tanner, H.G.: Flocking, formation control, and path following for a group of mobile robots. IEEE Trans. Control Syst. Technol. **23**(4), 1268–1282 (2015)

18. Wu, W., Liu, B., Zhang, H.T.: Model predictive flocking control for the cucker-smale multi-agent model. In: International Conference on Control (2017)

19. Yazdani, S., Haeri, M., Su, H.: Sampled-data leader-follower algorithm for flocking of multi-agent systems. IET Control Theory Appl. **13**(5), 609–619 (2019). https://doi.org/10.1049/iet-cta.2018.5533

20. Yu, P., Ding, L., Liu, Z.W., Guan, Z.H., Hu, M.X.: Flocking with a virtual leader based on distributed event-triggered hybrid control. In: Control Conference (2013)

21. Zhan, J., Li, X.: Decentralized flocking protocol of multi-agent systems with predictive mechanisms. In: Proceedings of the 30th Chinese Control Conference, pp. 5995–6000, July 2011

22. Zhan, J., Li, X.: Flocking of multi-agent systems via model predictive control based on position-only measurements. IEEE Trans. Ind. Inform. **9**(1), 377–385 (2013)

Detection of High Voltage Transmission Lines: A Survey and Perspective

Xiaoyuan Wang, Cheng Jin$^{(\boxtimes)}$ ⓘ, and Weijie Huang

University of Jinan, Jinan 250022, People's Republic of China
cse_chengj@ujn.edu.cn

Abstract. With the development of the national economy, the demand for electricity in various industries is expanding. It is necessary to ensure the safe operation of the high voltage transmission line. How to prevent and detect natural disasters and accidents that endanger transmission lines in a timely manner has become an important basic work to ensure power supply. Identifying high-voltage transmission lines first requires mathematical modeling of high-voltage transmission lines. Based on the mathematical model constructed, the image processing method is used to remove the blurred images in the images and restore the true background of the images. The establishment of mathematical models for high-voltage transmission lines has been relatively complete. This paper focuses on the analysis of existing methods for automatic identification and localization of foreign bodies on transmission lines and the existing research on deblurring, de-fogging, image denoising, image enhancement, etc. method. With the rapid development of deep learning, there are more and more methods for identifying high-voltage transmission lines and image restoration. More people will be engaged in this research in the future.

Keywords: Transmission line · Image restoration · Deep learning

1 Introduction

Images are the foundation of human face vision, and they give people a concrete and intuitive effect. Data digitization includes two parts: sampling and quantization. Digital image processing is the process of converting image signals into digital formats and using computers for processing and processing. Image restoration is an important issue in image processing, and it is of particular significance for improving image quality. The key to solving this problem is to establish a corresponding mathematical model for the degradation process of the image, and then obtain the restoration model of the image by solving the inverse problem and make a reasonable estimate of the original image.

Supported by National Nature Science Foundation under Grant 6120333 and 603150.

Y.-D. Zhang et al. (Eds.): ICMTEL 2020, LNICST 327, pp. 320–326, 2020.
https://doi.org/10.1007/978-3-030-51103-6_28

Image artificial intelligence processing can solve a series of hidden danger identification problems such as drones, video surveillance, and image surveillance. It is of great significance to improve the level of transmission management and personnel quality and efficiency. It also responds to the national and national network artificial intelligence development plans. Based on the analysis of this paper and the summary of existing methods, the foreign line and fault detection methods of transmission lines based on deep convolutional neural networks are analyzed and summarized.

2   Target Detection Methods

2.1   R-CNN

R-CNN uses AlexNet's network architecture and uses the Selective Search technology to generate Region Proposal. R-CNN is pre-trained on ImageNet, then fine-tune is performed on the PASCAL VOC dataset using mature weight parameters, then features are extracted using CNN, and then a series of SVMs are used for category prediction. Finally, R-CNN's bbox position regression was inspired by DPM and trained a linear regression model. The semantic segmentation of R-CNN uses CPMC to generate Region. Since RCNN comes earlier, it is not the best target detection model and semantic segmentation model.

2.2   SPP-net

The biggest improvement of SPP-net for R-CNN is that the feature extraction step has been modified, and other modules are still the same as R-CNN [12]. Feature extraction no longer needs to pass through the CNN for each candidate region. It only needs to input the entire image to the CNN. The ROI feature is directly obtained from the feature map. Compared with R-CNN, the speed has increased by a hundred times [5]. The shortcomings of SPP-net are also obvious. The conv layer in CNN cannot continue training when fine-tuning. It is still the framework of R-CNN.

2.3   Fast R-CNN

Sometimes, good results are not necessarily all original [17]. Fast R-CNN is a good illustration. SPPnet's pooling thought has been simplified and promoted on Fast, and author rbg further based on R-CNN [8]. The detection frame regression is integrated into the neural network, which makes the training test rate of Fast greatly improved.

2.4   Yolo

The core idea of YOLO is to use the whole graph as the input of the network, directly returning to the output layer and the position of the bounding box and its associated category [11]. The faster-RCNN also uses the entire graph as input directly, but the fast-RCNN uses the idea of RCNN's proposal + classifier as a whole, except that the steps to extract the proposal are implemented in CNN, while YOLO uses direct The idea of returning YOLO is not good for objects that are close to each other, and there is a small group detection [15]. This is because only two boxes are predicted in one grid and belong to only one category. The generalization ability is weak when new uncommon aspect ratios and other conditions occur for the same type of object. Due to the loss function problem, the positioning error is the main reason that affects the detection effect. Especially the handling of large and small objects has yet to be strengthened.

2.5 SSD

By comparing the network structure of SSD and YOLO, we can find that the advantage of SSD is that the default box generated by it is multi-scale [9]. This is because the feature map of the default box generated by SSD is not only the last layer of CNN output, but also the utilization. Compare the default box generated by the shallow feature map. Therefore, SSD detection of small targets will certainly be better than YOLO v1 (small targets almost disappear after high-level convolution). At the same time, because the multi-scale default box generated by SSD must have a higher probability to find a candidate box that is closer to Ground Truth, the stability of the model is definitely stronger than YOLO (YOLO's bounding box is very few, only 98, if Far from the GT, then the linear regression of the modified bounding box is not established, and the model may run when training). However, the number of candidate frames for SSDs is the highest among the three classic networks, with 8732, so it should be slower when training [9].

2.6 Yolov2

Although YOLOv1 has a fast detection speed, its detection accuracy is not as good as that of R-CNN. YOLOv1 is not accurate enough in object localization and has a low recall rate. YOLOv2 has proposed several improvement strategies to improve the positioning accuracy and recall rate of the YOLO model, thus improving mAP. YOLOv2 follows a principle in the improvement: maintaining the detection speed, which is also a big advantage of the YOLO model. It can be seen that most of the improved methods can significantly improve the mAP of the model.

2.7 Yolov3

Yolov3 solves the problem of difficult identification of small objects on the basis of v2, and is by far the most accurate framework for target detection.

2.8   Centernet

CenterNet's "anchor" only appears at the current target's position instead of the entire picture, so there is no such thing as a box anchor larger than a positive anchor, and there is no need to distinguish whether the anchor is an object or a background. Because each target corresponds to only one "anchor", this anchor is extracted from the heatmap, so no NMS is needed to filter the output resolution of CenterNet. The downsampling factor is 4, which is compared to other target detection frameworks. Smaller (Mask-Rcnn is at least 16 and SSD is at least 16). In general, the CenterNet structure is elegant and simple. It directly detects the center point and size of the target and is truly anchor-free.

3   Image Restoration Methods

Image restoration technology is a very important type of processing technology in the field of image processing. Similar to other basic image processing technologies such as image enhancement, it is also aimed at obtaining a certain degree of improvement in visual quality. The difference is that the image restoration process is actually An estimation process needs to restore the degraded image according to some specific image degradation models. In short, the process of image restoration is to improve the quality of degraded images, and to improve the visual improvement of images through the improvement of image quality. As there are many factors causing image degradation and their properties are different, there is no unified restoration method at present. Many researchers have adopted different degradation models, processing techniques and estimation criteria according to different corresponding physical environments, thus obtaining different restoration methods.

3.1   Deep Learning Method

Some scholars propose a six-layer convolutional neural network [14] for target recognition, and propose a localization algorithm based on output map information according to the characteristics of the transmission line itself. Firstly, the sliding window [3] method is used to make the convolutional neural network recognize each window of the input picture and obtain the output picture. Then, the output picture is binarized and opened, and finally the corresponding target is located according to the output picture information. In the process of network training, the image enhancement algorithm is used to expand the captured image of the transmission line fault model, and a training set consisting of 17,000 picture blocks and a detection set composed of 4000 picture blocks are obtained, which are used for network training and network detection. The set correct rate is 88.68% in the 125 position detection pictures, the transmission lines and related faults in 91.2% of the pictures can be detected. The experimental results show that in the picture with complex background, the deep learning algorithm can accurately identify [13] and locate the transmission line and its faults, and improve the versatility and applicability of the detection algorithm. It has practical significance for improving the detection efficiency of transmission line faults and improving the detection accuracy.

3.2 Method Based on Aerial Photography of Drone Control

Channel a priori is also a sparse priori that is manually designed by static statistically blurred images and sharp images. Different from the gray feature and the gradient feature of the direct statistical image, the channel prior is the pixel passing through the statistical channel. In 2009, He first proposed a dark channel prior and used it in image defogging, which made a breakthrough in image defogging [4]. An image dark channel means that the pixel value of at least one channel in the channel of the color image of the image block approaches zero. In 2016, Pan et al. first proposed applying the dark channel prior to the image deblurring field, pointing out that the clear image has a more sparse dark channel value than the blurred image, establishing dark channel sparsity.

3.3 Deblurring Algorithm Based on Edge Estimation

The core idea of blind deconvolution algorithm based on edge estimation is to estimate the image with large edge and small edge suppressed explicitly through some filtering and image enhancement algorithms, and then estimate the fuzzy kernel based on these significant edges [6]. The algorithm based on edge estimation is fast and has been proved to be very effective in practical application, but it is difficult to analyze, because it is not based on the model through optimization, but through some combination of heuristic steps [10].

The reason why the blind deconvolution algorithm based on edge estimation can successfully estimate the fuzzy kernel is that it can estimate the large scale edge in the image and suppress the small edge. The large scale edge is more beneficial to the estimation of the real fuzzy kernel, while the small edge plays an opposite role [7]. A bilateral filter is used to extract the large scale edge of the image and the extracted edge image is used for fuzzy kernel estimation.

3.4 Image Deblurring Based on Richardson-Lucy

Aiming at the ringing effect of RL algorithm, a new image deblurring algorithm based on RL was proposed, and Yuan's Gain Map was introduced in the iterative process, which effectively suppressed the ringing effect of the flat area and further amplified the image noise, and kept the details of the image [19]. At the same time, the effect of different parameter selection of the gain graph on the defuzzing result is discussed [16]. Experimental results show that the algorithm is effective in suppressing the ringing effect, preserving the image details, and recovering the fuzzy image with noise [20].

3.5 Image Deblurring Combined with Total Variation and Fractional Total Variation Model

In order to recover more details and texture information from fuzzy images, a digital image deblurring method based on combined total variation (TV) and fractional order total variation (FOTV) models is proposed [2]. The fuzzy image is decomposed into smooth region, convex edge and texture by the global gradient extraction method, the smooth region and convex edge are constrained by the full variation model, the details are constrained by the fractional order full variation model, the de-fuzzy convex optimization model is established, and the variable splitting and alternating direction method is used to quickly solve the model [1]. Experimental results verify the validity and rapidity of the model and algorithm [18].

4 Discussion

The safe operation of transmission lines is closely related to our daily lives. The country also invests a lot of money every year to ensure the normal operation of the power grid, but the transmission lines are high-voltage lines, and the cost of manual operations continues to increase and risks continue to increase. The perspective of artificial intelligence analyzes methods that can reduce human risks and provide guarantees for the development of the national economy.

References

1. Chen, F., Jiao, Y., Lin, L., Qin, Q.: Image deblurring via combined total variation and framelet. Circuits Syst. Signal Process. **33**(6), 1899–1916 (2014)
2. Gao, C., Zhang, J., Li, D.: Fractional-order total variation combined with sparsifying transforms for compressive sensing sparse image reconstruction. J. Vis. Commun. Image Represent. **38**(C), 407–422 (2016)
3. Golab, L.: Processing sliding window multi-joins in continuous queries over data streams. In: International Conference on Very Large Data Bases (2003)
4. He, K., Jian, S., Tang, X.: Single image haze removal using dark channel prior. In: IEEE Conference on Computer Vision & Pattern Recognition (2009)
5. He, K., Zhang, X., Ren, S., Sun, J.: Spatial pyramid pooling in deep convolutional networks for visual recognition. IEEE Trans. Pattern Anal. Mach. Intell. **37**(9), 1904–1916 (2015)
6. Hongbo, Z., Liuyan, R., Lingling, K., Xujia, Q., Meiyu, Z.: Single image fast deblurring algorithm based on hyper-laplacian model. IET Image Proc. **13**(3), 483–490 (2018)
7. Lee, D.B., Jeong, S.C., Lee, Y.G., Song, B.C.: Video deblurring algorithm using accurate blur kernel estimation and residual deconvolution based on a blurred-unblurred frame pair. IEEE Trans. Image Process. **22**(3), 926–940 (2013)
8. Li, J., Liang, X., Shen, S.M., Xu, T., Feng, J., Yan, S.: Scale-aware fast r-cnn for pedestrian detection. IEEE Trans. Multimedia PP(99), 1 (2015)
9. Liu, W., Anguelov, D., Erhan, D., Szegedy, C., Reed, S., Fu, C.Y., Berg, A.C.: Ssd: Single shot multibox detector. In: European Conference on Computer Vision (2016)

10. Mazinan, A.H., Karimi, A.: Block-based noise variance estimation algorithm in blurred and noisy images with its application to motion deblurring. Evol. Syst. **8**(2), 95–108 (2015). https://doi.org/10.1007/s12530-015-9134-4
11. Montgomery, M.J., Thiemann, T., Macedo, P., Brown, D.A., Scott, T.W.: Blood-feeding patterns of the culex pipiens complex in sacramento and yolo counties, california. J. Med. Entomol. **48**(2), 398–404 (2011)
12. Purkait, P., Zhao, C., Zach, C.: SPP-Net: deep absolute pose regression with synthetic views. arXiv, abs/1712.03452 (2017)
13. Rosenfeld, N., Aharonov, R., Meiri, E., Rosenwald, S., Spector, Y., Zepeniuk, M., Benjamin, H., Shabes, N., Tabak, S., Levy, A.: Micrornas accurately identify cancer tissue origin. Nat. Biotechnol. **26**(4), 462–469 (2008)
14. Schmidhuber, J.: Deep learning in neural networks: an overview. Neural Netw. **61**, 85–117 (2015)
15. Sommer, T., et al.: California's yolo bypass: evidence that flood control can be compatible with fisheries, wetlands, wildlife, and agriculture. Fisheries **26**(8), 6–16 (2001)
16. Wang, H., Miller, P.C.: Scaled heavy-ball acceleration of the richardson-lucy algorithm for 3D microscopy image restoration. IEEE Trans. Image Process. Publicat. IEEE Sig. Process. Soc. **23**(2), 848–54 (2014)
17. Xiang, W., Ma, H., Chen, X.: Salient object detection via fast R-CNN and low-level cues. In: IEEE International Conference on Image Processing (2016)
18. Xuan, Z., Ning, W., Lin, E., Li, Q.: Image decomposition model combined with sparse representation and total variation. In: IEEE International Conference on Information & Automation (2013)
19. Yang, H., Huang, P.: ShangHong: a novel gradient attenuation richardson–lucy algorithm for image motion deblurring. Signal Processing **103**, 399–414 (2014)
20. Zhao, B., Zhang, W.: Novel image deblurring algorithm based on richardson-lucy. Comput. Eng. Appl. (2011)

Image Processing-Based Electronic Fence: A Review

Xiaoyuan Wang, Weijie Huang$^{(\boxtimes)}$, and Qinjun Zhao

School of Electrical Engineering, University of Jinan,
Jinan 250022, People's Republic of China
{cse_huangwj,cse_zhaoqj}@ujn.edu.cn

Abstract. With the development of science and technology, using electronic fence to replace traditional physical fence becomes a trend. Research in the image processing-based electronic fence is gaining more popularity due to its low cost, low power consumption, and intelligence. In this paper, we are interested in the study of three types of scenarios with an image processing-based electronic fence. The scenarios contains national border protection, safety management in manufacturing field and vehicles safety plans. The state of the art frameworks of the three scenarios are reviewed, and the trend of these fields is also discussed.

Keywords: Image processing · Electronic fence · National border protection · Safety management in manufacturing field · Vehicles safety plans

1 Introduction

Nowadays, physical fence has the following problems:

- The physical fence height is not high enough so that it is easily to be crossed by intruder.
- Physical fences do not have monitoring character so that it is impossible to record which intruders are cross the fences.
- Physical fences have no automatic alarm character so that the offender can not stop the behavior of intrusion.

These problems have led to safety precautions that can not meet its requirements. With the development of electronic technology, sensing technology and computer technology, security precaution has gradually developed into a specialized public security technology discipline [48]. Among them, image processing-based electronic fence is a high-tech way that can not be ignored in security technology. Intrusion detection and alarm systems based on electronic fence have been widely used in national border, banks, financial departments, museums, office

Supported by key research and development projects of Shandong province (2019GNC106093, 2019QYTPY043).

buildings, manufacturing field, important workshops, office buildings, hotels and other areas [21]. Video surveillance system [22] has been widely used in electronic fence, and the core of video surveillance is image processing which can perform intelligent analysis based on images so that a large number of manpower is reduced.

Image processing-based electronic fence has five main advantages compared with traditional physical fence:

- First, the key technology of image processing-based electronic fence is to analyze the images in the scene of the monitoring area. This analysis can be used to analyze the specific area to achieve regional intrusion detection based on the algorithm without changing the physical device.
- Second, image processing-based electronic fence does not need significant intrusion detection devices to establish monitoring points while it only needs to analyze the collected images. The camera can be installed in a secret place so that the risk of equipment damage is reduced and remote detection is achieved.
- Third, image processing-based electronic fence analyses images captured in target area. No matter which direction the intruder enters the monitoring area, the alarm rings when the system detects the intrusion. This way can eliminate any form of bypassing the detector and achieve no omission alarm.
- Fourth, the monitoring area can be arbitrarily specified in image processing-based electronic fence so that it is easy to be modified without any additional cost.
- Finally, image processing-based electronic fence can not only detect the intrusion but also implement tracing objects compared with traditional physical fence.

Nowadays, image processing-based electronic fence with flexible settings, less false alarms, strong anti-interference ability and compatibility with each interface has become one of the hottest research areas.

The most important part of image processing-based electronic fence is objects detection [8]. Objects detection is to find all the objects (objects) of interest in the image and to determine their position and size. Objects detection has always been the most challenging problem in machine vision because of the different appearances, shapes, poses, and the interference of illumination and occlusion during processing. Objects detection focus on the following problems: a) objects may appear anywhere in the image; b) objects are available in a large number of different sizes from an image; c) objects may express a variety of different shapes in the image. Some research works on object detection based on Convolution Neural Network (CNN) [26] are recorded in Table 1. Based on the results of objects detection, the intruders are detected whether they exceeds the limits specified by the electronic fence. Next section we are interested in the study of the framworks of image processing-based electronic fence in national border protection [18], safety management in manufacturing field [36] and vehicles safety plans [1].

Table 1. Some results of objects detection are summarized.

Detector	VOC07 (mAP@IoU = 0.5)	VOC12 (mAP@IoU = 0.5)	COCO (mAP)
R-CNN [14]	58.5	–	–
OverFeat [37]	–	–	–
SPP-Net [17]	59.2	–	–
Fast R-CNN [13]	70.0	68.4	–
Faster R-CNN [35]	73.2	70.4	–
YOLO v1 [32]	66.4	57.9	–
AZNet [27]	70.4	–	22.3
ION [2]	80.1	77.9	33.1
OHEM [39]	78.9	76.3	22.4
SSD [25]	76.8	74.9	–
R-FCN [5]	79.5	77.6	29.9
DSSD [11]	81.5	80.0	–
YOLO v2 [33]	78.6	73.4	–
DeNet [43]	77.1	73.9	33.8
CoupleNet [52]	82.7	80.4	34.4
YOLO v3 [34]	–	–	33.0
RefineDet [51]	83.8	83.5	41.8

2   The Frameworks of Image Processing-Based Electronic Fence

2.1   National Border Protection

Border security refers to the protection of national borders from the illegal movement of goods, drugs, weapons and humans. It makes that trade and legal travel can be maintained and anti-terrorism protection can be provided globally. National border protection system [45] is employed to monitor the intrusion events around the fence and determine whether suspicious activities are carried out. If any suspicious event occurs, a set of scheduled tasks, such as warning or combat systems, are performed. Image processing-based electronic fence is an integral part of national border protection system. It aims to monitor hostilities, detect and track intruders, analyse their behaviors based on serial image sequences. This electronic fence insteads watching and monitoring 24×7 surveillance task by humans and the framework can be proved very useful to generate Intruder Detection System (IDS) [38] automatic alarms.

Vittal et al. [46] propose an intrusion detection and automatic shooting device controlled by computer to improve border security. Camera images are transfered from Video to computer in real-time through the USB bus, and then color images are convert to gray images. The current gray image is compared with the reference image to find out the changes between object and background so that the intruders are searched out based on these changes. It also intergrates an automatic shooting mechanism for automatic positioning and shooting targets. This framework has strong ability to suit to worse environment and saves a lot of manpower. KeremIrgan et al. [18] propose a cost-effective approach for dynamic (on-the-fly) prioritization of image macro-blocks. Data blocks are labeled as "important" and "not-important" to use impel encoding scheme at the source node. Various prioritization measurements and labeling method based on threshold can successfully improve efficiency and save time when the framework performs JPEG encoding. Raheja et al. [31] propose an approach of detecting instruders in hilly region. The instrusion can be noted no matter daytime or in night because the kinect sensor integrates infrared camera and color camera. The framework can easily distinguish animal and human with skeletal labeling approach provided by Kinect. Standing, walking, crawling, and bending etc. can be recognized based on the HMM. Deshmukh et al. [6] propose a three layer image processing-based electronic fence which each layer runs parallelly and the results of all layers are merged by a module. Two cameras are embedded into between first and second layer and provided image information for the framework. This electronic fence can effectively monitoring evil activities. Kim et al. [19] propose an image processing-based electronic fence with optical and thermal cameras. The electronic fence is generated by curve fitting and human-computer interactive processing. A convolutional neural network (CNN) is used to distinguish humans and animals and the results of classification is accurate. A long-term recurrent convolutional network (LRCN) model is also used in this framework to record five-types-behavior. The proposed method can be considered as an automatic protection system.

2.2 Safety Management in Manufacturing Field

Intrusion is an unauthorized entrance into dangerous areas without being aware of the potential hazards. Intrusion not only causes unauthorized workers to be suffered from accidents, but also disturbs or even injures other authorized workers in dangerous areas [4]. Traditional safety management relies mainly on the inspection of security officials and it is very time-consuming [28]. Image processing-based electronic fence solves this problem by integrating and processing information about the behaviors of workers and the scene environment [50]. The main aspects of image processing-based electronic fence building safety monitoring include: (1) on-site work behavior monitoring; (2) on-site environmental monitoring, and (3) information integration, analysis and early warning [40].

Naticchia et al. [29] propose an intelligent video surveillance system to address conflict issues that may arise among teams operating on building-sites. Actions in scene is tracked by image processing and the analyzing result is transformed by ZigBee. Real-time site state visualization and remote interaction with security inspectors are also designed. In this approach, the framework is considered to be cover a wide range of areas while maintaining its non-intrusive functions because a new low-power method is implemented. Revaud et al. [36]

propose a novel approach for detecting workers or other equipments with significant occlusions. The optical flow field is estimated by motion model. Edge-preserving interpolation and variational energy minimization are combined to obtain geodesic distance. This approach used in electronic fence can deal with the occlusion problem effectively. Fang et al. [9] develops a cognitive model of construction workers' unsafe behaviors (CM-CWUB). This module contains five steps: getting information, understanding information, perceiving response, choosing response and taking action. Works and fences are located by Building Information Modeling (BIM) approach. The framework can provide early warning and active protection. Tang et al. [42] propose that indoor industrial production space is divided into three different semantics (structure, connectivity and volume). Semantic information is generated by image segmentation, and then the unsafety areas are defined. Combining with action recognition, the framework can easily find the intruders.

2.3 Vehicles Safety Plans

There are two main types of uses of Image processing-based electronic fence in the field of vehicles safety plans. One is dockless bike-sharing services, the other is high-speed railway instrusion detection.

Bike-sharing is becoming a more popular form of transport in the world. With the increasing popularity of various types of shared bicycles, the problem of random parking has become more and more serious. These uncivilized actions have some adverse effects on the social environment and atmosphere [7]. It is impossible to manage 24 h a day by human resources testing alone. Despite a series of policies, the sharing of bicycles has not been alleviated. Image processing-based electronic fence can push reliable detection and warning information to traffic management, which greatly reduces the manual workload. Lee et al. [23] propose a static objects tracking and detection approach based on foreground model. The time for a pixel to remain foreground in this foreground model is determined by the brightness of the image. The longer the time means the brighter the position of the current pixel in the model. The framework also use $1-D$ transformation and image projection to reduce the dimensionality of image data. Thus bicycles and parking areas are detected in real-time. Xie et al. [49] also propose a foreground bicycles tracking approach based on CNN. The vehicle in the input image is detected by SSD+ROI (Region of Interest), and all vehicles in the ROI are tracked and counted once they stop moving. However, this method has higher algorithm complexity and poor real-time performance. Bock et al. [3] propose an objects detection approach based on static corner matching. The basic idea is to select the Harris corner according to the characteristics of the bicycles on each frame in the video. Then the static corner points is extracted according to the motion state of the Harris corners. However, this method requires a large number of training sets to learn in advance to achieve the interference of removing the static corners in the image and the calculation amount is too large.

Electronic fence used in high-speed railway instrusion detection needs to provide many features, which contains day and night surveillance, high-definition,

Table 2. Some researches to deal with various environmental effects are summarized.

Approach	Environment	Day or night	Features
[16]	Haze	Day	Dark channel priority Minimum filtering estimate for propagation
[20]	Rain	Day	Detecting the gradient of the brightness of the raindrops pixel
[24]	Rain	Day	Presenting a Dynamic Routing Residue Recurrent Network The context information is embedded into the network
[30]	Snow	Day	Using saturation and visibility characteristics
[44]	Sunny	Night	Glare-suppressed inter-channel fusion A deep KNN framework for joint classification Corlor and infrared information fusion
[41]	Sunny	Night	A hierarchical selforganizing network No-reference-based performance evaluation metrics
[12]	Sunny	Night	Fusing saliency map and colour information A proving approach for virtual tail-lamp
[10]	Sunny	Night	Using Deep Convolutional Neural Networks and Kernelized Correlation Filters

full coverage, system linkage, foreign object intrusion monitoring, intelligent identification and analysis, effective discovery, drive-out, prevention of climbing over the invasion [7]. It is the most direct, effective and credible party for railway perimeter prevention. There are many researches in various environmental effects and some are summarized in Table 2. After removing the impact of the special environment, Electronic fence detecting instruders can be handled. Guo et al. [15] propose an accurate and effective intrusion detection. They improve the average accuracy of Single Shot multibox Detector (SSD) with the fusion between high-level features and low-level features. Semantic information are enhanced based on the fusion results. The network are also clipped using the L1 norm so that the real-time performance is improved. Wang et al. [47] propose an adaptive segmentation algorithm to implement the function of the electronic fence on the high-speed railway. They first use hough-transform to generate adaptive direction of each Gaussian kernel. Then, according to the boundary weight and size of the region, a new clustering approach is used to merge small fragments to candidate regions. Last, the boundary of electronic fence is classified by a deep network. The framework can execute seamlessly on memory-constrained, tiny embedded devices.

3 The Trends of Image Processing-Based Electronic Fence

Image processing-based electronic fence must deal with massive video surveillance data, which brings huge challenges to intelligent system applications. These challenges can be summarized as cross-scenario, cross-media, and cross-space. The trends of image processing-based electronic fence contains high efficiency, high coordination and high initiative.

High Efficiency. The new electronic fence can be combined and adjusted automatically at any time. An orderly docking is formed between the dynamic monitoring tasks and the ubiquitous camera resources so the framework need to use distributed monitoring system. On the one hand, it can enhance the scalability of the monitoring network and can increase the monitoring terminal of different functions, on the other hand, it can improve the working efficiency and response speed.

High Coordination. The data from the physical space and the network space are fused. A three-dimensional perception and early warning mode of dual spatial cooperative sensing and active modulation are established. This can help the new electronic fence to better detect intruders.

High Initiative. Extracting various characteristics of public security events from different media data, including time, place, people, timing, relationship, etc., improves the integrity of semantic expression, and it is more reliable for event detection. The new electronic fence can learn massive cross-media data through robust structure, and the related information can express the multi-faceted features of the same event, thus this fence improves the reliability of monitoring network early warning.

4 Conclusions

This paper reviews some works of image processing-based electronic fence. We discuss research status of objects detection, the framworks in three scenarios and some technologies for eliminating the effects of bad weather. The development trend of electronic fence is forecasted and some advices are put forward with summarizing the characteristics of these studies.

References

1. Althoff, M., Althoff, D., Wollherr, D., Buss, M.: Safety verification of autonomous vehicles for coordinated evasive maneuvers. In: 2010 IEEE Intelligent Vehicles Symposium, pp. 1078–1083 (2010)
2. Bell, S., Lawrence Zitnick, C., Bala, K., Girshick, R.: Inside-outside net: detecting objects in context with skip pooling and recurrent neural networks. In: Proceedings of the IEEE Conference on Computer Vision and Pattern Recognition, pp. 2874–2883 (2016)
3. Bock, F., Liu, J., Sester, M.: Learning on-street parking maps from position information of parked vehicles. In: Sarjakoski, T., Santos, M.Y., Sarjakoski, L.T. (eds.) Geospatial Data in a Changing World. LNGC, pp. 297–314. Springer, Cham (2016). https://doi.org/10.1007/978-3-319-33783-8_17
4. Cocca, P., Marciano, F., Alberti, M.: Video surveillance systems to enhance occupational safety: a case study. Saf. Sci. **84**, 140–148 (2016)
5. Dai, J., Li, Y., He, K., Sun, J.: R-FCN: object detection via region-based fully convolutional networks. In: Advances in Neural Information Processing Systems, pp. 379–387 (2016)
6. Deshmukh, R., Kamdi, S., Pingle, M., Rajebhosale, S., Bhosale, A.: Intelligent surveillance system using energy efficient intrusion detection and tracking techniques. In: 2018 Second International Conference on Electronics, Communication and Aerospace Technology, pp. 1214–1218 (2018)
7. Dong, L., Qin, L., Xu, W., Jiang, H.: Detection method for vehicles in tunnels based on surveillance images. In: 2017 4th International Conference on Transportation Information and Safety, pp. 836–843 (2017)
8. Druzhkov, P., Kustikova, V.: A survey of deep learning methods and software tools for image classification and object detection. Pattern Recognit. Image Anal. **26**(1), 9–15 (2016)
9. Fang, D., Zhao, C., Zhang, M.: A cognitive model of construction workers' unsafe behaviors. J. Constr. Eng. Manag. **142**(9), 04016039 (2016)
10. Fernando, H., Perera, I., de Silva, C.: Real-time human detection and tracking in infrared video feed. In: 2019 Moratuwa Engineering Research Conference, pp. 111–116 (2019)
11. Fu, C.Y., Liu, W., Ranga, A., Tyagi, A., Berg, A.C.: DSSD: deconvolutional single shot detector. arXiv preprint arXiv:1701.06659 (2017)
12. Gao, F., Ge, Y., Lu, S., Zhang, Y.: On-line vehicle detection at nighttime-based tail-light pairing with saliency detection in the multi-lane intersection. IET Intell. Transport Syst. **13**(3), 515–522 (2018)
13. Girshick, R.: Fast R-CNN. In: Proceedings of the IEEE International Conference on Computer Vision, pp. 1440–1448 (2015)
14. Girshick, R., Donahue, J., Darrell, T., Malik, J.: Rich feature hierarchies for accurate object detection and semantic segmentation. In: Proceedings of the IEEE Conference on Computer Vision and Pattern Recognition, pp. 580–587 (2014)
15. Guo, B., Shi, J., Zhu, L., Yu, Z.: High-speed railway clearance intrusion detection with improved ssd network. Appl. Sci. **9**(15), 2981 (2019)
16. He, K., Sun, J., Tang, X.: Single image haze removal using dark channel prior. IEEE Trans. Pattern Anal. Mach. Intell. **33**(12), 2341–2353 (2010)
17. He, K., Zhang, X., Ren, S., Sun, J.: Spatial pyramid pooling in deep convolutional networks for visual recognition. IEEE Trans. Pattern Anal. Mach. Intell. **37**(9), 1904–1916 (2015)

18. Irgan, K., Ünsalan, C., Baydere, S.: Low-cost prioritization of image blocks in wireless sensor networks for border surveillance. J. Netw. Comput. Appl. **38**, 54–64 (2014)
19. Kim, S.H., Lim, S.C., et al.: Intelligent intrusion detection system featuring a virtual fence, active intruder detection, classification, tracking, and action recognition. Ann. Nucl. Energy **112**, 845–855 (2018)
20. Krishnan, S., Venkataraman, D.: Restoration of video by removing rain. Int. J. Comput. Sci. Eng. Appl. **2**(2), 19 (2012)
21. Krüger, J., Nickolay, B., Heyer, P., Seliger, G.: Image based 3d surveillance for flexible man-robot-cooperation. CIRP Ann. **54**(1), 19–22 (2005)
22. Lao, W., Han, J., De With, P.H.: Automatic video-based human motion analyzer for consumer surveillance system. IEEE Trans. Consum. Electron. **55**(2), 591–598 (2009)
23. Lee, J.T., Ryoo, M.S., Riley, M., Aggarwal, J.K.: Real-time detection of illegally parked vehicles using 1-D transformation. In: 2007 IEEE Conference on Advanced Video and Signal Based Surveillance, pp. 254–259 (2007)
24. Liu, J., Yang, W., Yang, S., Guo, Z.: D3R-NET: dynamic routing residue recurrent network for video rain removal. IEEE Trans. Image Process. **28**(2), 699–712 (2018)
25. Liu, W., et al.: SSD: single shot multibox detector. In: Leibe, B., Matas, J., Sebe, N., Welling, M. (eds.) ECCV 2016. LNCS, vol. 9905, pp. 21–37. Springer, Cham (2016). https://doi.org/10.1007/978-3-319-46448-0_2
26. Liu, W., Wang, Z., Liu, X., Zeng, N., Liu, Y., Alsaadi, F.E.: A survey of deep neural network architectures and their applications. Neurocomputing **234**, 11–26 (2017)
27. Lu, Y., Javidi, T., Lazebnik, S.: Adaptive object detection using adjacency and zoom prediction. In: Proceedings of the IEEE Conference on Computer Vision and Pattern Recognition, pp. 2351–2359 (2016)
28. Melzner, J., Hollermann, S., Kirchner, S., Bargstädt, H.J.: Model-based construction work analysis considering process-related hazards. In: 2013 Winter Simulations Conference, pp. 3203–3214 (2013)
29. Naticchia, B., Vaccarini, M., Carbonari, A.: A monitoring system for real-time interference control on large construction sites. Autom. Constr. **29**, 148–160 (2013)
30. Pei, S.C., Tsai, Y.T., Lee, C.Y.: Removing rain and snow in a single image using saturation and visibility features. In: 2014 IEEE International Conference on Multimedia and Expo Workshops, pp. 1–6 (2014)
31. Raheja, J.L., Deora, S., Chaudhary, A.: Cross border intruder detection in hilly terrain in dark environment. Optik-Int. J. Light Electron. Opt. **127**(2), 535–538 (2016)
32. Redmon, J., Divvala, S., Girshick, R., Farhadi, A.: You only look once: unified, real-time object detection. In: Proceedings of the IEEE Conference on Computer Vision and Pattern Recognition, pp. 779–788 (2016)
33. Redmon, J., Farhadi, A.: YOLO9000: better, faster, stronger. In: Proceedings of the IEEE Conference on Computer Vision and Pattern Recognition, pp. 7263–7271 (2017)
34. Redmon, J., Farhadi, A.: YOLOV3: an incremental improvement. arXiv preprint arXiv:1804.02767 (2018)
35. Ren, S., He, K., Girshick, R., Sun, J.: Faster R-CNN: towards real-time object detection with region proposal networks. In: Advances in Neural Information Processing Systems, pp. 91–99 (2015)

36. Revaud, J., Weinzaepfel, P., Harchaoui, Z., Schmid, C.: EpicFlow: edge-preserving interpolation of correspondences for optical flow. In: Proceedings of the IEEE Conference on Computer Vision and Pattern Recognition, pp. 1164–1172 (2015)
37. Sermanet, P., Eigen, D., Zhang, X., Mathieu, M., Fergus, R., LeCun, Y.: OverFeat: integrated recognition, localization and detection using convolutional networks. arXiv preprint arXiv:1312.6229 (2013)
38. Sharma, R., Athavale, V.A.: Survey of intrusion detection techniques and architectures in wireless sensor networks. Int. J. Adv. Netw. Appl. **10**(4), 3925–3937 (2019)
39. Shrivastava, A., Gupta, A., Girshick, R.: Training region-based object detectors with online hard example mining. In: Proceedings of the IEEE Conference on Computer Vision and Pattern Recognition, pp. 761–769 (2016)
40. Skibniewski, M.J.: Information technology applications in construction safety assurance. J. Civil Eng. Manag. **20**(6), 778–794 (2014)
41. Soumya, T., Thampi, S.M.: Self-organized night video enhancement for surveillance systems. SIViP **11**(1), 57–64 (2017). https://doi.org/10.1007/s11760-016-0893-6
42. Tang, L., Li, L., Ying, S., Lei, Y.: A full level-of-detail specification for 3D building models combining indoor and outdoor scenes. ISPRS Int. J. Geo-Inf. **7**(11), 419 (2018)
43. Tychsen-Smith, L., Petersson, L.: DeNet: scalable real-time object detection with directed sparse sampling. In: Proceedings of the IEEE International Conference on Computer Vision, pp. 428–436 (2017)
44. Ulhaq, A., Yin, X., He, J., Zhang, Y.: FACE: fully automated context enhancement for night-time video sequences. J. Vis. Commun. Image Represent. **40**, 682–693 (2016)
45. Vijayanandh, R., Kumar, J.D., Kumar, M.S., Bharathy, L.A., Kumar, G.R.: Design and fabrication of solar powered unmanned aerial vehicle for border surveillance. In: Proceedings of International Conference on Remote Sensing for Disaster Management, pp. 61–71 (2019)
46. Vittal, K., Pai, A., Shenoy, A., Rao, C.S.: Computer controlled intrusion-detector and automatic firing-unit for border security. In: 2010 Second International Conference on Computer and Network Technology, pp. 289–293 (2010)
47. Wang, Y., Zhu, L., Yu, Z., Guo, B.: An adaptive track segmentation algorithm for a railway intrusion detection system. Sensors **19**(11), 2594 (2019)
48. Wei, C., Yang, J., Zhu, W., Lv, J.: A design of alarm system for substation perimeter based on laser fence and wireless communication. In: 2010 International Conference on Computer Application and System Modeling, vol. 3, pp. V3–543 (2010)
49. Xie, X., Wang, C., Chen, S., Shi, G., Zhao, Z.: Real-time illegal parking detection system based on deep learning. In: Proceedings of the 2017 International Conference on Deep Learning Technologies, pp. 23–27 (2017)
50. Zhang, M., Cao, T., Zhao, X.: Applying sensor-based technology to improve construction safety management. Sensors **17**(8), 1841 (2017)
51. Zhang, S., Wen, L., Bian, X., Lei, Z., Li, S.Z.: Single-shot refinement neural network for object detection. In: Proceedings of the IEEE Conference on Computer Vision and Pattern Recognition, pp. 4203–4212 (2018)
52. Zhu, Y., Zhao, C., Wang, J., Zhao, X., Wu, Y., Lu, H.: CoupleNet: Ccupling global structure with local parts for object detection. In: Proceedings of the IEEE International Conference on Computer Vision, pp. 4126–4134 (2017)

Tightly INS/UWB Combined Indoor AGV Positioning in LOS/NLOS Environment

Peisen Li, Shuhui Bi$^{(\boxtimes)}$ ⓘ, Tao Shen, and Qinjun Zhao

School of Electrical Engineering, University of Jinan,
Jinan 250022, Shandong, China
574640363@qq.com, {cse_bish,cse_st,cse_zhaoq}@ujn.edu.cnj

Abstract. In view of the defects and shortcomings of traditional Automated Guided Vehicle (AGV) robots in the localization mode and working scene, this paper studies the tightly-coupled integrated localization strategy based on inertial navigation system (INS) with ultra wide band (UWB). This paper presents an interactive multi-model (IMM) to solve the influence of non-line-of-sight (NLOS) on positioning accuracy. In IMM framework, two parallel Kalman filter (KF) models are used to filter the measured distance simultaneously, and then IMM distance is obtained by weighted fusion of two KF filtering results. This paper adopts the tightly-coupled combined method, and performs indoor positioning by extending Kalman filter (EKF). Experiments show that the method can effectively suppress the influence of NLOS error and improve the localization accuracy.

Keywords: Kalman filter (KF) · Extending kalman filter (EKF) · Tightly-coupled

1 Introduction

Automated Guided Vehicle (AGV) is the key equipment of self-flow transportation system and flexible manufacturing system, and it has been applied more and more widely (Li et al. 2017). Therefore, how to obtain accurate localization information is the core of AGV positioning technology. Inertial navigation system (INS) is an autonomous navigation system which does not depend on external information or radiate energy to the outside world (Liu et al. 2018). The disadvantage is that the localization error accumulates with the increase of time. Compared with INS, ultra wide band (UWB) technology is a kind of wireless technology rising in recent years. But in the complex indoor environment with more obstacles, the accuracy will be reduced because of the influence of

This work was supported by the Shandong Key R&D Program under Grants 2019GGXI04026 and 2019GNC106093.

Y.-D. Zhang et al. (Eds.): ICMTEL 2020, LNICST 327, pp. 337–344, 2020.
https://doi.org/10.1007/978-3-030-51103-6_30

signal transmission (Sobhani et al. 2017; Xu et al. 2017). Therefore, this paper combines INS with UWB, and the two localization methods complement each other, which can provide a more comprehensive and reliable localization solution.

When the AGV moves in the indoor environment, there are a large number of obstacles. The linear communication path between the beacon node and the reference nodes (RNs) is prone to obstruction of obstacles. The non-line-of-sight (NLOS) error produces a large deviation, and the localization accuracy is drastically reduced (Morelli et al. 2007). So, this paper proposes an interactive multi-model (IMM) algorithm. In IMM framework, two parallel Kalman filter (KF) (Shi and Fang 2010) models are used to filter the measured distance simultaneously. The distance estimation of IMM is obtained by weighted fusion of the two filtering results. When the state is switched, it can still get a more accurate distance estimation. To a certain extent, it avoids the error caused by the inaccuracy of the algorithm for identifying LOS/NLOS, thus better weakening the influence of LOS/NLOS error. In the integrated localization stage, the INS/UWB integrated localization can be divided into loosely-coupled integrated localization and tightly-coupled integrated localization. The accuracy of navigation results depends on the accuracy of UWB solution, and the sub-localization technology alone completes the localization, which also introduces the calculation error for the integrated localization system. The tightly-coupled integrated proposed in this paper is to reduce the influence of NLOS error by using UWB's original ranging information through IMM algorithm, and then fuse the data with INS's estimated distance information, which can reduce the calculation error of sub-localization system's independent localization and improve the overall localization accuracy.

2   The Tightly INS/UWB Combined AGV Positioning Strategy

When the AGV moves indoors, the channel state between the AGV and the RNs is switched between LOS and NLOS, and the measurement models of LOS and NLOS are different. Therefore, it is difficult for a single filtering model to effectively filter the measured distance in real time. In this paper, an IMM algorithm is proposed to estimate the measurement distance. Two parallel KF are used to adapt to the LOS and NLOS ranging models. In the distance filtering phase, two KF simultaneously estimate the measurement distance, and the model likelihood probability can be calculated according to the distance measurement value input by the system and the model filtered observation value, and then the model probability can be calculated. Models with higher model matching can obtain higher model probabilities, so the filtered results of the model will dominate the output at the weighted output stage. The IMM algorithm is a multi-basic interaction algorithm based on the second-order Markov chain model. According to the characteristics of the given model parameters, given the input data, the input data can be adaptively distinguished from which model is closer

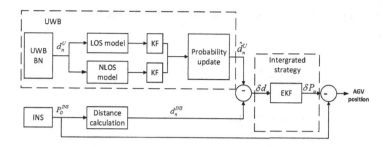

Fig. 1. The tightly-coupled strategy.

to, and expressed by the model probability. The Combination strategy framework is shown in Fig. 1.

The space equation of tightly-coupled combined localization system includes state equation and measurement equation. Tightly-coupled combined localization is more original than loosely-coupled combined localization. So when the number of Reference nodes (RNs) of UWB visible tags is small, the tightly-coupled combined localization still keeps a high localization accuracy. Figure 1 shows the tightly-coupled strategy for AGV in complex work environments. Firstly, d_n represent the range information obtained by UWB positioning system, $\hat{d}_n$ is the distance information after the NLOS error is weakened by the IMM algorithm. P^{INS} is the location information solved by INS, d_n^{INS} is the distance information of the INS to the RNs and is calculated from the UWB and INS position information. The last, the optimal navigation solution of AGV is obtained by making a difference between P^{INS} and optimal error δP.

2.1 IMM Model

We define the equation of state of the system as:

$$\underbrace{\begin{bmatrix} d_{n,k+1}^U \\ \dot{d}_{n,k+1}^U \end{bmatrix}}_{\mathbf{x}_{n,k|k-1}} = \underbrace{\begin{bmatrix} 1 & T \\ 0 & 1 \end{bmatrix}}_{\mathbf{F^I}} \underbrace{\begin{bmatrix} d_{n,k}^U \\ \dot{d}_{i,k}^U \end{bmatrix}}_{\mathbf{x}_{n,k-1}} + \mathbf{W}_{n,k}, \tag{1}$$

$d_{n,k}^U$ represents distance measured in the nth UWB at k time, $\dot{d}_{n,k}^U$ represents the rate of change of distance measured at the nth UWB at k time, system noise $\omega_{n,k} \sim N(0, \mathbf{Q}_{n,k})$.

In this paper, the observation equation of IMM model in LOS environment is defined as:

$$\underbrace{\tilde{d}_{n,k}^U}_{\mathbf{y}_{n,k}^U} = \underbrace{\begin{bmatrix} 1 & 0 \end{bmatrix}}_{\mathbf{H}^L} \mathbf{x}_{n,k|k-1} + \omega_{n,k}^L, \tag{2}$$

where noise $\omega_{n,k}^L \sim N\left(0, R_{n,k}^{los}\right)$.

In this paper, the observation equation of IMM model in NLOS environment is defined as:

$$\underbrace{\bar{d}_{n,k}^U}_{\mathbf{y}_{n,k}^U} = \underbrace{\begin{bmatrix} 1 & 0 \end{bmatrix}}_{\mathbf{H}^N} \mathbf{x}_{n,k|k-1} + \omega_{n,k}^N, \tag{3}$$

where noise $\omega_{n,k}^N \sim N\left(0, R_{n,k}^{nlos}\right)$.

2.2 INS/UWB Tightly-Coupled Model

At this stage, The INS/UWB tightly-coupled Combined positioning strategy space equation includes a state equation and a measurement equation. The equation of state can be defined as:

$$\underbrace{\begin{bmatrix} \delta P_{E,k+1} \\ \delta P_{N,k+1} \\ \delta V_{E,k+1} \\ \delta V_{N,k+1} \end{bmatrix}}_{\mathbf{x}_{k+1}} = \underbrace{\begin{bmatrix} 1 & 0 & T & 0 \\ 0 & 1 & 0 & T \\ 0 & 0 & 1 & 0 \\ 0 & 0 & 0 & 1 \end{bmatrix}}_{\mathbf{F}} \underbrace{\begin{bmatrix} \delta P_{E,k} \\ \delta P_{N,k} \\ \delta V_{E,k} \\ \delta V_{N,k} \end{bmatrix}}_{\mathbf{x}_k} + \mathbf{V}_k \tag{4}$$

where $\delta P_{E,k}$ is the eastward position error of AGV and $\delta P_{N,k}$ is the error in the north direction at k time. $\mathbf{F}$ represents the state transition matrix and noise $\mathbf{V}_n \sim N\left(0, \mathbf{Q}_k\right)$.

$$\delta d_n = (d_n^{INS})^2 - (d_n^I)^2 \tag{5}$$

It can be concluded that:

$$\begin{aligned} (d_n^{INS})^2 - (d_n^I)^2 = 2(P_E^{INS} - P_E^i)\delta P_E \\ + 2(P_N^{INS} - P_N^i)\delta P_N - (\delta P_E^2 + \delta P_N^2) \end{aligned} \tag{6}$$

where δd_n is distance measurement information, d_n^{INS} is the distance between the BN estimated by the INS and the nth RN reference station. d_n^I is the ideal position of the BN is difficult to obtain in the measurement. P_E^I and P_N^I represent the ideal position coordinates of BN in the east and north directions. Detailed formula derivation can be found in the literature (Xu and Chen 2016).

So the measurement equation is defined as:

$$\underbrace{\begin{bmatrix} \delta d_{1,k} \\ \delta d_{2,k} \\ \vdots \\ \delta d_{i,k} \end{bmatrix}}_{\mathbf{Z}_k} = \underbrace{\begin{bmatrix} h_1(\delta P_{E,k}, \delta P_{N,k}) \\ h_2(\delta P_{E,k}, \delta P_{N,k}) \\ \vdots \\ h_i(\delta P_{E,k}, \delta P_{N,k}) \end{bmatrix}}_{f(\mathbf{x}_k)} + \mathbf{V}_k \tag{7}$$

where $\mathbf{V}_k$ is system observation noise matrix.

In the integrated localization phase, AGV localization uses an extended Kalman filter (EKF) that can handle nonlinear dynamic problems. The basic idea of the EKF is to perform a first-order Taylor function expansion on a non-linear system, linearize the first two terms, and then use the KF algorithm for state estimation.

3   Test

During the experiment, AGV traveled around two obstacles, and the route size was $3.5\,\text{m} \times 9.5\,\text{m}$. The Fig. 2 displays the test platform. We set the sampling time to: $T_i = T = 0.02\,\text{s}$. Moreover, the initial values of the parameters are set to:

$$\mathbf{\Pi} = \begin{bmatrix} 0.5 & 0.5 \\ 0.5 & 0.5 \end{bmatrix}, \tag{8}$$

and

$$\theta_k = \begin{bmatrix} 0.5 \\ 0.5 \end{bmatrix}, \tag{9}$$

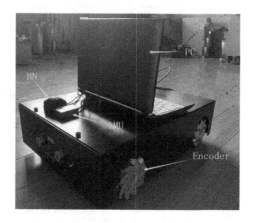

Fig. 2. The test platform.

Figure 3 shows the INS and UWB localization trajectories. As INS error gradually accumulates with time, the deviation between localization trajectory and reference trajectory is large. Although INS is an autonomous localization system, it can not be used alone. The accuracy of UWB localization trajectory is higher than INS, but due to the obstacles, it can be seen that there are many burrs in the track, which will also lead to the increase of localization error.

Figure 4 shows the loosely-coupled trajectory, the traditional tightly-coupled trajectory and the proposed tightly-coupled trajectory. The error between the positioning trajectory and the reference trajectory of each positioning system is shown in Table 1. The proposed tightly-coupled combined approach's performance is better than the traditional tightly-coupled combined and loosely-coupled combined localization. It can be seen that tightly-coupled combined has

higher localization accuracy than loosely-coupled combination, which depends on
that tightly-coupled combined uses the original data of each sensor and reduces
the error brought by positioning algorithm. The proposed tightly-coupled combined method is using the IMM algorithm weakens the LOS/NLOS error and
improves the positioning accuracy. Therefore, through the analysis of Fig. 5 and
Fig. 6, we can draw a conclusion that the IMM algorithm studied in this paper
can effectively reduce the LOS/NLOS error and the proposed tightly-coupled
combined method has smaller error than the other two combination methods.

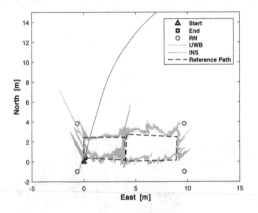

Fig. 3. Trajectory of single positioning system.

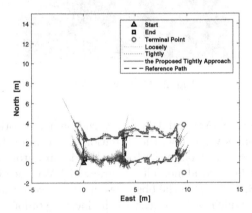

Fig. 4. The localization trajectory of INS/UWB tight combination method and other
traditional combination location methods.

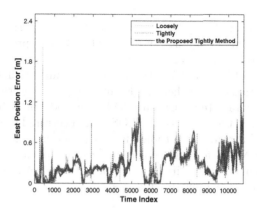

Fig. 5. Error in the east.

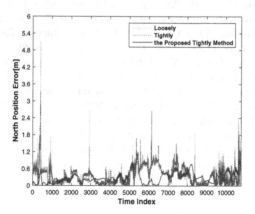

Fig. 6. North position error.

Table 1. Statistics of positioning errors in the east and north directions.

Method	East Mean/m	East Std./m	North Mean/m	North Std./m
Loosely	0.2959	0.3566	0.3922	0.4266
Tightly	0.2912	0.3453	0.3846	0.3986
IMM-tightly	0.2887	0.3404	0.3733	0.3644

4   Conclusion

We propose indoor AGV tightly INS/UWB combined positioning method. In this combined positioning method, an IMM algorithm is proposed for the location of moving targets in indoor environment. Through error analysis of off-line measurement data of LOS and NLOS, the error parameters under two propagation states are obtained, and then the Kalman distance filtering models for the two channel error parameters are established. The model probability is calculated adaptively to weaken the influence of NLOS ranging error. The simulation results show that: the IMM algorithm studied in this paper can effectively weaken the influence of LOS/NLOS for the navigation accuracy of indoor moving targets, and the tightly combined positioning method can make AGV achieve higher positioning accuracy when moving indoors.

References

Li, B., Liu, H., Xiao, D., et al.: Centralized and optimal motion planning for large-scale AGV systems: a generic approach. Adv. Eng. Softw. **106**, 33–46 (2017)

Liu, B., Wei, S., Su, G., et al.: An improved fast self-calibration method for hybrid inertial navigation system under stationary condition. Sensors **18**(5), 1303 (2018)

Sobhani, B., Paolini, E., Giorgetti, A., et al.: Target tracking for UWB multistatic radar sensor networks. IEEE J. Sel. Topics Signal Process. **8**(1), 125–13603 (2017)

Morelli, C., Nicoli, M., Rampa, V.: Hidden Markov models for radio localization in mixed LOS/NLOS conditions. IEEE Trans. Signal Process. **55**(4), 1525–1542 (2007)

Xu, Y., Chen, X.: Range-only UWB/INS tightly-coupled integrated navigation method for indoor pedestrian. Chin. J. Sci. Instrum. **37**(8), 142–148 (2016)

Shi, Y., Fang, H.: Kalman filter-based identification for systems with randomly missing measurements in a network environment. Int. J. Control **83**(3), 538–551 (2010)

Xu, Y., Shmaliy, Y.S., Li, Y., Chen, X.: UWB-based indoor human localization with time-delayed data using efir filtering. IEEE Access **5**, 16676–16683 (2017)

Prediction Analysis of Soluble Solids Content in Apples Based on Wavelet Packet Analysis and BP Neural Network

Xingwei Yan, Shuhui Bi[⊠], Tao Shen, and Liyao Ma

School of Electrical Engineering, University of Jinan, Jinan 250022, China
497754391@qq.com, {cse_bish,cse_st}@ujn.edu.cn

Abstract. Considering Fuji apple, the relationship between the near infrared spectrum and the soluble solids content (SSC), which is one of the important indexes to measure the internal quality of apple, is studiedin this paper. In order to reduce the computational complexity and to improve the accuracy of modeling, this paper adopts the wavelet packet threshold denoising method for spectral spectrum processing, and usesthe method of wavelet packet analysis (WPA) to filter the characteristic wavelength of the spectrum. Moreover, a prediction model of SSC is proposed based on BP neural network due to its characteristics of anti-noise, anti-interference, strong nonlinear conversion abilityand the good capacity in handling nonlinear measured data with uncertain causality. Finally, the simulation results show that wavelet packet analysis can not only reduce the calculation of modeling variables, but also Improve modeling accuracyof the BP neural network model. The proposed method can make a better prediction of the SSC of apple.

Keywords: Apple · Wavelet packet analysis · Near infrared spectrum · BP neural network

1 Introduction

China is a major fruit production country. After years of development, China's apple industry has ranked among the major countries in the development of the world's fruit industry. China's apple output, output value and planting area rank among the top in the world, and apple industry has developed into one of the pillar industries of China's agriculture. Against the background of fierce competition in the domestic market, China's apple industry still has problems

This work is supported by the National Natural Science Foundation of China No. 61473135, Shandong Agricultural Machinery Research and Development Innovation Project Grant No. 2018YF011, and Shandong Provincial Key Research and Development Project 2017GGX10116, and Shandong Provincial Natural Science Foundation ZR2018PF009.

Y.-D. Zhang et al. (Eds.): ICMTEL 2020, LNICST 327, pp. 345–354, 2020.
https://doi.org/10.1007/978-3-030-51103-6_31

such as relatively excessive market, weak market competitiveness and less foreign exchange from export Shu (2018). The main reason lies in the lag of apple postpartum quality inspection and safety commercialization. Therefore, improving the detection level of the internal quality of Chinese apples can not only promote the research and development of Chinese apple sorting equipment, but also have positive significance for improving the international competitiveness of Chinese apples. NIRS detection technology has the advantages of rapid, non-destructive and environmental protection, and has become the fastest developing and most widely used modern food analysis and detection technology Chen (2019). Shang Jing uses near-infrared spectroscopy to identify apple varieties. The results show that the rapid and non-destructive identification of apple varieties can be achieved by using near-infrared spectroscopy Shang (2019). Guo Zhiming et al. corrected the intensity of hyperspectral images, it predicted the sugar content and distribution of apples quickly and without loss Guo (2015). Sanaz Jarolmasjed et al. used spectroscopy to classify healthy apples and bitter apples, and proved that near-infrared spectroscopy can be used as an indicator of apple bitter nuclear development Sanaz (2017).

The near-infrared spectrum obtained by instrument scanning will inevitably have noise interference, which will affect the accuracy of spectral analysis. In this paper, we study the optimization problem of modeling the SSC of red Fuji apples. Wavelet analysis was used to pretreat the spectrum and feature wavelength screening. The robustness and applicability of the prediction model of near-infrared spectroscopy were improved by establishing a neural network model.

2   The Experimental Part

Experimental Materials. In this study, Red Fuji apples from Yantai Qixia were selected, and 200 apples without defects and damage were selected. Among them, 130 were randomly selected as the correction set, and the remaining 70 were used as prediction sets, and they were stored in a $0°$ cold storage. Before the experiment, it was taken out of the cold storage two hours in advance, so that the sample temperature was consistent with the laboratory temperature, and the experimental medium temperature was basically unchanged.

Spectral Collection and Determination of SSC. The near-infrared spectrum of the sample was acquired by an Antaris II Fourier transform near-infrared spectrometer. From Fig. 1, the InGaAs detector was used, and the integrated ball diffuse reflection acquisition method was used to set the number of scans of the sample to 32 times' resolution is $8 \, cm^{-1}$, the collection range is $4000–10000 \, cm^{-1}$, and 1557 variables are obtained for each spectrum. At the time of sampling, three different positions of the equator of the apple were collected, take the average as the experimental spectral data.

After the spectrum acquisition is completed, the soluble solid content is measured at the position of the collected spectrum. The Japanese Atago Brix meter is used to peel the apple, crush the pulp and take 1–2 drops of juice, and drop it in the center of the prism plane to read. The SSC in apples, the average of the soluble solids content of each of the three spectral collection sites of apples was used as a reference value for the apple sample.

Fig. 1. Near infrared spectrometer.

3   Data Processing

The near-infrared spectrum acquisition process may be affected by various factors such as the state of the spectrometer and the detection conditions, which leads to the noise interference in the spectrum affecting the accuracy of the modeling. Therefore, it is necessary to preprocess the acquired spectrum.

3.1   Spectral Pretreatment

In engineering applications, the noise signal is usually a high-frequency signal, and the low-frequency signal that is relatively stable is a real signal. Therefore, the denoising process mainly has the following steps:

(1) Import raw spectral data.
(2) Wavelet decomposition of the acquired spectral signal. Specifically, it is divided into the following steps:

(a) Select the order of the wavelet function and the wavelet function;

(b) Determine the wavelet decomposition scale;

(c) Select reasonable parameters for wavelet packet transform.

(3) Select the corresponding threshold quantization method for the high and low frequency coefficients of wavelet decomposition.

(4) Wavelet reconstruction. Wavelet reconstruction of the decomposition coefficients of the optimal wavelet packet base and the wavelet packet coefficients after threshold quantization.

As an improvement of wavelet analysis technology, wavelet packet analysis uses a multi-scale analysis tower algorithm to decompose each frequency band of the signal. Each layer of wavelet after decomposition contains all frequencies of the signal, which can completely reproduce the original signal Xiong (2005). The wavelet packet decomposition tree is shown in Fig. 2.

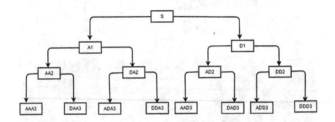

Fig. 2. Three-layer wavelet packet decomposition tree diagram.

Commonly used to measure signal denoising effects are:

(1) Root mean square error (RMSE):

$$RMSE = \sqrt{\frac{1}{N} \sum (\tilde{X} - X_0)^2} \tag{1}$$

Where N is the size of the signal, X_0 is the original signal, and $\tilde{X}$ is the signal after the wavelet packet is denoised. The smaller the RMSE, the higher the approximation of the denoised signal and the original signal, and the better the effect.

(2) Signal to noise ratio (SNR):

$$SNR = 10 * \log 10 * (power_{signal}/power_{noise})$$

$$power_{signal} = \frac{1}{N} \sum_N X_0^2 \tag{2}$$

$$power_{noise} = \frac{1}{N} \sum (\tilde{X} - X_0)^2$$

$power_{signal}$ is the real data energy, $power_{noise}$ is the energy of the noise, and the higher the signal-to-noise ratio, the better the denoising effect.

The premise of wavelet analysis is to choose the appropriate wavelet basis function. After experimental comparison, this paper adopts the db4 wavelet decomposition widely used in engineering to perform three-layer decomposition. How to perform threshold quantization is also the key to the denoising quality of the signal. The spectral denoising is performed by the four thresholds of sqtwolog, minmaxi, rigrsure and heursure respectively. From Table 1, we can see its denoising effect.

Table 1. Comparison of SNR and RMSE results for four threshold denoising methods

Threshold selection	SNR	RMSE
sqtwolog	39.8248	0.3275
minmaxi	35.0180	0.5712
heursure	35.4934	0.6044
rigrsure	30.5389	0.9522

In general, the larger the SNR and the smaller the RMSE, the better the denoising effect. From Table 1, it can be seen that the sqtwolog denoising effect is better than the other three threshold selection methods, so the sqtwolog threshold principle is used in this paper. The comparison chart before and after spectral preprocessing is shown in Fig. 3. A partial effect diagram of one of the spectra is shown in Fig. 4. The blue spectrum is the original spectrum, and the red spectrum is the spectrum after denoising.

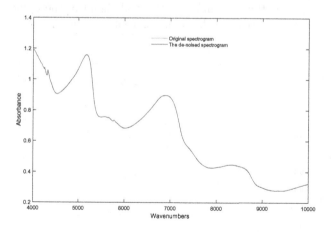

Fig. 3. Spectrum before and after wavelet packet denoising. (Color figure online)

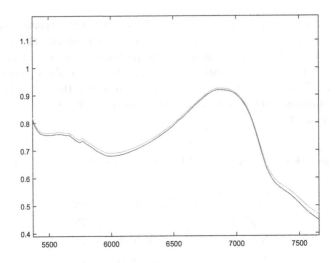

Fig. 4. Partial detail enlarged view. (Color figure online)

3.2   Spectral Feature Screening

Full-spectrum modeling, and some spectral regions lack correlation with sample properties, and modeling results are poor. The variable group with the lowest information redundancy is selected from the full-band spectral information to ensure that the selected feature band has the least collinearity, which can greatly reduce the complexity in the model building process. The wavelet transform is used to screen the characteristic wavelength of the sample spectrum. After the wavelet decomposition, the coefficients of each node contain some details of the signal. In this paper, the near-infrared spectral signal of the apple is collected by using the Matlab wavelet packet analysis toolbox. The signal is decomposed

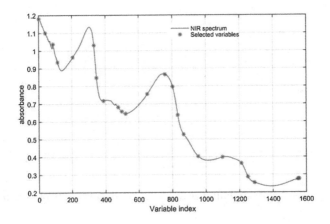

Fig. 5. Screened characteristic wavelength points.

in three layers and the spectral signals are analyzed in each frequency band. Calculate the coefficient standard deviation of the spectrum in the band at each wavelength point. The larger the standard deviation is, the larger the dispersion is, and the wavelength corresponding to the position where the coefficient dispersion of each band is the largest. Figure 5 shows the wavelengths screened based on wavelet packet analysis. Using the selected 24 characteristic wavelengths for modeling analysis greatly reduces the complexity of modeling.

4   BP Neural Network Modeling Analysis

Artificial neural network is suitable for processing non-linear measurement data with uncertain cause and result relationship. It simulates the working principle of neuron to build a model. It has the characteristics of anti-noise, anti-interference and strong nonlinear conversion ability. As shown in the Fig. 6, the BP neural network consists of three parts: the input layer, the hidden layer, and the output layer. It can realize arbitrary nonlinear mapping of input and output subtraction, and has good nonlinear mapping approximation ability and predictive ability Li (2019).

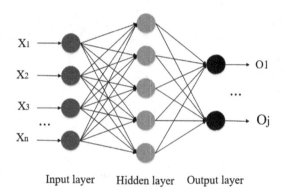

Fig. 6. Artificial neural network model.

The model is comprehensively evaluated by the correlation coefficient R between the measured value and the predicted value of the soluble solid content and the predicted root mean square error (RMSEP). The calculation formula is as shown in Eqs. (3) and (4):

$$R = \sqrt{1 - \frac{\sum\limits_{i=1}^{n} (\hat{y}_i - y_i)^2}{\sum\limits_{i=1}^{n} (y_i - \bar{y}_i)^2}} \tag{3}$$

Where y_i and $\hat{y}_i$ are the measured and predicted sets of the i_{th} sample in the sample set (including the training set and the prediction set), respectively, and $\bar{y}_i$ is the average of the measured values of all samples in the sample set.

$$RMSEP = \sqrt{\frac{\sum\limits_{i=1}^{n} (y_i - \hat{y}_i)^2}{n}} \tag{4}$$

Where y_i and $\hat{y}_i$ are the measured and predicted sets of the i_{th} sample in the prediction set, respectively, and n is the number of samples in the prediction set.

The collected apple spectrum has a total of 1557 wavelength points. Using all collected spectral points to build a model will greatly increase the amount of calculation, and the lack of correlation between some spectral bands and the nature of the sample, resulting in poor modeling results. Figure 7 shows the full spectrum modeling forecast result graph.

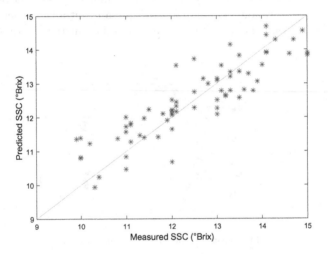

Fig. 7. Full spectrum.

In theory, increasing the number of layers in the network can reduce the model's error and improve the accuracy of the network, but it will complicate the network structure, increase the training time, and reduce the training efficiency. Generally, we first consider designing a hidden layer. When we need to improve the accuracy, we can achieve this by increasing the number of nodes in the hidden layer. Therefore, this paper designs a hidden layer and builds a three-layer BP network model. The number of nodes in the hidden layer has a great influence on the modeling effect. Too much will make the learning time too long and the generalization ability will decrease. Too little will make the learning time shorter and the fault tolerance performance of the network will decrease.

Generally speaking, we choose a network with different numbers of hidden layer nodes to train the sample set. The smaller the error, the better the network model. The number of hidden layer nodes is generally determined based on the empirical formula (5).

$$l = \sqrt{m+n} + a \qquad (5)$$

Where m is the number of input layer nodes, n is the number of output layer nodes, and a is a constant of 1–10.

A three-layer BP neural network was selected to establish a prediction model for soluble solids content in apples. The 24 wavelength points obtained by wavelet packet analysis were used as the input of the BP neural network. Through experimental tests, the parameters of BP neural network are selected as follows: the maximum number of training is 2000; the number of hidden nodes is 30; learning rate is 0.3; the target error is 0.04, the hidden layer selects the tansig activation function, and when the output layer selects the purein activation function, the network structure is more stable. The model prediction results after screening the characteristic wavelengths are shown in Fig. 8. It can be seen from Table 2 that the use of wavelet packet analysis to screen out the spectral characteristic wavelengths not only improves the modeling efficiency, but also the effect of the model is better than the model established by the full spectrum.

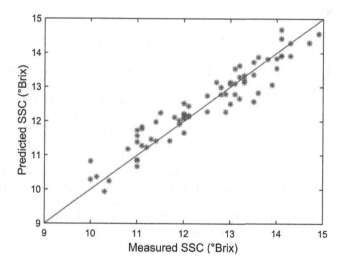

Fig. 8. After wavelet packet analysis and screening.

Table 2. Screening characteristics before and after spectral spectroscopy

Modeling variable	R	RMSEP
Full spectrum	0.863	0.639
Screening characteristic wavelength	0.936	0.471

5   Conclusion

This article takes red Fuji apples as the research object, and studies the relationship between near-infrared spectrum and apple internal quality. The original spectra were pretreated and the characteristic wavelengths of the spectra were screened by wavelet packet analysis. The selected wavelengths were used in BP neural network to establish apple soluble solids. The quantitative analysis model of the content, the R and the PRMSE were 0.936 and 0.471, respectively. The wavelet packet analysis method and BP neural network modeling method eliminate a large number of bands that are not related to soluble solids, which greatly reduces the complexity of the model and improves the prediction accuracy and stability of the model. The results show that it is feasible to use wavelet packet analysis and BP neural network to predict the soluble solid content of apple.

References

Shu, H., Chen, X.: The current task of the development of fruits industry in China. Chin. Fruit Tree **02**, 1–3 (2018)

Chao, C., Qi, L., Fan, L., et al.: Research and application of infrared spectroscopy technology in food safety testing. Food Res. Dev. **40**(14), 219–224 (2019)

Jing, S., Yan, Z., Qinglong, M.: Nondestructive identification of apple varieties by VIS/NIR spectroscopy. Preserv. Process. **19**(03), 8–14 (2019)

Guo, Z., Zhao, C., Huang, W., et al.: Intensity correction of visualized prediction for sugar content in apple using hyperspectral imaging. Agric. Mach. **46**(07), 227–232 (2015)

Sanaz, J., Carlos, Z.E., Singdhuja, S.: Near infrared spectroscopy to predict bitter pit development in different varieties of apples. Food Meas. **11**(3), 987–993 (2017)

Xiong, Y., Wen, Z., Chen, G., et al.: Study on spectral identification based on wavelet packet analysis. Spectrosc. Spectr. Anal. **25**(8), 1332–1335 (2005)

Li, J., Liu, Y., Liu, Z.: Soil total iron content hyperspectral inversion based on BP neural network. Technol. Econ. Guide **27**(22), 93–94 (2019)

Research on Positioning Accuracy of Indoor and Outdoor Pedestrian Seamless Navigation

Kailong Wang, Huixia Li$^{(\boxtimes)}$, and Hang Guo

Nanchang University, Nanchang 330031, China
lihuixia0601@163.com

Abstract. The accuracy of pedestrian positioning is helpful to ensure the pedestrian safety in both indoor and outdoor environments. Improve the accuracy of pedestrian positioning is a key research issue. In order to solve the problem that the indoor and outdoor pedestrian navigation is not continuous and the accuracy is low, a pedestrian seamless navigation and positioning method based on BDS/GPS/IMU is proposed. In outdoor environment, in order to improve the availability of dynamic positioning when the single-system observation geometry is not ideal, the key techniques such as the differential coordinates and time benchmark in BDS/GPS positioning are studied, and a method of eliminating time difference by the independent combination difference in the system is proposed. This method simplifies the operation steps and overcomes the current compatible positioning difficulties without the time difference between BDS and GPS. It can be seen that the more the number of visible satellites, the better the space geometric distribution. In the combined positioning experiment results of BDS and GPS, the number of visual satellites are about 6–8 when GPS is used alone. When using the combined system, the number of visible satellites increased to about 16. The increase in the number of visible satellites greatly improves the observation geometry. For the Position Dilution of Precision (PDOP), the maximum PDOP of the dual system is 2.7, which is significantly lower than that of the single system, and the observation geometry performance is greatly improved. In the effective positioning time, for BDS/GPS, the positioning accuracy of elevation (U) direction is better than 4 cm, and the positioning accuracy of North (N) and East (E) direction is better than 2 cm. Based on the analysis of pedestrian gait characteristics, a multi-condition constrained zero-velocity detection algorithm is proposed. For the error of the inertial sensor error is accumulated over the time, the zero velocity update (ZUPT) algorithm is implemented to correct the cumulative errors by using to the designed extended Kalman filter (EKF) with the velocity and angular velocity information as the measurements. The results show that the accuracy of dual-mode positioning system of BDS compatible with GPS is better than the single-mode GPS positioning, the outdoor position accuracy can reach centimeter level, and under ZUPT compensation the indoor error ratio is 1%, which can achieve more accurate pedestrian seamless navigation.

Keywords: Pedestrian seamless navigation · BDS/GPS · Zero Velocity Update (ZUPT) · Extended Kalman Filter (EKF)

© ICST Institute for Computer Sciences, Social Informatics and Telecommunications Engineering 2020
Published by Springer Nature Switzerland AG 2020. All Rights Reserved
Y.-D. Zhang et al. (Eds.): ICMTEL 2020, LNICST 327, pp. 355–367, 2020.
https://doi.org/10.1007/978-3-030-51103-6_32

1 Introduction

In a closed environment without satellite signals, the pedestrian navigation system can achieve independent positioning. It can be effectively used in emergency rescue, jungle adventure and military fields, and in recent years it has become a research hotspot in the field of navigation.

At present, many scholars have studied the indoor and outdoor positioning technology. The positioning technology in document [1–10] requires pre-installation of the sensor network, using time difference of arrival (TDOA), angle of arrival (AOA), signal strength and other algorithms to achieve indoor positioning, but the navigation range is limited. The document [11] proposed an indoor navigation system based on GPS/IMU/Zigbee, using Zigbee to establish a fingerprint database fusion with IMU, and combined with GPS to achieve positioning, but single GPS signal is easy occluded, it needs to establish complex Zigbee nodes fingerprint database in advance, and the IMU solution error diverges over time, the positioning accuracy is not high. Document [9–12] studied the pseudolite-based positioning technology, and used the pseudolite to launch the satellite-like signals to achieve indoor navigation, however, this method is influenced by multipath effect, it is not suitable for indoor complex environment. In document [13], a monocular vision indoor positioning technology is studied. It collects images, extracts feature points, and gets position through the coordinate transformation, but the computation is large, and the image extraction algorithm is complex, which requires the computer with fast speed and high power consumption.

Compared with the above algorithms, a navigation and positioning method based on the BDS/GPS/IMU is studied. The dual-mode positioning of BDS and GPS not only extends the advantages of single-satellite, but also achieves the common coverage of the dual-mode navigation system at the same time and the same place, relieves the influence of factors such as terrain, obstacles, and so on. In indoor positioning, uses the MEMS-IMU equipment with small size, low cost and autonomous navigation module, improves the pedestrian dead reckoning algorithm (PDR). At same time, by increasing the constraint conditions, the pedestrian zero speed state is accurately determined, and adopts ZUPT and zero angular velocity update (ZARU) to auxiliary correct the angular velocity update further eliminates the cumulative error of the inertial device. The system has strong autonomous positioning and better navigation continuity even in complex environments. This method can achieve high-accuracy indoor and outdoor pedestrian navigation.

2 The Positioning Method of BDS/GPS

Because of the difference between the BDS and the GPS in the definition of coordinate and the time system, in order to realize the dual-mode positioning, it is necessary to solve the transformation and unification of the space-time coordinate. Ignoring the coordinate changes in the two system coordinate system, it only needs to solve the deviation problem between different systems [14–16]. In order to simplify the calculation, the time deviation can be eliminated by the form of the difference algorithm between the systems. In the process of observation, the base station receiver and the rover station receiver simultaneously observed multiple BDS satellites and GPS satellites. Among them, two BDS satellite and two GPS satellites were selected to further operation, obtained the respective carrier phase observation equations between the satellites, and then calculated the difference equation between the satellites and the receivers. Therefore, we can get two equations about the respective satellites, the simultaneous equation is the combined positioning double difference observation equations [17].

$$\begin{bmatrix} \nabla\Delta\Phi_{AB}^{G} \\ \nabla\Delta\Phi_{AB}^{C} \end{bmatrix} = \begin{bmatrix} \nabla\Delta\rho_{AB}^{G} - \lambda^{G}\nabla\Delta N_{AB}^{G} + \nabla\Delta Trp_{AB}^{G} + \varepsilon_{\nabla\Delta}^{G} \\ \nabla\Delta\rho_{AB}^{C} - \lambda^{C}\nabla\Delta N_{AB}^{C} + \nabla\Delta Trp_{AB}^{C} + \varepsilon_{\nabla\Delta}^{C} \end{bmatrix} \tag{1}$$

Which $\nabla\Delta\Phi_{AB}^{G}$, $\nabla\Delta\Phi_{AB}^{C}$ represent GPS and BDS carrier phase double-difference observations, $\nabla\Delta\rho_{AB}^{G}$, $\nabla\Delta\rho_{AB}^{C}$ represent GPS and BDS pseudo-range double-difference observations, λ is the carrier wavelength, $\nabla\Delta N$ is the double-difference full-circumference ambiguity vector, and Trp z is the tropospheric delay error, $\varepsilon_{\nabla\Delta}$ is the observation noise, and subscripts A and B represent the base station and the rover station, respectively.

The double-difference observation equation composed of BDS and GPS, not only eliminates most of the error such as the time deviation, but also as the distance between the two stations is relatively close, the error caused by multipath is small and can be ignored. According to Eq. (1), using the least squares method or Kalman filter to solve the double-difference full-circumference ambiguity $\nabla\Delta N$, and then the position coordinates can be solved.

3   Indoor Pedestrian Positioning Method

3.1   PDR Algorithm

The purpose of the pedestrian positioning system is to achieve positioning by using the MTI IMU device strapping onto a foot/shoe. The proposed PDR positioning method is implemented under the framework of the Extended Kalman Filter (EKF). The main idea of this project is to use these filtering algorithms to estimate the cumulative errors (biases) of the IMU sensors. When the foot is on the floor, the EKF is updated by the velocity and angular velocity measurements respectively by the Zero-Velocity-Update (ZUPT) and Zero-Angular-Rate-Update (ZARU). Then the sensor biases are compensated with the estimated errors. Therefore, frequent use of the ZUPT/ZARU measurements can assist in correcting the errors, so even relatively low cost sensors can provide useful navigation performance. The PDR system prepared in the Matlab environment consists of five algorithms:

1) Initial alignment, using the static data of the accelerometer and magnetometer for the first few minutes to calculate the initial attitude.
2) IMU mechanization algorithm for computing the navigation parameters (position, velocity and attitude).

3) Zero velocity detection algorithm for determining when the foot is on the ground, the velocity and angular velocity of the IMU are zero.
4) ZUPT and ZARU feed the EKF with the measured errors when pacing is detected.
5) EFK estimates the errors and gives feedback to the IMU mechanization algorithm.

3.2 Initial Alignment

The initial static alignment provides the initial attitude information for the navigation system. The initial alignment of the IMU sensor is accomplished by two steps: leveling and gyro compassing. Leveling refers to obtain the roll and pitch by using the acceleration, and gyro compassing refers to obtain the heading by using the angular velocity. However, the bias and noise of gyroscopes are larger than the value of the Earth's rotation rate for the micro-electronic mechanical system (MEMS) IMU, so the heading error is large. In this paper, the initial alignment of the MEMS IMU is completed by using the static data of accelerometer and magnetometer during the first few minutes, and a method of heading measurement using a magnetometer is proposed.

3.3 Zero Velocity Detection Process

The movement of the foot-mounted IMU can be divided into two phases. The first one is the swing phase that means the foot-mounted IMU is moving. The second one is the stance phase which means the foot-mounted IMU is on the ground. The angular velocity and linear velocity of the foot-mounted IMU must be very close to zero in the stance phase. Therefore, the angular velocity and linear velocity of the foot-mounted IMU can be zero as the measurements of the Extended Kalman Filter (EKF) [18–20]. This is the main idea of the ZUPT and ZARU method.

1) Acceleration amplitude detection. In the zero speed phase, the acceleration vector value of the pedestrian's foot changes around the gravitational acceleration. Therefore, the logical value of the moment is obtained by judging the relationship between the acceleration amplitude and the set threshold.

$$\left| a_k^b \right| = \sqrt{ {a_{x(k)}^b}^2 + {a_{y(k)}^b}^2 + {a_{z(k)}^b}^2 } \tag{2}$$

$$c_1 = \begin{cases} 1, & T_{\min} < |a_k| < T_{\max} \\ 0, & others \end{cases} \tag{3}$$

2) Acceleration amplitude variance detection. Variance is an important condition for measuring the degree of dispersion of variables and averages. By setting the sliding window w, the variance of the data is obtained to determine the dispersion of the zero-speed point.

$$\sigma_{a_k}^2 = \frac{1}{2w+1} \sum_{j=k-w}^{k+w} (a_j - \bar{a}_k)^2 \tag{4}$$

$$c_2 = \begin{cases} 1, & |\sigma_{a_k}^2| < T_{\sigma\max} \\ 0, & others \end{cases} \tag{5}$$

3) Angular velocity amplitude detection. When in the zero speed interval, the angular velocity of the foot changes to be close to 0, and the logical value is obtained by judging the magnitude relationship between the modulus value of the angular velocity at time k and the threshold value.

$$|\omega| = \sqrt{\omega_x^2 + \omega_y^2 + \omega_z^2} \tag{6}$$

$$c_3 = \begin{cases} 1, & |\omega_k| < T_{\omega\max} \\ 0, & others \end{cases} \tag{7}$$

4) Peak acceleration detection. To determine the acceleration peak of Δt t between the two times before and after the vertical direction, Δt t is the zero speed interval.

$$a_{\max}(k) = \begin{cases} a_z(k) < T_{\max} \\ a_z(k) < a_z(k-1) \\ a_z(k) < a_z(k+1) \end{cases} \tag{8}$$

$$C_4(k + \Delta t_1, k + \Delta t_2) = \begin{cases} 1, & a_{\max}(k) \to true \\ 0, & others \end{cases} \tag{9}$$

According to the above four constraints, the conditional threshold is set respectively, and the obtained judgment result is calculated by the relationship of logical "AND", the equation is as follows:

$$C = C_1 \, \& \, C_2 \, \& \, C_3 \, \& \, C_4 \tag{10}$$

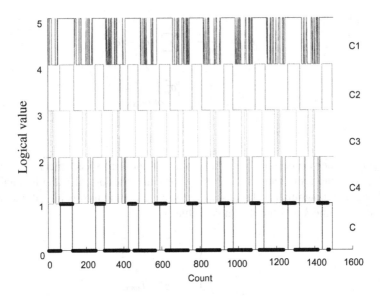

Fig. 1. The logical value of zero velocity detection

When C is 1, the zero speed interval. Figure 1 shows the pedestrian zero velocity detection experiment. The figure captures the data of 9 steps of pedestrian walking. It shows that the algorithm results are consistent with the actual number of steps, which indicates that the algorithm accurately determines the zero velocity phase of pedestrians.

4 Experiment and Analysis

The receiver equipment used in the experiment can receive GPS and BDS signals, and the IMU includes a three-axis gyroscopes, accelerometers, and magnetometers, technical specifications of these three sensors are show in Table 1. The sampling frequency of the receiver is 1 Hz, the cut-off angle is 10°, and the IMU sampling frequency is 100 Hz. In the experiment, we selected a path from outdoor to indoor, as shown in Fig. 2, it is the experimental environment for aerial photography. Fixed the base station antenna on the rooftop, fixed IMU on the pedestrian's foot, and handheld rover station receiver at the same time, after initial alignment, follow the route and walk normally to collect data.

Table 1. Sensor specification

Technical specification	Gyroscopes	Accelerometers	Magnetometers
Standard full range	±300 deg/s	±50 m/s^2	±700 mGauss
Zero stability	5 deg/s	0.02 m/s^2	0.5 mGauss
Noise density	0.1 deg/s/$\sqrt{Hz}$	0.02 m/s^2 $\sqrt{Hz}$	0.5 mGauss

When the pedestrian is outside, the satellite signal is good and the BDS/GPS positioning is effective. In order to verify the superiority of BDS/GPS dual-mode positioning to the single-satellite system, the BDS/GPS fusion data and GPS data were calculated by GrafNav software respectively, and the Kinpos software developed by Wuhan University was used to process the data of the single BDS system. The standard deviations of the East, north, and up components of the three solutions are given in Figs. 3, 4 and 5. Table 2 summarizes the root mean square error of each component of the three schemes.

Table 2. RMS of GPS/BDS

Constellation	E/m	N/m	U/m
GPS	0.011	0.012	0.022
BDS	0.007	0.018	0.024
BDS/GPS	0.008	0.0067	0.018

Through three kinds of solving methods, the observation data is dynamically located by each epoch, and the epochs participating in the solution are all fixed. The positioning accuracy of each component is less than 3 cm, the average error of the horizontal position of the BDS/GPS system is less than 1.1 cm, and the accuracy is improved by 5 mm to 8 mm. During the experiment, the walking route gradually approaches the indoor environment from the outdoor. Due to the obstruction of obstacles such as walls, the error fluctuates when walker is entering into the indoor environment. The maximum horizontal error of GPS and BDS single system is 0.04 m and 0.03 m, respectively, while the maximum horizontal error of combined system is 0.028 m. It indicates that under the influence of obstacles and other factors, the BDS/GPS dual-system can reduce the error and improve the positioning accuracy compared with the single system (Table 2).

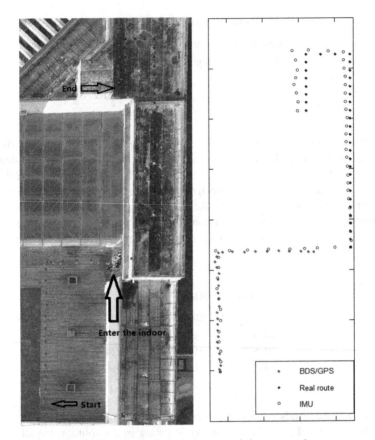

Fig. 2. Experimental environment and data processing

The average error of the BDS and GPS combination solution is 1.08 cm, and the positioning accuracy can reach the centimeter level. The main reason for the improved positioning accuracy of the combined dual-mode BDS and GPS positioning system is that the joint solution increases the number of spatial visible satellites, improves the geometric distribution of satellite and reduces the value of PDOP that affects the positioning accuracy. As shown in Table 3, the PDOP value is reduced due to the increase of the number of satellites. The smaller the PDOP, the better the accuracy of the positioning, therefore, the BDS/GPS dual-mode positioning accuracy is higher than that of the single system.

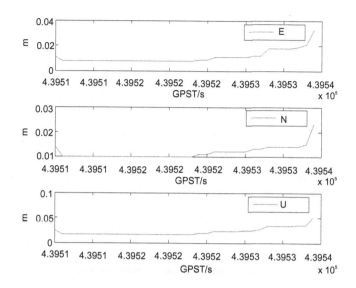

Fig. 3. GPS position error estimation

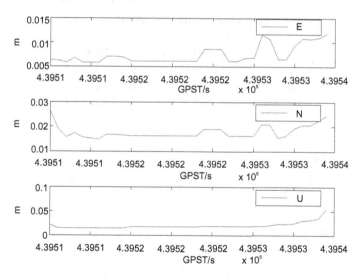

Fig. 4. BDS position error estimation

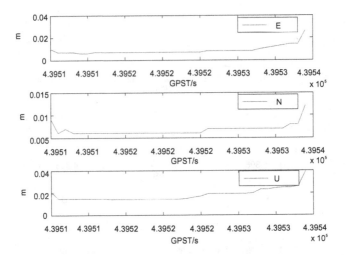

Fig. 5. GPS/BDS position error estimation

Table 3. Visual satellite number and PDOP value of each system

Constellation	Visual satellite number			PDOP		
	MAX	MIN	AVG	MAX	MIN	AVG
GPS	9	5	8.3	3.3	1.3	1.73
BDS	9	5	7.7	3.64	2.11	2.41
BDS/GPS	18	10	16	2.7	1.1	1.4

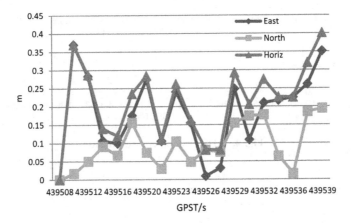

Fig. 6. IMU positioning error in the outdoor

At the same time, the outdoor IMU data is processed separately and compared with the satellite combination positioning. As shown in Fig. 7, the maximum absolute position error of the PDR algorithm is less than 0.4 m, which indicates that the algorithm achieves good results.

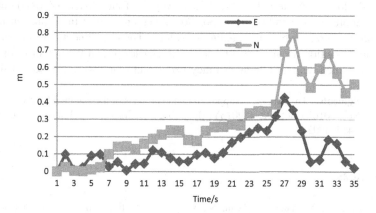

Fig. 7. PDR indoor positioning error

When the pedestrian entered the room, the BDS and GPS signal were interrupted, PDR algorithm was used to fix position. Table 4 shows the result of indoor navigation. The indoor algorithm results show that the average position errors in the East and the North are 0.102 m and 0.285 m respectively. The standard deviation in the East and the North are 0.123 m and 0.220 m, respectively. It can be seen that the zero velocity detection algorithm under multiple constraint conditions can improve the accuracy of step detection. Therefore, the zero velocity correction algorithm based on the EKF can significantly reduce the positioning divergence with time and reduce the positioning error, and realize seamless navigation between indoor and outdoor environments.

Table 4. Indoor positioning error

	East/m	North/m
MIN	0	0
MAX	0.427	0.799
AVG	0.102	0.285
STD	0.123	0.220

5 Conclusion

A higher precision method for seamless pedestrian navigation is proposed in this paper. Combined with the dual-mode positioning of BDS and GPS, the number of visible satellites can still be achieved in the relatively poor outdoor shelter, which makes the geometric distribution of the observation satellites good, reduces the influence of multipath effect and other factors, and improves the positioning accuracy. In the environment where satellite signals cannot be received or mismatches, proposed the zero velocity multi-constraint conditions detection algorithm based on the Extended Kalman Filter (EKF), applied and updated with velocity and angular rate measurements by Zero-Velocity-Update (ZUPT) and Zero-Angular-Rate-Update (ZARU) solutions, greatly suppresses the influence of position divergence over time, and ensures the positioning accuracy when the satellite loses lock. The experimental results show that the outdoor position accuracy can reach centimeter level, and the indoor error ratio is about 1%, which realizes a high accuracy pedestrian seamless navigation method.

Acknowledgments. The paper was supported by the projects of the National Key R&D Program of China (Nos. 2016YFB0502204, 2016YFB0502002), National Natural Science Foundation of China (No. 41764002).

References

1. Deng, Z., Yu, Y., Yuan, X., Wan, N., Yang, L.: Situation and development tendency of indoor positioning. China Commun. **10**(3), 42–55 (2013)
2. Popleteev, A.: Indoor Positioning using FM Radio Signals. University of Trento, Trento (2011)
3. Xu, Y., Ahn, C.K., Shmaliy, Y.S., et al.: Adaptive robust INS/UWB-integrated human tracking using UFIR filter bank. Measurement **123**, 1–7 (2018)
4. Xu, Y., Shmaliy, Y.S., Li, Y., Chen, X.: UWB-based indoor human localization with time-delayed data using EFIR filtering. IEEE Access **5**(1), 16676–16683 (2017)
5. Li, H.X., Wen, X., Guo, H., et al.: Research into kinect/inertial measurement units based on indoor robots. Sensors **18**(3), 839 (2018)
6. Xu, Y., Karimi, H.R., Li, Y.Y., Zhou, F.Y., Bu, L.L.: Real-time accurate pedestrian tracking using EFIR filter bank for tightly coupling recent inertial navigation system and ultra-wideband measurements. Proc. Inst. Mech. Eng. Part I-J. Syst. Control Eng. **232**(4), 464–472 (2018)
7. Guo, H., Li, H., Xiong, J., Yu, M.: Indoor positioning system based on particle swarm optimization algorithm. Measurement **134**, 908–913 (2019)
8. Xu, Y., Shmaliy, Y.S., Li, Y., Chen, X., Guo, H.: Indoor ins/lidar-based robot localization with improved robustness using cascaded fir filter. IEEE Access **7**(1), 34189–34197 (2019)
9. Xu, Y., Tian, G., Chen, X.: Enhancing INS/UWB integrated position estimation using federated EFIR filtering. IEEE Access **6**, 64461–64469 (2018)
10. Xu, Y., Chen, X.: Online cubature Kalman filter Rauch–Tung–Striebel smoothing for indoor inertial navigation system/ultrawideband integrated pedestrian navigation. Proc. Inst. Mech. Eng. Part I-J. Syst. Control Eng. **232**(4), 390–398 (2018)
11. Pany, T., Winkel, J., Riedl, B., Niedermeier, H., Eissfeller, B., Wörz, T., et al.: Experimental results from an ultra-tightly coupled GPS/Galileo/WiFi/ZigBee/MEMS-IMU indoor navigation test system featuring coherent integration times of several seconds. In: Satellite Navigation Technologies & European Workshop on GNSS Signals & Signal Processing, pp. 1–8. IEEE (2011)

12. Stone, J.M., Powell, J.D., Powell, P.J.D.: Precise positioning with GPS near obstructions by augmentation with pseudolites. In: IEEE Position Location and Navigation Symposium, pp. 562–569 (1998)
13. Sakamoto, Y., Niwa, H., Ebinuma, T., Fujii, K., Sugano, S.: Indoor positioning with pseudolites—the effect of the number of receivers and transmitters. In: Sice Conference, pp. 398–399 (2010)
14. Huang, J., Millman, D., Quigley, M., Stavens, D., Thrun, S., Aggarwal, A.: Efficient, generalized indoor WiFi GraphSLAM. In: 2011 IEEE International Conference on Robotics and Automation, pp. 1038–1043. IEEE (2011)
15. Junyong, C.: On the establishment of Chinese modern geodetic coordinate system. Board Geomat. Inf. Sci. Wuhan Univ. 27(5), 441–44 (2002)
16. Li, J., Yang, Y., Xu, J., He, H., Guo, H.: GNSS multi-carrier fast partial ambiguity resolution strategy tested with real BDS/GPS dual- and triple-frequency observations. GPS Solut. 19 (1), 5–13 (2013)
17. Liang, W., Shen, L.Z., Hong, Y., Kai, Z.: Validation and analysis of the performance of dual-frequency single-epoch BDS/GPS/GLONASS relative positioning. Chin. Sci. Bull. 60 (9), 857 (2015). Opto-Electronics, A.O.
18. Caron, F., Duflos, E., Pomorski, D., Vanheeghe, P.: GPS/IMU data fusion using multisensor Kalman filtering: introduction of contextual aspects. Information Fusion 7(2), 221–230 (2016)
19. Wendel, J., Meister, O., Schlaile, C., Trommer, G.F.: An integrated GPS/MEMS-IMU navigation system for an autonomous helicopter. Aerosp. Sci. Technol. 10(6), 527–533 (2006)
20. Jiménez, A.R., Seco, F., Prieto, J.C., Guevara, J.: Indoor pedestrian navigation using an INS/EKF framework for yaw drift reduction and a foot-mounted IMU. In: 2010 7th Workshop on Positioning, Navigation and Communication, pp. 1–9. IEEE (2010)

Intelligent Technology and Design
for Special Education/Rehabilitation

Sign Language Video Classification Based on Image Recognition of Specified Key Frames

Zhaosong Zhu[1]⑩, Xianwei Jiang[1,2], and Juxiao Zhang[1(✉)]⑩

[1] Nanjing Normal University of Special Education, Nanjing 210038, China
zzs2019@foxmail.com, 3301611@qq.com
[2] Department of Informatics, University of Leicester, Leicester LE1 7RH, UK

Abstract. This paper is based on the Chinese sign language video library, and discusses the algorithm design of video classification based on handshape recognition of key frames in video. Video classification in sign language video library is an important part of sign language arrangement and is also the premise of video feature retrieval. At present, sign language video's handshape classification work is done manually. The accuracy and correctness of the results are quite erroneous and erroneous. In this paper, from the angle of computer image analysis, the definition and extraction of key frames are carried out, and then the region of interest is identified. Finally, an improved SURF algorithm is used to match the area of interest and the existing hand image, and the classification of the video is completed. The entire process is based on the actual development environment, and it can be used for reference based on the classification of video image features.

Keywords: Classification of videos · Classification of sign language · Key frame extraction · Image matching · Handshape matching

1 Introduction

With the development of information technology, Internet technology and multimedia technology have been greatly improved, In particular, the emergence of "we media", such as YouTube, Facebook, tiktok and so on, has led to the explosive growth of videos on the Internet, in this case, the manual annotation has become impossible, and the subjectivity of artificial tagging cannot meet the needs of users. In order to facilitate the management and retrieval of massive video, automatic video classification is particularly important, Automatic video classification is also widely used in video monitoring, network supervision, medicine and other fields: For example, Johnson et al. proposed a multi-mode monitoring method, which extracts the static features of human body from multiple angles to realize the detection, separation and recognition of human beings at a distance [1]; Through video classification, video on the Internet is regulated to filter out undesirable videos (pornography, violence, etc.) [2]. The video classification method was applied to the video library obtained by the wireless endoscope, and all the videos was classified according to the different organs diagnosed [3]. In addition to specific fields, there are also some general video classification methods: Fischer et al. proposed in 1995 that video could be divided into news, sports, business, advertising,

© ICST Institute for Computer Sciences, Social Informatics and Telecommunications Engineering 2020
Published by Springer Nature Switzerland AG 2020. All Rights Reserved
Y.-D. Zhang et al. (Eds.): ICMTEL 2020, LNICST 327, pp. 371–381, 2020.
https://doi.org/10.1007/978-3-030-51103-6_33

cartoons, etc. [4]; Huang et al. proposed a classification algorithm based on text features, extracted user-generated text features, and used a classifier for classification [5]. Jiang et al. proposed a method of video classification using support vector machine (SVM) based on visual features (color, motion, edge, etc.) [6]; Subashini et al. proposed a machine learning algorithm based on audio features and image histograms [7].

However, the current video classification methods generally have two problems:

1) Insufficient universality. Some prior knowledge is needed to design the classification rules, which can only be targeted at certain fields;
2) Complex algorithms. It needs a lot of computing resources to use multi-level deep learning algorithm to deal with video library with a large amount of video.

In order to solve the above problems, this paper proposes a classification method for key frame images of sign language video. The main steps are as follows:

1) Extraction of key frames from sign language video;
2) Image visual feature preprocessing and hotspot extraction;
3) Feature matching with the designated image to achieve video classification.

2 Datasets and Problems

The object of this paper is video library of Chinese sign language (csl-lib, Project NO. zda125-8, 2016). The library contains 57,531 sign language vocabulary videos from nine specific regions in China. At present, the copyright of this dataset belongs to the Chinese language and script commission, and some contents will be released later. In the retrieval operation of video library, the handshape index of video is an important retrieval method, which is also the only retrieval method based on video image features, and has important video analysis and research value. The classification is mainly based on the 60 hand shapes in sign language (see Fig. 1). It will consume a lot of human, material and time resources by manual classification. In view of this problem, a set of practical classification methods is proposed from the perspective of computer image processing, Furthermore, SURF algorithm [9] which is based on Lowe D G's SIFT algorithm [8], is improved by plane angle rotation. Meanwhile, the key frame extraction algorithm of literature [10, 11] is applied. Finally, a classification method based on key frame matching is proposed which can be applied to sign language handshape classification and has the characteristics of batch and high efficiency.

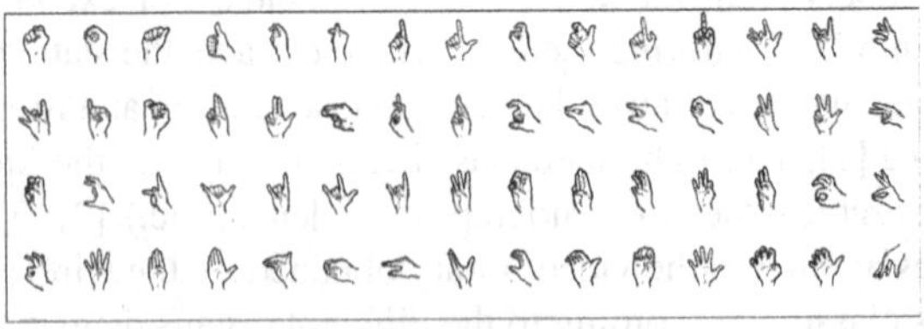

Fig. 1. The handshape index for Chinese sign language

3 Extraction of Video Key Frames

The key frame of video retrieval is defined as the image of the handshape used by the gesture in the sequence formed by the video stream. Take the sign language video 'lightning/electricity' as an example. This sign language is the standard sign language. In most parts of mainland China, it has the same or similar stroke (the left hand does not move, and the right hand draws the shape of lightning in the air), so it is typical of cases.

Key frame extraction is mainly divided into two steps: Firstly, video serialization and graying. The grayscale processing adopts the general formula (1) proposed in literature [12] to form the video sequence as shown in Fig. 2. Secondly, extract key frames according to the algorithm.

$$gray = 0.299 \times R + 0.587 \times G + 0.114 \times B \tag{1}$$

Fig. 2. The grayscale sequence of sign language video 'lightning/electricity'

The extracted key frame image shoulsd have two characteristics: 1) it should have a certain duration of stability; 2) clear edges can be preprocessed into hand recognition materials.

According to the two characteristics, the key frames are described as follows during the processing: The current frame has a small difference with the preceding m frames and the following m frames, The value of m is a natural number, so it should not exceed the number of frames, In the method of this paper, the efficiency of video library processing is considered. So m is equal to 2, that is, the difference is calculated with the first two frames and the second two frames. The difference coefficient is calculated by the gaussian function, and the difference is summarized at last.

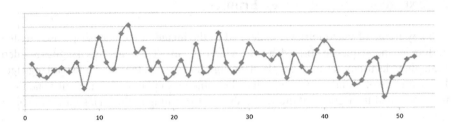

Fig. 3. The mean grayscale image sequence of sign language 'lightning/electricity', the X-axis is the video image sequence, and the Y-axis shows the change of grayscale mean

Assuming that the number of frames in the video is n, and the key frame is i, V_i is defined as the grayscale matrix of the sequence image i, The vector (i, V_i) forms a discrete function: $V_i = f(i)$, Fig. 3 shows the mean value function of gray square of "lightning" sign language video sequence. Key frame image extraction is completed according to the following formula:

$$V_{key} = f\left(\int Min_p(|f'(x)|)\right) \tag{2}$$

But it's hard to compute the derivative of the function formed by the discrete sequence directly, So $f'(x)$ is calculated using the series of coefficient operators S. S is normalized by the gaussian function formula (3), (4), (5) (A = 1, σ = 1.5)

$$G(x) = Ae^{-\frac{x^2}{2\sigma^2}} \tag{3}$$

$$S = |-0.135, -0.365, 1, -0.365, -0.135| \tag{4}$$

$$f'(x) = |V_{i-2}, V_{i-1}, V_i, V_{i+1}, V_{i+2}| \times S^T \tag{5}$$

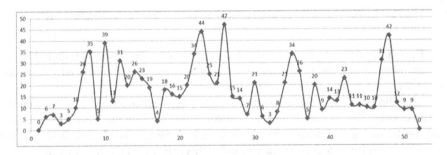

Fig. 4. The change rate of the overall grayscale pixel in the image sequence

Generate $|f'(x)|$ data sequence (Fig. 4) to reflect the changes between images. According to the actual situation of the video, the head subsequence and the tail subsequence are removed, because these two sub-sequences may contain useless information. In this project, the front part is about 10 frames for the video model gesture preparation stage, and the back part is about 10 frames for her gesture homing stage, which needs to be removed. Then find the p frames in the sequence where the rate of change is from small to large, $Min_p(|f'(x)|)$, When $p = 3$, three Ordinal Numbers conforming to the conditions are obtained, which are 17, 32 and 37 respectively. The corresponding images are frame 17, 32 and 37 (Fig. 5).

Fig. 5. The three images with the lowest rate of change in the video sequence

The video key frame is extracted, and then the hand shape matching recognition is carried out.

4 Image Visual Feature Preprocessing and Thermal Region Extraction

The interference of hand image matching mainly comes from the following two points: 1) Differences between the matched images. That is, the source of the key frame image and the matching handshape image are different, and the difference of the image feature is relatively large, so it is not easy to form a match. 2) Spatial difference. Although it is the same handshape, the spatial position transformation gap is large and it is not easy to match.

For the second point, there is no valid three-dimensional transpose algorithm for plane space, Therefore, in order to reduce the first kind of interference and form an effective matching feature, the key frame image is preprocessed before matching to improve the matching accuracy. Preprocessing is divided into two steps: 1) Image marginalization and binarization to reduce pixel interference; 2) image hotspot extraction to reduce the interference of non-hot region.

4.1 Image Marginalization and Binarization

High pass filter is used to process the image. The improved regional edge detection algorithm proposed by J Canny [13] and wang [14] et al. was referred to and simplified to meet the needs of this project. The kernel matrix of Sobel operator is used as filter (Fig. 6).

-1	0	1
-2	0	2
-1	0	1

-1	-2	-1
0	0	0
1	2	1

Fig. 6. Sobel operator for edge detection

The horizontal filtering and vertical filtering were performed respectively, and after filtering, the matrices sobel_X and sobel_Y were formed, Then the L1 normal form is used to obtain filtering results.

$$L1 = |sobel_X| + |sobel_Y| \qquad (6)$$

Finally, the low threshold is used for binarization processing to retain the basic contour information. In the case of this project, the gray intermediate value 128 is adopted as the threshold. The result is shown in Fig. 7. The selection of threshold value should be optimized according to the specific situation of the image.

Fig. 7. key frame images that has been marginalized

4.2 Extraction of Image Hot Area

The extraction process of hot area is as follows: Suppose the key frame image is Img_i. It's previous frame is Img_{i-1}, and next frame is Img_{i+1}, The extraction formula is as follows:

$$Img_{key} = |Img_i - Img_{i-1}| + |Img_i - Img_{i+1}| \qquad (7)$$

By using the difference of adjacent images to extract the hot areas, the background interference can be minimized and only the dynamic region can be concerned, Finally, the results are filtered by low threshold, and the remaining part is the hot areas, The final results are shown in Fig. 8, and the position of hot areas is shown in Table 1.

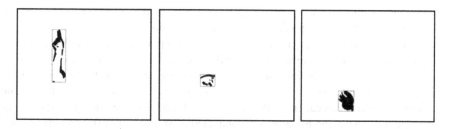

Fig. 8. Sign language video 'lightning/electricity' key frame image hot areas

Table 1. Coordinates of hot areas in three images of Fig. 8

Upper left and lower right coordinates	X_0	Y_0	X_1	Y_1
Left image	177	104	275	104
Middle image	218	337	297	337
Right image	200	419	283	419

5 Matching the Hot Areas of Key Frame with Handshape Image

In the process of matching the key frame hot areas with the feature handshape images, the scale-invariant feature matching is needed. The accelerated version of SIFT algorithm [8], SURF algorithm [9], was used and the results were further filtered, SURF algorithm has the following 6 steps:

(1) Construct the Hessian matrix of the image through formula (8) and calculate the eigenvalue. The convolution window of gaussian filter is used in the calculation process, and the simplified matrix of 3 * 3 is adopted $\begin{bmatrix} 1 & 2 & 1 \\ 2 & 4 & 2 \\ 1 & 2 & 1 \end{bmatrix}$.

(2) Construct gaussian difference scale space of the image. It is generated by convolution of gaussian difference kernel with image at different scales. Core formula (8), (9) and (10) are as follows:

$$L(x,y,\sigma) = G(x,y,\sigma) * I(x,y) \tag{8}$$

$$G(x,y,\sigma) = \frac{1}{2\pi\sigma^2} e^{\frac{-(x^2+y^2)}{2\sigma^2}} \tag{9}$$

$$D(x,y,\sigma) = (G(x,y,k\sigma) - G(x,y,\sigma))I(x,y) = L(x,y,k\sigma)L(x,y,\sigma) \tag{10}$$

Since the image size is small, the space is set as 3 scales. Each scale contains five levels. The template size of the filter increases gradually from 1 to 3 at different scales. In 1 scale, the filter's fuzzy coefficient increases gradually, forming a 3 * 5 scale space.

(3) Feature point location. Compare each pixel processed by Hessian matrix with 26 points in the neighborhood of 2d image space and scale space. The key points were preliminarily located, and then the weak key points and the wrong key points were filtered out to screen out the final stable characteristic points.

(4) Main direction distribution of feature points. The harr wavelet transform in the circular neighborhood of the statistical feature points is used. The sum of the horizontal and vertical Haar-like features of all points in the 60-degree sector was calculated. After the rotation was conducted at an interval of 0.2 radians and the harr small baud value in the region was calculated again, the direction of the sector with the largest value was finally taken as the main direction of the feature point.

(5) generate feature point descriptors. Take a rectangle of 4 * 4 around the feature point. The direction of the rectangle region is the main direction along the feature point. Each subregion counts the horizontal and vertical Haar-like features of 25 pixels. The Haar-like features contains four sum operations, that is, the sum of the horizontal value, the sum of the vertical value, the sum of the horizontal absolute value, and the sum of the vertical absolute value.

(6) Feature point matching The Euclidean distance between two feature points is calculated to determine the matching degree. The shorter the Euclidean distance is, the better the matching degree of the two feature points is. At the same time, the Hessian matrix trace was judged. The matrix trace of the feature points had the same sign, which represented the contrast change in the same direction. If different, it was the opposite, even if the Euclidean distance was 0, it was directly excluded.

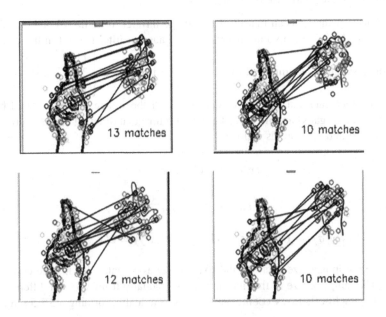

Fig. 9. The matching result of the key frame hot area and handshape image

This algorithm is used to match the hot areas of the key frames with each hand-shape image in turn, Finally, the percentage matching value is formed, Not all of the 60 handshape feature matching results are listed, but some typical matching results are listed as shown in Fig. 9.

As can be seen from Fig. 9, although the handshape with the highest matching degree can be analyzed as the top left figure, with the matching degree of about 14.44% among the 60 hand shapes, the difference between the matching degree of other handshapes is not large enough to form matching results, such as figure 13.33% in the bottom left figure, figure 11.11% in the top right figure, and figure 11.11% in the bottom right figure.

From the matching results, it can be seen that the matched feature point pairs need to be filtered due to more interference. Inspired by the geometric characteristics proposed by Liu et al. [15], The filtering method is as follows:

If the matching image is placed in the same plane, the two images will have relatively fixed positions in the plane coordinate system, as shown in Fig. 10.

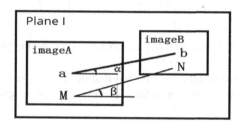

Fig. 10. Image A is the hot area of key frame, and image B is the handshape image. They are in the same plane I.

The connecting line segment of the center point of imageA and imageB is ab, The Angle between ab and the horizontal line is α. Let the Angle increment be $\Delta\alpha$, and assume that the points M and N are a pair of matching points. The Angle between the line MN and the horizontal line is β, If β satisfies formula (11), then MN is the matching point, otherwise it is deleted from the matching point pairs. You can make some adjustments by setting the size of the $\Delta\alpha$.

$$\beta\varepsilon\left[\alpha - \Delta\alpha,\ \alpha + \Delta\alpha\right] \tag{11}$$

The matching point pairs formed after filtering are shown in Fig. 11. The matching degrees of handshapes in the four cases were 21.43% in the top left, 17.86% in the bottom left, 5.36% in the top right, and 5.36% in the bottom right. The difference is large enough for classification, and A good result distribution was formed in 60 handshapes matches.

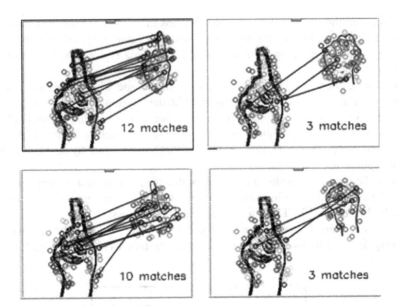

Fig. 11. The result of the Angle operation on matching point pairs

Based on the above results, we can find that the handshapes in the sign language video "lightning" have a good matching degree with the seventh and eighth hand shapes in Fig. 1.

6 Conclusion

In order to classify the videos in the sign language database by specified handshapes, this paper proposes an algorithm combination flow. Firstly, extract the key frames of digital video. The video image sequence matrix difference is used to calculate the sequence change rate and the smaller frame is taken as the key frame; Secondly, the feature extraction and hot areas extraction of the key frame images are carried out. Feature extraction uses Sobel operator to extract the contour and conduct binarization, and hot areas extraction is to use the difference between adjacent images to separate the image area with larger values. Finally, handshapes matching was performed, mainly using SURF algorithm. In addition, the point pairs generated by SURF algorithm were filtered into the plane at an Angle to form a handshape-matching distribution satisfying the requirements. Compared with the supervised learning algorithm, this process avoids the stage of sample learning and the complicated classification calculation, saves the computing resources, and has a certain efficiency and practicability. But because the video key frame image has spatial transformation, it can't match the specified hand-shape completely. The next research direction focuses to solve this problem with depth information [16], and at the same time, deep learning algorithm and more graphical features [17] is introduced to apply this algorithm flow to video classification and other practical applications.

Acknowledgement. This work was supported by Surface Project of Natural Science Research in Colleges and Universities of Jiangsu China (No.16KJB520029), The Major Programs of Natural Science Foundation of the Jiangsu Higher Education Institutions of China (No. 19KJA310002.) and The Natural Science Foundation of the Jiangsu Higher Education Institutions of China (No. 17KJD520006).

References

1. Johnson, A.Y., Bobick, A.F.: A multi-view method for gait recognition using static body parameters. In: Bigun, J., Smeraldi, F. (eds.) AVBPA 2001. LNCS, vol. 2091, pp. 301–311. Springer, Heidelberg (2001). https://doi.org/10.1007/3-540-45344-X_44
2. Jiang, C., Jiang, X., Sun, T.: Video filtration for content security based on multimodal features. Inf. Secur. Communi. Privacy **3**, 76–77 (2012). (in Chinese)
3. Feng, W., Gao, J., Bill, P.B., et al.: Wireless capsule endoscopy video classification using an unsupervised learning approach. J. Image Graph. **16**(11), 2041–2046 (2011). (in Chinese)
4. Fischer, S., Lienhart, R., Effelsberg, W.: Automatic recognition of film genres. In: Proceedings of the 3rd ACM International Conference on Multimedia, pp. 295–304. ACM Press, New York (1995)
5. Huang, C.N., Fu, T.J., Chen, H.C.: Text-based video content classification for online video-sharing sites. J. Am. Soc. Inf. Sci. Technol. **61**(5), 891–906 (2010)
6. Jiang, X.H., Sun, T.F., Wang, S.L.: An automatic video content classification scheme based on combined visual features model with modified DAGSVM. Multimedia Tools Appl. **52**(1), 105–120 (2011)
7. Subashini, K., Palanivel, S., Ramalingam, V.: Audio-video based classification using SVM. IUP J. Sci. Technol. **7**(1), 44–53 (2011)
8. Lowe, D.G.: Distinctive image features from scaleinvariant keypoints. Int. J. Comput. Vis. **60**(2), 91–110 (2004)
9. Bay, H., Tuytelaars, T., Van Gool, L.: SURF: speeded up robust features. In: Leonardis, A., Bischof, H., Pinz, A. (eds.) ECCV 2006. LNCS, vol. 3951, pp. 404–417. Springer, Heidelberg (2006). https://doi.org/10.1007/11744023_32
10. Zhen, E., Lin, J.: Unordered image key frame extraction based on image quality constraint. Comput. Eng. **43**(11), 210–215 (2017)
11. Wang, Y., Sun, S., Ding, X.: A self-adaptive weighted affinity propagation clustering for key frames extraction on human action recognition. J. Vis. Commun. Image Representation **33**(3), 193–202 (2015)
12. Yi, R., Tomasi, C., Guibas, L.J.: Mover's distance as a metric for image retrieval. Int. J. Comput. Vis. **40**(2), 99–121 (2000)
13. Canny, J.: A computational approach to edge detection. IEEE Trans. Pattern Anal. Image Understand. **18**(6), 679–698 (1986)
14. Wang, X., Liu, X., Guan, Y.: Image edge detection algorithm based on improved Canny operator. Comput. Eng. **34**(14), 196–198 (2012)
15. Wu, L., Yadong, C.Y.: Gesture recognition based on geometric features. Comput. Eng. Des. **35**(2), 636–640 (2014). (in Chinese)
16. Binjue, Zhang, Liaoyin, Zhao, Yixuan, Wang: Fingeritip detection and gesture recognition based on kinect depth data. IEEE Trans. Comput. Sci. Technol. **3**(1), 9–14 (2014)
17. Mitra, S., Acharya, T.: Gesture recognition: a servey. IEEE Trans. Syst. Man Cybern. Part C Appl. Rev. **37**(3), 311–324 (2007)

Chinese Fingerspelling Recognition via Hu Moment Invariant and RBF Support Vector Machine

Ya Gao[1], Ran Wang[1], Chen Xue[1], Yalan Gao[1], Yifei Qiao[1], Chengchong Jia[1], and Xianwei Jiang[1,2(✉)]

[1] School of Mathematics and Information Science, Nanjing Normal University of Special Education, Nanjing 210038, China
jxw@njts.edu.cn
[2] Joint Accessibility Key Laboratory, China Disabled Persons' Federation, Nanjing 210038, China

Abstract. Sign language plays a significant role in smooth communication between the hearing-impaired and the healthy. Chinese fingerspelling is an important composition of Chinese sign language, which is suitable for denoting terminology and using as basis of gesture sign language learning. We proposed a Chinese fingerspelling recognition approach via Hu moment invariant and RBF support vector machine. Hu moment invariant was employed to extract image feature and RBF-SVM was employed to classify. Meanwhile, 10-fold across validation was introduced to avoid overfitting. Our method HMI-RBF-SVM achieved overall accuracy of 86.47 ± 1.15% and was superior to three state-of-the-art approaches.

Keywords: Hu moment invariant · RBF · Support Vector Machine · Chinese fingerspelling recognition

1 Introduction

Around the world, hundreds of thousands of people suffer from hearing impairment [1]. Deafness is a problem that affects people's lives in many aspects because all communication between individuals is through language. Nowadays, continuous advances in mobile technology and new forms of user interaction have made it possible to overcome the communication problems between deaf and healthy people. Even so, the use of sign language is necessary. Sign language (SL) refers to the communication tool that people use in the deaf environment to express a specific meaning according to certain grammatical rules in accordance with the gestures, movements, positions and orientations of their hands. Chinese sign language (CSL) can generally be divided into gesture sign language and fingerspelling language [2]. Gesture sign language is a major supplement of spoken language and plays an auxiliary role in communicating information and feelings with the outside world. Gesture sign language often uses both hands at the same time, supplemented by facial expressions, three main features were extracted from the gesture signature gestures, namely the position of the hand, the

Y.-D. Zhang et al. (Eds.): ICMTEL 2020, LNICST 327, pp. 382–392, 2020.
https://doi.org/10.1007/978-3-030-51103-6_34

direction of the hand signature, and the shape of the hand [3]. It emphasizes the situation, relatively complex. While the latter fingerspelling language emphasizes 30 basic finger languages, including 26 basic pinyin letters and 4 raised tongues, and expresses pinyin or some special meanings through their combination. Fingerspelling language is relatively simple and definite. In addition, the origin of gesture sign language is early, which has experienced three main stages, namely the germination of ancient times, the disunity of modern times and the perfection and unification of modern times. Fingerspelling language originated from the west, through the introduction and learning, gradually realize the Chinese practice. Currently, refinement and standardization are two basic tasks in the development of Chinese sign language.

Sign language recognition (SLR) is the use of computer technology to translate sign language information into text, natural language, audio and other information to facilitate understanding and communication [4]. Moreover, sign language recognition is a significant part of intelligent human-computer interaction. According to the data output, SLR technology can be divided into two categories: sensor-based sign language recognition technology and computer vision-based sign language recognition technology. The former employs wearable devices such as data gloves and EMG signal arms, while the latter mainly adopts depth cameras. Due to practical factors, sign language recognition based on computer vision is more popular. The features commonly used in sign language recognition are hand shape, direction, position and movement track. Most experimental studies have revolved around these features.

There are many methods of sign language recognition, one of which utilizes statistical analysis techniques to derive the various eigenvectors of a sample and then classify it. Hidden markov model (HMM) [5] is a typical representative of this. The second method employs template matching technology, that is, first construct a defined template, then match the original data with the template, and use the similarity as a reference to complete the identification. Another method uses the self-learning and organizational functions of the network to build on a new neural network. However, these methods have shortcomings. HMM relies on complex initialization process and huge computation workload. It requires the existence of successfully detected gesture region and gesture movement, which determines the robustness of the algorithm. The accuracy of traditional template matching methods is not satisfactory. New neural network techniques require large amounts of data and consume more training sessions. To solve these problems, a Chinese sign language recognition method based on wavelet entropy and support vector machine was proposed [6]. Gray-level co-occurrence matrix (GLCM) was also employed to identify sign language [7]. In addition, many experts try to use the depth information of gestures to solve the issue of recognition accuracy. Taiwan sign language recognition system based on Kinect extracted three main features from sign language gestures, namely gesture position, gesture direction and gesture shape [8]. The position of the hand was obtained by input sensor. HMM was introduced to determine the direction of the gesture, and a trained SVM was employed to identify the shape of the gesture, the recognition rate of this system reached 85.14% [9]. Convolutional neural network (CNN) and its variant were also tested to identify fingerspelling [10]. For the problem of hand posture recognition, hand contour is usually adopted because of the simplification of real-time. All of the current related work can be divided globally into software-based and hardware-based solutions.

In this study, a suitable method for Chinese fingerspelling recognition via Hu moment invariant and RBF support vector machine was proposed. Hu moment invariant was employed to extract the image features and accelerate the training. RBF was employed to enhance classification due to its effective performance. Meanwhile, the experiment was carried out on 10-fold cross-validation to prevent overfitting.

2   Method

2.1   Subjects and Dataset

We acquired 1320 Chinese fingerspelling language samples to construct experiment materials, which were from 44 volunteers. Following 30 categories in Chinese fingerspelling language, every volunteer simulates action one time, thus totally 1320 images were gained. All these images were preprocessed by software and normalized to size of 256×256. Figure 1 showed a sample from one volunteer.

Fig. 1. A sample of Chinese fingerspelling language from one volunteer

2.2   Hu Moment Invariant

Hu moment was proposed by Hu. M.K in 1962, which has translation, rotation and scale invariance. For images with gray distribution of $f(x, y)$, the $(p+q)$ order invariant moment is defined as:

$$M_{pq} = \int_{-\infty}^{\infty} \int_{-\infty}^{\infty} x^p y^q f(x, y) dx dy \qquad p, q = 0, 1, 2, \ldots \qquad (1)$$

The central moment of order $(p+q)$ is:

$$C_{pq} = \int_{-\infty}^{\infty} \int_{-\infty}^{\infty} (x-x_1)^p (y-y_1)^q f(x,y) dx dy \qquad p,q = 0,1,2,\ldots \qquad (2)$$

where moment center is (x_1, y_1), and centroid of the component is as follows:

$$x_1 = M_{10}/M_{00} \qquad y_1 = M_{01}/M_{00}$$

For the digital image, discretization means that in the discrete state, the formula of $(p+q)$ order ordinary moment and central moment of $f(x,y)$ is as follows:

$$M_{pq} = \sum\sum x^p y^q f(x,y) \qquad (3)$$

$$C_{pq} = \sum\sum (x-x_1)^p (y-y_1)^q f(x,y) \qquad p,q = 0,1,2,\ldots \qquad (4)$$

When the image changes, M_{pq} also changes. Although C_{pq} has translation invariance, it is still sensitive to rotation [11–13]. To address these problems, we normalize the central moment,

$$N_{pq} = C_{pq}/C_{00}^{1+p+q/2}, \qquad p+q = 2,3,4\ldots \qquad (5)$$

Directly represented by ordinary moments or central moments, features can't be translated, rotated, bonus scaled invariant at the same time. If the normalized center distance is used, the feature not only has translation invariance but also proportional invariance.

The Hu moment constructs seven invariant moments using the second and third order center distances. They keep the translation, zooming and rotation unchanged in consecutive pixels, making the image more efficient to optimize. This is the seven invariant moments of the Hu moment:

$$I_1 = N_{20}N_{02} \qquad (6)$$

$$I_2 = (N_{20} + N_{02})^2 = 4N_{11}^2 \qquad (7)$$

$$I_3 = (N_{30} + N_{12})^2 + (3N_{21} - N_{03})^2 \qquad (8)$$

$$I_4 = (N_{30} + N_{12})^2 + (N_{21} - N_{03})^2 \qquad (9)$$

$$I_5 = (N_{30} - 3N_{12})(N_{30} + N_{12})\left[(N_{30} + N_{12})^2 - 3(N_{21} + N_{03})^2\right] \\ + (3N_{21} - N_{03})(N_{21} + N_{03})\left[3(N_{30} + 3N_{12})^2 - (N_{21} + N_{03})^2\right] \qquad (10)$$

$$I_6 = (N_{20} - N_{02})\left[(N_{30} + N_{12})^2 - (N_{21} + N_{03})^2\right] + 4N_{11}(N_{30} + N_{12})(N_{21} + N_{03}) \qquad (11)$$

$$I_7 = (3N_{21} - N_{03})(N_{30} + N_{12})\left[(N_{30} + N_{12})^2 - 3(N_{21} + N_{03})^2\right]$$
$$- (N_{30} - 3N_{12})(N_{21} + N_{03})\left[3(N_{30} + N_{12})^2 - (N_{21} + N_{03})^2\right] \qquad (12)$$

In short, the Hu moment is the center (x_1, y_1) of the image obtained from the ordinary moment M_{pq}, and then defines the center distance C_{pq}, normalizes the center distance to obtain the normalized center distance N_{pq}. In addition, together with the normalized second-order moment of the center distance and the third moments, seven invariant moments are formed.

2.3   Support Vector Machine

Among supervised classification methods, Support Vector Machine (SVM) [14] is the latest classification method based on machine learning theory. But it is a linear classifier which just uses one hyperplane to divide samples in D-dimension into two categories, thus it has a poor fitting effect [15–19]. For the purpose of improving the fitting effect, kernel-SVM was proposed. It is an innovation of SVM which can solve the problem of the disability of separating the practical data with complex distribution in supervised learning by combining several hyperplanes. Kernel-SVM extends original linear SVM to nonlinear SVM classifier which represents many machine learning algorithms by using the dot product between samples [20–24].

Based on the above theory, the linear function of SVM can be redefined as follows:

$$w^T x + m = m + \sum_{i=1}^{n} \gamma_i x^T x^{(i)} \qquad (13)$$

Here, γ_i denotes coefficient vector, $x^{(i)}$ indicates training sample. At the same time, we can substitute $\varphi(x)$ with the output of the eigenfunction x, and introduce a kernel function to replace the dot product. The equation is as follows:

$$k\left(x, x^{(i)}\right) = \varphi(x) \cdot \varphi\left(x^{(i)}\right) \qquad (14)$$

Where $\cdot$ denotes the dot product. Thus, we can use the substitution function to predict.

$$f(x) = m + \sum_i \gamma_i k\left(x, x^{(i)}\right) \qquad (15)$$

The function $f(x)$ is nonlinear and it is completely equivalent to preprocessing inputs using $\varphi(x)$ and then learning a linear model in the new transformation space. After transformation, the kernel-SVM is allowed to fit the maximum-margin hyperplane in a transformed feature space. The transformation can be nonlinear and the transformed space can be high dimensional, but after a series of transformations and replacements, although the classifier is a hyperplane in the high-dimensional feature space, it may be nonlinear in the original input space.

Based on the advantages that kernel trick has, it guarantees the effective convergence of optimization techniques for learning nonlinear models and the implementation of kernel function. $k(x, x^{(i)})$ is much more efficient than constructing $\varphi(x)$ and then computing the dot product, so we introduced the kernel function into the kernel-SVM.

2.4 Radial Basis Function

Among all of kernel functions, Gaussian kernel is the most popular used kernel which usually be utilized to solve two-dimensional time-independent convection-diffusive-reaction equations with inhomogeneous boundary conditions. It is also called as Radial Basis Function (RBF) [25], which can be expressed as follow:

$$k_G(x, x^{(i)}) = T(x - x^{(i)}; 0; \sigma^2 I) \tag{16}$$

Here T represents standard normal density. RBF means that its value decreased in the direction in which $x^{(i)}$ radiates outward from x. RBF also can be defined as the following formula in details.

$$k_G(x, x^{(i)}) = \exp(-\beta \|x - x^{(i)}\|^2) \tag{17}$$

Where β is a parameter which needs to be tuned.

To conclude, kernel-SVM has a plenty of advantages which are summarized as follows: (1) it guarantees the effective convergence of optimization techniques for learning nonlinear models; (2) the implementation of kernel function is much more effective than constructing and computing the dot product; (3) it has the faster speed to training samples and classifying samples, besides, it's memory requirement is relatively low; (4) it has less adjustable parameters; (5) convex quadratic optimization is utilized in training, which provides global and unique solution and avoids convergence to the local minimum. We do not use deep learning in this paper, since our dataset is small and not suitable for training a deep neural network model [26–30].

2.5 Implementation

We carried out the experiment on a platform of the personal computer installed Windows 10 operating system with Core i7 CPU and 16 GB memory. Overall accuracy (OA) was introduced to evaluate the results, which denotes the value of correct prediction over all test sets divide the total numbers.

In order to obtain the parameters of top optimal kernel-SVM, trial-and-error algorithm is commonly used. Meanwhile, 10-fold cross validation [31] is introduced to validate the performance. Cross validation is a practical method for cutting data samples into smaller subsets. In the 10-fold cross validation, the obtained data samples are randomly divided into 10 equal-sized subsamples. Among the 10 sub-samples, one sample is retained as validation data for testing the model, and the remaining 9 samples are used as training data. Then, the cross validation was repeated 10 times, and each sample was used as a test sample. The average of the 10 results is used as an estimate of the accuracy of the algorithm. (See Fig. 2) The advantage of 10-fold cross validation is that all data samples are trained and verified, and the subset is verified only once, and the error is greatly reduced during the experiment.

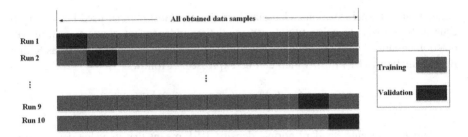

Fig. 2. An illustration of 10-fold cross validation

3   Experiment and Results

3.1   Statistical Results of Proposed Method

In this experiment, we utilized this proposed method, called HMI-RBF-SVM, in which Hu moment invariant and radial basis function support vector machine were employed. The results of 10 × 10-fold cross validation are listed in Table 1. It can be observed that the highest accuracy in column Total (Overall Accuracy) achieves 88.79% that has been highlighted with bold font. Another marked number represents the highest accuracy of single run, which reaches 93.18%. Finally, the value of means and standard deviation based on 10 runs is 86.47 ± 1.15%, which indicates the results are effective and accepted.

Table 1. Results of 10 × 10-fold cross validation

Run	F1	F2	F3	F4	F5	F6	F7	F8	F9	F10	Total
1	85.07	88.37	87.59	83.08	86.15	87.88	87.79	84.85	86.67	86.92	86.44
2	85.71	81.82	86.57	79.55	86.03	88.97	84.50	89.23	83.70	83.08	84.92
3	83.72	81.10	88.06	86.36	84.62	88.37	87.22	84.78	85.19	86.47	85.61
4	83.85	87.50	85.61	91.67	88.15	**93.18**	88.72	89.15	89.31	90.77	**88.79**
5	79.10	86.76	84.13	88.15	85.50	80.80	87.88	88.81	87.97	83.58	85.30
6	83.85	85.07	84.21	86.05	82.31	86.26	85.07	88.81	90.91	87.22	85.98
7	85.19	85.83	85.38	85.38	85.94	87.41	86.13	87.97	89.15	86.76	86.52
8	87.22	89.31	90.98	83.46	82.68	83.70	90.15	90.91	89.47	89.31	87.73
9	87.79	87.88	84.33	88.81	86.67	87.31	89.15	84.85	84.96	85.71	86.74
10	87.41	89.06	84.09	88.49	82.95	87.31	87.88	87.60	84.96	86.82	86.67
Mean ± SD	86.47 ± 1.15										

3.2 SVM Versus RBF-SVM

We compared RBF-SVM with traditional SVM and polynomial SVM. All the parameters were chosen by trial-and-error method. As can be seen from Table 2, Mean ± SD of traditional SVM, polynomial SVM (PSVM) and RBF-SVM are 82.39 ± 1.13%, 85.50 ± 1.09%, and 86.47 ± 1.15%, respectively. In term of accuracy, RBF-SVM gains about four percentage points ahead of traditional SVM and also gets about one point ahead of PSVM. It denotes that the higher accuracy is benefit from radial basis function which guarantees the effective convergence and speeds up the training.

Table 2. Comparison with other SVM variants

Run	SVM	PSVM	RBF-SVM (Ours)
1	83.18	86.06	86.44
2	82.42	87.12	84.92
3	83.71	85.45	85.61
4	82.27	85.76	88.79
5	80.00	86.06	85.30
6	82.35	83.64	85.98
7	82.58	83.71	86.52
8	81.59	85.15	87.73
9	83.94	85.91	86.74
10	81.89	86.14	86.67
Mean ± SD	82.39 ± 1.13	85.50 ± 1.09	86.47 ± 1.15

3.3 Comparison of State-of-the-Art Approaches

In this experiment, three state-of-the-art approaches were compared with our HMI-RBF-SVM method. As can be seen from Table 3, WE-SVM [6], GLCM-MGSVM [7], and HMM-SVM [9] achieved overall accuracy of 85.69 ± 0.59%, 85.3% and 85.14%, respectively. Different feature extraction methods, such as wavelet entropy, gray-level co-occurrence matrix, hidden markov model and Hu moment invariant were employed in these four approaches. Kernel SVM was applied in GLCM-MGSVM and HMI-RBF-SVM, and another two approaches utilized traditional SVM. Our method is superior others about one percentage point, which indicates that Hu moment invariant maintains image feature when extracting and radial basis function provides the effective convergence to enhance performance of classification.

Table 3. Comparison of four state-of-the-art approaches

Approach	Overall accuracy
WE-SVM [6]	85.69 ± 0.59
GLCM-MGSVM [7]	85.3
HMM-SVM [9]	85.14
HMI-RBF-SVM (Ours)	86.47 ± 1.15

4 Discussions

In our study, three advanced techniques: Hu moment invariant, RBF-SVM and 10-fold cross validation were applied. As the picture is identified by the feature quantity composed of the Hu moment invariant, the advantage is that the speed can be accelerated, and the disadvantage is that the recognition rate may drop relatively. Therefore, Hu moments invariant are generally used to identify large objects in an image. It describes the shape of the object better, but the texture features of the image cannot be too complicated. Thus, it will be relatively suitable for identifying the hand shape preprocessed. Due to excellent performance of RBF, the Gaussian kernel was chosen. It needs few parameters to tune and employs convex quadratic optimization to train. Particularly, kernel SVM provides unique and global solutions, preventing the convergence to local minima. When compared with traditional SVM, its superiority is highlighted distinctly. The application of 10-fold cross-validation is determined by its easy-to-use properties and the use of all data for training and validation, which will not improve the accuracy of the final classification, but will make the classifier reliable, so it can be generalized to other independent data sets. All these techniques contribute to improving of performance.

5 Conclusions

This study proposed a novel Chinese fingerspelling recognition via Hu moment invariant and RBF-SVM, carrying out on 10-fold cross validation. Hu moment invariant adapts to the feature traction of fingerspelling images, which accelerates the speed of recognition. Based on radial basis function, kernel SVM guarantees the effective convergence and improves the classification. 10-fold cross validation provides the sufficient confirmation of data samples, reducing the error. This approach gained overall accuracy of 86.47 ± 1.15%, which indicates it has superiority in all four state-of-the-art approaches.

In the future, some contributions shall be tried as follows: (1) other advanced methods such as particle swarm optimization, principal component analysis, deep neural network and transfer learning may be applied in this theme. (2) We shall try to apply our approach to other fields.

Acknowledgement. This work was supported from Jiangsu Overseas Visiting Scholar Program for University Prominent Young & Middle-aged Teachers and Presidents of China, The Natural Science Foundation of Jiangsu Higher Education Institutions of China (19KJA310002), The Philosophy and Social Science Research Foundation Project of Universities of Jiangsu Province (2017SJB0668), The Natural Science Foundation of Jiangsu Province (16KJB520029).

References

1. Maharani, A., et al.: Hearing impairment, loneliness, social isolation, and cognitive function: longitudinal analysis using english longitudinal study on ageing. Am. J. Geriatr. Psych. **27**(12), 1348–1356 (2019)
2. Eifring, H.: Language contact across time: classical chinese on modern public signs. J. Chin. Linguist. **47**(2), 562–614 (2019)
3. Deriche, M., et al.: An intelligent arabic sign language recognition system using a pair of LMCs with GMM based classification. IEEE Sens. J. **19**(18), 8067–8078 (2019)
4. Kumar, E.K., et al.: 3D sign language recognition with joint distance and angular coded color topographical descriptor on a 2-stream CNN. Neurocomputing **372**, 40–54 (2020)
5. El Afia, A., et al.: Adaptive cooperation of multi-swarm particle swarm optimizer-based hidden Markov model. Progress Artif. Intell. **8**(4), 441–452 (2019)
6. Jiang, X., Zhu, Z.: Chinese sign language identification via wavelet entropy and support vector machine. In: Li, J., Wang, S., Qin, S., Li, X., Wang, S. (eds.) ADMA 2019. LNCS (LNAI), vol. 11888, pp. 726–736. Springer, Cham (2019). https://doi.org/10.1007/978-3-030-35231-8_53
7. Jiang, X.: Isolated Chinese sign language recognition using gray-level co-occurrence matrix and parameter-optimized medium gaussian support vector machine. In: Satapathy, S.C., Bhateja, V., Nguyen, B.L., Nguyen, N.G., Le, D.-N. (eds.) Frontiers in Intelligent Computing: Theory and Applications. AISC, vol. 1014, pp. 182–193. Springer, Singapore (2020). https://doi.org/10.1007/978-981-13-9920-6_19
8. Kumar, P., et al.: A position and rotation invariant framework for sign language recognition (SLR) using Kinect. Multimedia Tools Appl. **77**(7), 8823–8846 (2018)
9. Lee, G.C., et al.: Kinect-based Taiwanese sign-language recognition system. Multimed Tools Appl. **75**, 261–279 (2016)
10. Jiang, X.: Chinese sign language fingerspelling recognition via six-layer convolutional neural network with leaky rectified linear units for therapy and rehabilitation. J. Med. Imaging Health Inform. **9**(9), 2031–2038 (2019)
11. Sun, P.: Pathological brain detection based on wavelet entropy and Hu moment invariants. Bio-Med. Mater. Eng. **26**(s1), 1283–1290 (2015)
12. Yang, J.: Pathological brain detection in MRI scanning via Hu moment invariants and machine learning. J. Exp. Theor. Artif. Intell. **29**(2), 299–312 (2017)
13. Hou, X.-X.: Alcoholism detection by medical robots based on Hu moment invariants and predator-prey adaptive-inertia chaotic particle swarm optimization. Comput. Electr. Eng. **63**, 126–138 (2017)
14. Tharwat, A.: Parameter investigation of support vector machine classifier with kernel functions. Knowl. Inf. Syst. **61**(3), 1269–1302 (2019)
15. Yang, J.: Identification of green, Oolong and black teas in China via wavelet packet entropy and fuzzy support vector machine. Entropy **17**(10), 6663–6682 (2015)

16. Chen, S., et al.: Magnetic resonance brain image classification based on weighted-type fractional Fourier transform and nonparallel support vector machine. Int. J. Imaging Syst. Technol. **25**(4), 317–327 (2015)

17. Liu, G.: Pathological brain detection in MRI scanning by wavelet packet Tsallis entropy and fuzzy support vector machine. SpringerPlus, **4**(1) (2015). Article ID. 716

18. Yang, M.: Dual-tree complex wavelet transform and twin support vector machine for pathological brain detection. Appl. Sci. **6**(6) (2016). Article ID. 169

19. Chen, M.: Morphological analysis of dendrites and spines by hybridization of ridge detection with twin support vector machine. PeerJ **4** (2016). Article ID. e2207

20. Zhou, X.-X.: Comparison of machine learning methods for stationary wavelet entropy-based multiple sclerosis detection: decision tree, k-nearest neighbors, and support vector machine. Simulation **92**(9), 861–871 (2016)

21. Gorriz, J.M., et al.: Wavelet entropy and directed acyclic graph support vector machine for detection of patients with unilateral hearing loss in MRI scanning. Front. Comput. Neurosci. **10** (2016). Article ID. 160

22. Lu, H.M.: Facial emotion recognition based on biorthogonal wavelet entropy, fuzzy support vector machine, and stratified cross validation. IEEE Access **4**, 8375–8385 (2016)

23. Li, Y.: Detection of dendritic spines using wavelet packet entropy and fuzzy support vector machine. CNS Neurol. Disorders - Drug Targets **16**(2), 116–121 (2017)

24. Chen, Y.: Pathological brain detection by wavelet-energy and fuzzy support vector machine. In: 8th International Symposium on Computational Intelligence and Design (ISCID), Hangzhou, China, pp. 409–412. IEEE (2015)

25. De Mulder, W., et al.: An interpretation of radial basis function networks as zero-mean Gaussian process emulators in cluster space. J. Comput. Appl. Math. **363**, 249–255 (2020)

26. Li, Z.: Teeth category classification via seven-layer deep convolutional neural network with max pooling and global average pooling. Int. J. Imaging Syst. Technol. (2019). https://doi.org/10.1002/ima.22337

27. Muhammad, K.: Image based fruit category classification by 13-layer deep convolutional neural network and data augmentation. Multimedia Tools and Applications **78**(3), 3613–3632 (2019)

28. Sangaiah, A.K.: Alcoholism identification via convolutional neural network based on parametric ReLU, dropout, and batch normalization. Neural Comput. Appl. (2019). https://doi.org/10.1007/s00521-018-3924-0

29. Zhao, G.: Polarimetric synthetic aperture radar image segmentation by convolutional neural network using graphical processing units. J. Real-Time Image Proc. **15**(3), 631–642 (2018)

30. Huang, C.: Multiple sclerosis identification by 14-layer convolutional neural network with batch normalization, dropout, and stochastic pooling. Front. Neurosci. **12** (2018). Article ID. 818

31. Gilanie, G., et al.: Automated and reliable brain radiology with texture analysis of magnetic resonance imaging and cross datasets validation. Int. J. Imaging Syst. Technol. **29**(4), 531–538 (2019)

Multi-touch Gesture Recognition of Braille Input Based on RBF Net

Zhang Juxiao[1,2] (ID), Zeng Xiaoqin[2], and Zhu Zhaosong[1(✉)] (ID)

[1] College of Information and Mathematics Science, Nanjing Normal University of Special Education, Nanjing 210038, Jiangsu, China
`330161l@qq.com, zzs2019@foxmail.com`
[2] College of Computer and Information, Hohai University, Nanjing 210038, Jiangsu, China

Abstract. One challenging task for the blind is to input Braille while by no way could they sense the location information on touch screens. The existing Braille input methods are suffering from problems including inaccurate positioning and lack of interactive prompts. In this paper, touch gestures are recognized by trained RBF network while combined gestures are modelled. By doing so, the Braille input concerning multi-touch gesture recognition is then implemented. The experimental results show that the method is effective and blind people can friendly input Braille with almost real-time interaction.

Keywords: Input method of braille · RBF net · Multiple touch gesture recognition

1 Introduction

In recent years, although many improvements have been made in the availability of smartphones and other electronic devices for the visually impaired, there is still a huge gap between blind people and touch-screen devices in human-computer interaction.

1.1 Research Status

At present, the output stage mostly uses Text to Speech (TTS) technology or external point input devices [1], the main input methods of touch-screen mobile phones include Pinyin [2], stroke [3], speech recognition [4] and on-line handwriting recognition input methods [5]. E.g: Multi-touch input method for the blind [1], Single-Finger touch gesture input method [6], Touch action recognition input method [7].

Other common Braille input methods on touch-screen include Perkinput [8], TypeInBraille [9] and BrailleTouch [10], etc.

1.2 Deficiencies

Blind touch screen input method has two major shortcomings: one is precise localization; the other is an interactive prompt barrier. The biggest difficulty for blind people to use the touch screen is that they cannot get the location information on the screen,

Y.-D. Zhang et al. (Eds.): ICMTEL 2020, LNICST 327, pp. 393–399, 2020.
https://doi.org/10.1007/978-3-030-51103-6_35

and they cannot obtain a series of operations such as visual focus [11]. The keys of the virtual keyboard used in the input method are too close to each other, which also lead to an increase in the rate of input error [12]. Interactive prompt accessibility barrier means that the time and focus of fingers sliding on the touch screen cannot be fed back to blind persons.

1.3   Innovations

A Braille Input concerning Multi-Touch Gesture Recognition based on RBF Net is designed. This input method integrates Braille phonetic notation, which makes it possible to complete Braille input and output through voice input and Text to Speech (TTS) conversion. Besides, when the Braille gesture is inputting on the touch screen, it reduces the interference of orientation. It relates the high inherent logicality of Braille, which makes it easy for blind people to learn and remember. It is suitable for blind people to input Braille information on touch-screen devices, providing a new method for the human-computer interaction between blind people and touch screen.

2   Multi-touch Gesture Recognition Braille Input Module

The main algorithm modules include: multi-touch area loaded on electronic devices and the multi-touch area graphic information library module generated by this multi-touch area, multi-touch area graphic information library recognition module and Braille input module and so on.

2.1   Multi-touch Area Graphic Information Library Module

The basic graphics feature of multi-touch area is established based on the four-point position features at the top of Braille and the graphic information of multi-touch area. The graphic information library of multi-touch area is constructed by adding one or two-finger touches to the basic graphics feature, and the correspondence table between graphic information of the multi-touch area and the Braille points is generated.

2.2   Establish the Correspondence Table Between Graphic Information of the Multi-touch Area and Braille

Based on the inherent logic of Braille points, the Braille point positions composed of points 1, 2, 4 and 5 are established, with points 3 and 6 points kept empty. The graphics shapes of these four points are taken into account when establishing the Braille points, and the multi-touch area graphic information library is established based on the regional information composed by the above four points.

In order to reduce the interference of noisy graphical information caused by touch and improve the degree of distinction between graphical information in the multi-touch area, finger movement in touch area should follow the regulated order principle.

Rule 1: The principle of sequence uniqueness.

$$f(\chi_i, \chi_j) \neq f(\chi_j, \chi_i), i \neq j \wedge \chi_i, \chi_j \in X \tag{1}$$

Where X is a finite state set, $X = \{\chi_i, \chi_i \in C\}, C = \{1, 2, 4, 5\}$;

f is Transfer Function, $f(\chi_i, \chi_j)$ is sliding from point area χ_i to point area χ_j, $\{\chi_i, \chi_j \in X\}$.

Rule 2: The sequence of touch sliding is from top to bottom and from left to right.
Where X is a finite state set,

$$X = \{\chi_i, \chi_i \in C\}, C = \{1, 2, 4, 5\} \tag{2}$$

f is Transfer Function, $f(\chi_i, \chi_j)$ is sliding from point area χ_i to point area χ_j, $\{\chi_i, \chi_j \in X\}$;

$f(\chi_i, 0)$ denotes hitting point area χ_i ,Then

$$f(\chi_i, \chi_i) = f(\chi_i, 0) = f(\chi_i) \tag{3}$$

Continuous sliding:

$$f(\chi_i, \chi_k) = f(\chi_i, \chi_j) + f(\chi_j, \chi_k) \tag{4}$$

$$iff \ i < j < k \wedge \chi_i, \chi_j, \chi_k \in X \tag{5}$$

$$f(\chi_i) < f(\chi_j) < f(\chi_k) \Rightarrow f(\chi_i, \chi_k) = f((\chi_i, \chi_j), \chi_k) \tag{6}$$

$$f(\chi_i) < f(\chi_j), f(\chi_i) \langle f(\chi_k) \wedge f(\chi_j) \rangle f(\chi_k) \Rightarrow f(\chi_i, \chi_k) = f(\chi_i, \chi_j) \tag{7}$$

$$f(\chi_i) > f(\chi_j) \Rightarrow f(\chi_i, \chi_k) = f(\chi_i, 0) = f(\chi_i) \tag{8}$$

2.3 Multi-touch Area Graphic Information Recognition Module

When touching the screen with both left and right hands, pressing and holding or sliding over the multi-touch area to form graphical information, the module collects the input points and area to form a graphical structure, thus draws the features of multi-touch area graphical information. Moreover, it compares the graphical information features with those in the multi-touch graphics area information library and looks up the Braille comparison table to extract the Braille.

3 Multi-touch Gesture Recognition Algorithm

3.1 Multi-touch Gesture Action Data Structure

A touch gesture action is a description of a single point touch, which can be described by a quadruple data structure [13]:

$$Metagesture = \{ID, \ Touchstate, \ Coordinate, \ Time\} \tag{9}$$

Where: $ID = \{1, 2, 3, 4, 5\}$ respectively represent the touchpoint identification numbers of different fingers in one hand. Each touchpoint at a certain moment has a unique ID number, so as to record the trajectory and direction of the same touchpoint.

$Touchstate = \{0, 1\}$ represents the direct touch state between the finger and the touch screen, $Touchstate = 1$ represents touch state, and $Touchstate = 0$ represents the not-touch state.

$Coordinate = P(x, y)$ represents the coordinates of the touchpoint. When $Touchstate = 1$, the touch screen records a series of coordinate data by IDs according to touchpoint detection and touchpoint tracking. When $Touchstate = 0$, it will be considered as noise data and the coordinate information will be meaningless.

Time is the time data when gestures are recorded, which is acquired by reading the system time. It represents the time parameter when the state or position of the touchpoint changes.

Multi-touch gestures consist of touch gesture element actions, which can be described by a quintuple data structure [14]:

$$Multigesture = \{Number, ID, Touchstate, Coordinate, Time\} \tag{10}$$

Where: $Number = \{1, 2, \ldots\}$ represents the number of touchpoints tracked by touchpoint detection, usually, $(0 \leq Number \leq 5)$. The other parameters have the same meanings as Metagesture.

3.2 Data Preprocessing

Touch gestures on the touch-screen are recorded in the form of data, but it is impossible that the time and area scope of all actions are consistent, besides, there will be noise data, which need to be preprocessed. Preprocessing includes noise elimination, absorption of gesture graphics deformation, compression of redundant information and so on.

3.3 Normalization Processing

Network training cannot be carried out if the data volume of different touch gestures is inconsistent. Therefore, the data of touch gesture actions need to be normalized. Normalization is to take out the same number of points for all touch gestures according to different time differences. After repeated experiments, 15 sampling points are selected from A to B and are divided equally according to time. The obtained information of the 15 sampled points are stored as 14 two-dimensional vectors and are normalized to obtain the normalized feature vectors of multi-touch gestures, which are used as the input into the RBF network.

3.4   Input Vector of Network

The multi-touch gesture action is composed of several gesture element actions arranged and combined in chronological order. The feature matrix of the touch gesture of m touchpoints is as follows:

$$
\begin{bmatrix}
ID_1 & Touchstate_1 & (x_{10}, y_{10}) & \cdots & (x_{1n}, y_{1n}) & t_1 \\
ID_2 & Touchstate_2 & (x_{20}, y_{20}) & \cdots & (x_{2n}, y_{2n}) & t_2 \\
\cdots & \cdots & \cdots & \cdots & \cdots & \cdots \\
ID_m & Touchstate_m & (x_{10}, y_{10}) & \cdots & (x_{mn}, y_{mn}) & t_m
\end{bmatrix}
\tag{11}
$$

3.5   Feature Extraction

The collected multi-touch gesture information is classified and merged based on their features so as to extract the graphics with corresponding feature information. We first determined the number of touchpoints m; then we found the matching gestures in the database through the gesture features. Then we distinguished the actions of double-click and clicking twice as well as long-press and sliding actions by combining with the given threshold function. Sliding Displacement Threshold Expression:

$$
f(x, y) = \begin{cases} 0, \sqrt{(x_0 - x_1)^2 + (y_0 - y_1)^2} \geq a \\ 1, \sqrt{(x_0 - x_1)^2 + (y_0 - y_1)^2} < a \end{cases}
\tag{12}
$$

Time Threshold Expression:

$$
f(t) = \begin{cases} 0, t \geq a \\ 1, t < a \end{cases}
\tag{13}
$$

3.6   RBF Network Design and Training

If the Gaussian function is used as the radial basis function (RBF), the expression of the radial basis function of the ith hidden layer is as follows:

$$
\psi_i(x) = exp\left(-\frac{\|x - c_i\|^2}{2\sigma_i^2} \right), i = 1, 2, \ldots n
\tag{14}
$$

The Gaussian function is adopted as the radial basis function:

$$\psi_i(t) = exp\left(-\frac{t^2}{2\sigma_i^2}\right) \text{ where } t = \|x - c_i\|, \text{ then} \tag{15}$$

$$\psi_i\left(\|x - c_i\|^2\right) = exp\left(-\frac{(\|x - c_i\|)^2}{2\sigma_i^2}\right) \tag{16}$$

The width of the RBF network is determined by σ. Let d be the maximum distance of the samples and let N be the number of samples, then the width is determined by fixation method.

$$\sigma = \frac{d_{max}}{\sqrt{2N}} \tag{17}$$

3.7   Recognition of Touch Gesture Element Actions

Following the previous data preprocessing steps, the module sends the input vectors of touch gestures into the trained RBF Net for classification. If the output vector of RBF is closest to the standard output of a gesture in the gesture library, it is identified as the desired gesture [15].

4   Results and Discussion

Eight students were organized to form a user test group. The average time of multi-touch gesture recognition entering Braille is respectively s1–s8: 2.49 s/square, 3.21 s/square, 2.68 s/square, 2.88 s/square, 3.75 s/square, 3.38 s/square, 3.57 s/square, 3.99 s/square.

The Braille input method of multi-touch gesture recognition features the following advantages:

(1) Availability: the function of inputting Braille on the electronic device of touch-screen is realized. The issue of text information interaction between blind persons and touch-screen is addressed.
(2) Simplicity: points 1, 2, 4, 5 in Braille points constitute a graph which is highly similar to Braille points. There are fewer basic graphics, and the inherent logic in Braille points distribution is retained in the expanded graphics, therefore it facilitates blind persons to learn and remember.
(3) Efficiency: multiple touches are operated and one block of Braille is input in one operation. Braille Points 3 and 6 are totally dependent on increasing finger touches, which is highly similar to the internal logic structure of Braille. This reduces the time for transformation and thinking when inputting, thus improving the efficiency of inputting Braille.

Acknowledgement. This work was supported by The Major Programs of Natural Science Foundation of the Jiangsu Higher Education Institutions of China (No. 19KJA310002.) and The Natural Science Foundation of the Jiangsu Higher Education Institutions of China (No. 17KJD520006).

References

1. Zhang, J.X., Zeng, X.Q., Meng, C.H.H.: Design and implementation of multi-touch input method for blind usage. Comput. Appl. Software **10**, 231–235 (2015)
2. Liu, F., Wang, Y.H., Tang, B.Z.H., et al.: Intelligent Chinese input method based on android. Comput. Eng. **37**(07), 225–227 (2011)
3. Yu,T.Z.H.: The Design and Implementation of Cross-Platform Stroke Input Method Engine. Harbin Institute of Technology (2013)
4. Zheng, Y.D., Chen, Z.H.S., Xiao, L.S.H.: Research and implementation of a voice control audio system based on Android speech recognition. Mod. Electron. Tech. **42**, 93–96 (2019)
5. Yan, X.L., Wang, L.M.: Handwritten Chinese character recognition system based on neural network convolution depth. Comput. Eng. Appl. **53**(10), 246–250 (2017)
6. Chen, H.F., Xu, S.H., Wang, J.L.: A Braille Input Method Based on Gesture Recognition: JiangSu, CN102929394A, 13 February 2013
7. Hu, Y.P.: A Method and Device for Output and Input of Braille Characters on Touch Screen. BeiJing: CN103870008A, 18 June 2014
8. Fukatsu, Y., Shizuki, B., Tanaka, J.: No-look flick: single-handed and eyes-free japanese text input system on touch screens of mobile devices. human computer interaction with mobile devices and services, pp. 161–170 (2013)
9. Mascetti, S., Bernareggi, C., Belotti, M., et al.: TypeInBraille: a braille-based typing application for touchscreen devices. In: Conference on Computers and Accessibility (ASSETS 2011), pp. 295–296 (2011)
10. Frey, B., Southern, C., Romero, M., et al.: Brailletouch: mobile texting for the visually impaired. In: International Conference on Universal Access in Human Computer Interaction, pp. 19–25 (2011)
11. Fukatsu, Y., Shizuki, B., Tanaka, J., et al.: No-look flick: single-handed and eyes-free japanese text input system on touch screens of mobile devices. In: Human Computer Interaction with Mobile Devices and Services, pp. 161–170 (2013)
12. Nicolau, H., Guerreiro, T.J., Jorge, J.A., et al.: Proficient blind users and mobile text-entry. In: Proceedings of the 28th Annual European Conference on Cognitive Ergonomics. New York USA, pp. 19–22 (2010)
13. Ji, H.Y.: Human-Computer Interaction Research Based on Multi-touch Technology. Diss. East China Normal Unversity (2011)
14. Wang, X.Q., Chen, G., Wang, D., Wang, C.H.: Research on multi-touch gesture analysis and recognition algorithm. Comput. Sci. **39**(S1), 522–525 (2012)
15. Li, W.S.H., Deng, C.H.J., Lv, Y.: Interaction gesture analysis based on touch screen. Chin. J. Liquid Crystals Displ. **26**(2), 194–199 (2011)

Transfer Learning Methods Used in Medical Imaging and Health Informatics

Research on Early Warning Monitoring Model of Serious Mental Disorder Based on Multi-source Heterogeneous Data Sources

Xinlei Chen, Dongming Zhao, Wei Zhong$^{(\boxtimes)}$, Jiufeng Ye,
and Feng Gao

Suzhou Guangji Hospital, Suzhou 215000, Jiangsu, China
13862339400@163.com

Abstract. Patients with severe mental disorders are sudden and aggressive, and the means may be more cruel. The data shows that the number of serious mental disorders is increasing. In order to prevent the occurrence of accidents and disasters in patients with mental illness, active intervention should be carried out to design an early warning and monitoring system for serious mental disorders. By collecting administrative departments of health and family planning at all levels, it is necessary to cooperate with the political and legal, public security, civil affairs, human resources, social security, and the Disabled Persons' Federation. The established platform information is used to summarize multi-source heterogeneous data. Establish an early warning monitoring model, classify 10 risk factors from four levels, establish a risk factor assessment model, and set up different levels of treatment intervention programs. It is described from the perspectives of design ideas, design principles, and architecture design. The construction of an early warning and monitoring mechanism for serious mental disorders can effectively integrate the high-quality resources of mental health institutions at all levels, guide the rational allocation of resources, improve the management of serious mental disorders, detect the morbidity of patients with mental disorders early, and promptly intervene to reduce the risk of accidents.

Keywords: Severe mental disorders · Disease surveillance · Disease warning · Data fusion

1 Introduction

Psychiatric patients, who are dominated by hallucinations and delusions, often lose their sense of self-control and have uncontrollable consequences for their own and social safety. Severe mental disorders are classified into schizophrenia, schizoaffective disorder, paranoid psychosis, bipolar disorder, mental disorder caused by epilepsy, mental retardation accompanied by mental disorder [1]. Severe mental disorders are sudden and aggressive, and the means may be more cruel. Nearly 90% of the victims are close relatives; from 2013 to 2016, the provincial courts tried 304 cases of compulsory treatment, resulting in 243 cases of casualties. Pieces, accounting for 79.9%, of which 118 people were killed. According to the National Mental Health Epidemiological Survey data, the prevalence of severe mental disorders in China is about 1%,

© ICST Institute for Computer Sciences, Social Informatics and Telecommunications Engineering 2020
Published by Springer Nature Switzerland AG 2020. All Rights Reserved
Y.-D. Zhang et al. (Eds.): ICMTEL 2020, LNICST 327, pp. 403–410, 2020.
https://doi.org/10.1007/978-3-030-51103-6_36

and the prevalence of any one mental disorder is 9.32%. The situation of the country: February 18, 2017, Hubei Province The head of the store near Wuhan Wuchang Station was beheaded; on February 2, 2018, there were 3 deaths and 1 injury in Chenghai District, Shantou City, Guangdong Province; on July 26, 2018, the southeast gate of the US Embassy in Chaoyang District, Beijing The vicinity of the visa office exploded; in November 2019, a 9-year-old boy was killed in Changsha, Hunan Province.

Suzhou currently has 34,424 people with mental illness, 22,604 patients with stable disease; 28,364 patients with confirmed serious diseases; 851 with mild troubles; and 336 with accidents. See Fig. 1.

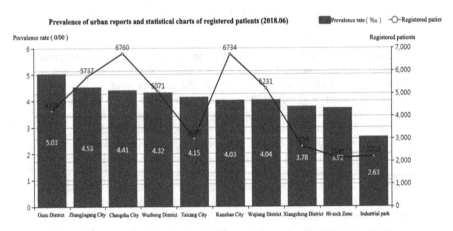

Fig. 1. Prevalence of urban reports and statistical charts of registered patients.

2   Purposes

Adhere to the work principle of prevention-oriented, prevention and treatment, key intervention, extensive coverage, and legal management [2, 3]. Through information technology, relying on the comprehensive management platform for serious mental disorders and the data sharing analysis mechanism of each system platform, establish a serious mental disease early warning monitoring system and establish Street leaders, relevant department cooperation, social participation work mechanism and mental illness prevention and control system based on mental disease prevention and control, community (village)-based, family-based, and strengthen psychological interventions for key mental illnesses and key populations To reduce the occurrence of accidents and illnesses in patients with mental illness [4, 5].

Recently, the National Health and Wellness Committee issued the "Regulations on the Management and Treatment of Severe Mental Disorders (2018 Edition)" (National Guardian Control (2018) No. 13) combined with the National Basic Public Health Service Regulations (Third Edition) [6–8].

The management of patients with severe mental disorders requires the implementation of new norms, requiring medical staff in primary health care institutions to timely standardize the management and treatment of patients with severe mental disorders [9].

At the same time, the administrative departments of health and family planning at all levels must establish information sharing mechanisms with the departments of politics, public security, civil affairs, human resources, social security, and the Disabled Persons' Federation to coordinate the comprehensive management of mental health [10–12]. This provides data support for the early warning system for severe mental disorders.

3 Design of Early Warning Monitoring Model

3.1 Classification of 10 Risk Factors in Four Levels: See Table 1

Table 1. Four levels and 10 categories.

Level	Serial number	Content
A	1	Weak guardian (care) family (caregiver or guardian age $\geq$ 65 years old, social agent guardian)
	2	No previous home visit record in previous years
	3	In the past year, outpatient clinics were all dispensed by family members
B	1	In the past year, he was treated repeatedly ($\geq$2 times) due to illness (six major categories);
	2	Have had a risk rating of Level 3 and above in the past 1 year;
	3	In the recent two consecutive follow-ups, the mode of administration was "dark medication" with unstable condition;
	4	There have been minor incidents in the past year
C	1	At the last follow-up, he was found to have not taken the drug or refused to take the drug and had a risk rating of Grade 3 or higher in the past 1 year;
	2	There were no visits or dispensing records for more than two months and there were more than three grades of dangerous behavior in the past 1 year
D	1	Persons who have historically recorded incidents of misconduct or are in the police officer's information base

3.2 Risk Factor Assessment Model

By collecting the multi-platform information established by the administrative departments of health and family planning at all levels and the departments of politics, public security, civil affairs, human resources, social security, and the Disabled Persons' Federation, data collection and analysis are carried out [13, 14]. Classified according to risk factors. Determine the level of risk for mentally ill persons for further treatment. See Fig. 2.

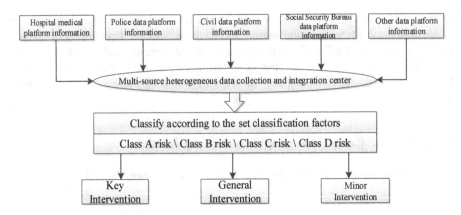

Fig. 2. Multi-source data collection and integration classification model.

3.3   Classification and Classification Event Handling Settings

Data collection, data processing, data classification, and data processing are performed in a closed loop operation [15].

Disposal settings for early warning of alarm objects at all levels:

a. a slight intervention in the subject. The grassroots psychiatrist completed the intervention within one month.
b. General intervention targets. The district led the Jingwei Office and the grassroots fine prevention doctors to complete the intervention under the guidance of the sugar candy office within 2 weeks.
c. Key intervention targets. City, county and township three-level fine prevention network, grassroots joint intervention, 72 h integrated management team completed the intervention.

Interventions include emergency treatment of patients with mental disorders who have suspected or diagnosed mental disorders who are harming themselves, harming the safety or safety of others, relapsed, acute or severe adverse drug reactions. Emergency prevention personnel, civilian police, members of the village (residential) committee, grid staff and other members of the care support team and psychiatrists, nurses and other emergency response teams to carry out emergency response.

4   System Overall Design

4.1   Design Ideas

According to the particularity of severe mental illness, the importance and necessity of early warning and monitoring of mental patients' accidents and accidents are carried out, and the design of early warning monitoring system is carried out. The system is designed primarily by following the classification of risk factors and the risk rating model [16, 17].

Data availability: The early warning and monitoring system for severe mental disorders is connected to various hospitals such as HIS, and the platform information established by departments such as politics, public security, civil affairs, human resources and social security, and the Disabled Persons' Federation, and other information systems from different vendors. There are multiple connections between the data link layer and the transport layer to the network layer. Different data format standards are adopted based on different data interconnection methods, which makes the interconnection relationship between systems complicated. So data governance starts with the underlying data structure and contains the type attribute of the data. Create a library element that contains assemblies, spaces, types, data structures, and so on, and the naming guidelines should use these elements to determine the naming rules: for example, capitalization conventions, identifier case styles, name conflict avoidance, and so on.

Follow the overall design philosophy of advanced, targeted, integrated, secure, expansive, normative, and practical.

This project uses JAVA technology, combined with a powerful oracle database. Strictly define all the business logic processing objects and unify the data storage rules. On this basis, the platform is divided into the background data processing layer, the intermediate business logic processing layer, the foreground business logic processing layer and the presentation layer. The business module adopts the B/S architecture, and the simple operation of WEB browsing query and business operation further enhances the practicability and ease of use of the software.

In order to adapt to the needs of interface changes caused by changes in data specifications in the later period, the expansion capability of this system is also a major feature. Based on the in-depth study and induction of relevant normative documents in the national fine defense field, we have formulated uniform data processing rules. The existing business logic objects are also uniformly encapsulated, and a simple and flexible access interface is provided for other government departments to access data in the future. In the later stage, the new functions are expanded without direct manipulation of the database, thereby ensuring the integrity of the database. Sex and safety.

4.2 Design Principles

Security principle: The system should be based on a large data center, a powerful information processing environment and a high-speed network. Since the entire system involves a large amount of confidential data, and part of the data sharing is based on the network environment, information security must be considered in the design process. The confidentiality measures ensure that the information resources in the system are not illegally stolen and tampered, the data center is not destroyed, and the user is required to use the shared resources in the system to provide the information services that should be provided. In order to ensure the security of the system, effective security and confidentiality technologies must be adopted in addition to establishing a sound security management system. The system adopts a set of scientific and convenient security management mode. The system strictly defines the permissions of the end users, and the permissions of the end users are refined to each function of each module.

On this basis, the system flexibly uses the group management mode. It greatly simplifies the work complexity of system administrators [15]. See Fig. 3.

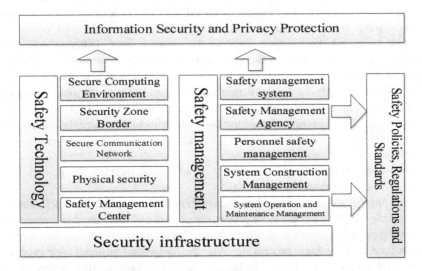

Fig. 3. Severe mental disorder early warning monitoring security system

Normative principle: In the system design process, the data structure and data coding settings conform to strict technical specifications, the data coding adopts the national standard, and the design is designed according to the "Minimum Data Set of Jiangsu Province Defense Platform Management Part".

Practicality principle: According to the comprehensive user demand research and expert analysis results, the business operation module adopts the WEB end form of the B/S architecture, and the data exchange level adopts the WEBSERVICE standard interface. Has powerful data processing and business management functions. From the beginning to the end, focusing on the integrity of the system, consider the links between Jiangsu Province's fine defense platform, Suzhou's fine defense platform, psychiatric hospital, other medical institutions, civil affairs, public security, social security, community, etc., and the internal relationship between their businesses. Make it an organic whole.

4.3 Architecture Design

According to the Ministry of Health's "Regulations on the Management and Treatment of Severe Mental Disorders (2018 Edition), "National Basic Public Health Service Regulations (Third Edition)", "Implementation Plan for the Rescue and Relief Work for Severe Mental Disorders in Suzhou City" and "The Spirit" The Health Law is a blueprint for business and builds a system. The system adopts the B/S framework to establish a city-level mental health data center business integration system. Each district and county security office, medical and health institutions, and grassroots health organizations access through a private network or VPN. The city's severe mental disorder early warning and monitoring system to carry out business management. The project system and other information platforms are in the form of interfaces for data interconnection.

5 Summary

Taking the three-level psychiatric hospital of Suzhou Guangji Hospital as the main body, the construction of early warning mechanism for serious mental disorders can effectively integrate the high-quality resources of mental health institutions at all levels, guide the rational allocation of mental health resources, and thus improve the management of serious mental disorders. Early detection of patients with mental disorders fluctuations, and timely intervention; obtain a large number of potential high-risk patients data, and use different strategies according to the level of early warning risk; improve work efficiency, reduce the unnecessary work of grassroots prevention doctors, timely and accurate positioning of risk patients; Reduce the occurrence of accidents, incidents, and accidents, reduce the risk of accidents and prevent social risks.

References

1. Swanson, J., Estroff, S., Swartz, M., Borum, R., Lachicotte, W., Zimmer, C.: Violence and severe mental disorder in clinical and community populations: the effects of psychotic symptoms, comorbidity, and lack of treatment. Psychiatry (1997)
2. Paz GarcíaPortilla, M., Sáiz, P.A., DíazMesa, E.M., Fonseca, E., Arrojo, M., Sierra, P., et al.: Psychometric performance of the oviedo sleep questionnaire in patients with severe mental disorder. Revista De Psiquiatría Y Salud Mental 2(4), 169–177 (2009)
3. Xinlei, C., Xiaogang, R., Yue, W., Jiufeng, Y.: Design and realization of a comprehensive management system for severe mental disorders based on FLUX mode. J. Med. Imaging Health Inform. ASP 10(2), 522–527(6) (2020)
4. Kessler, R.C., Barker, P.R., Colpe, L.J., Epstein, J.F., Gfroerer, J.C., Hiripi, E., et al.: Screening for serious mental illness in the general population. Int. J. Meth. Psychiatric Res. 19(S1), 4–22 (2003)
5. Thompson, P.E., Farah, N.M.: What is mental illness? BMJ (2019)
6. Ruggeri, M., Leese, M., Thornicroft, G., Bisoffi, G., Tansella, M.: Definition and prevalence of severe and persistent mental illness. Br. J. Psychiatry J. Ment. Sci. 177(2), 149 (2018)
7. Mueser, K.T., Bond, G.R., Drake, R.E., Resnick, S.G.: Models of community care for severe mental illness: a review of research on case management. Schizophrenia Bull. 24(1), 37–74 (1998)
8. Simmonds, S., Coid, J., Joseph, P., Marriott, S., Tyrer, P.: Community mental health team management in severe mental illness: a systematic review. Br. J. Psychiatry J. Ment. Sci. 178 (3), 497 (2001)
9. Agius, M., Biocina, S.M., Alptekin, K., Rotstein, V., Persaud, A.: Basic standards for management of patients with serious mental illness in the community. Psychiatria Danubina 17(1–2), 42–57 (2005)
10. Dalum, H.S., et al.: Illness management and recovery (IMR) in danish community mental health centres. Trials (2011)

11. Xiang, Y.F., Chao, L.Y., Hua, D.Z.: An overview of data cleaning. Appl. Res. Comput. **3**, 3–5 (2002)
12. Liu, Y., Liu, D.: Mining ordinal patterns for data cleaning. In: Liu, G., Tan, V., Han, X. (eds.) Computational Methods. Springer, Dordrech (2006). https://doi.org/10.1007/978-1-4020-3953-9_3
13. Galhardas, H., Lopes, A., Santos, E.: Support for user involvement in data cleaning. In: Cuzzocrea, A., Dayal, U. (eds.) DaWaK 2011. LNCS, vol. 6862, pp. 136–151. Springer, Heidelberg (2011). https://doi.org/10.1007/978-3-642-23544-3_11
14. Lijun, W., Yongfeng, H., Ji, C., Ke, Z., Chunhua, L.: Medoop: a medical information platform based on Hadoop In: 2013 IEEE 15th International Conference on e-Health Networking, Applications and Services, Piscataway, NJ, USA, 9–12 October 2013. IEEE (2013)
15. Wang, Q., Ma, C., Yu, Y., Zhu, G.: Research on the construction of regional medical information service platform. In: Li, S., Jin, Q., Jiang, X., Park, J.J.(Jong Hyuk) (eds.) Frontier and Future Development of Information Technology in Medicine and Education. LNEE, vol. 269, pp. 3519–3524. Springer, Dordrecht (2014). https://doi.org/10.1007/978-94-007-7618-0_460
16. Jansen, A., Bosch, J.: Software architecture as a set of architectural design decisions. In: Fifth Working IEEE/IFIP Conference on Software Architecture (WICSA 2005), 6–10 November 2005, Pittsburgh, Pennsylvania, USA. IEEE (2005)
17. Bosch, J., Molin, P.: Software architecture design: evaluation and transformation. In: IEEE Conference & Workshop on Engineering of Computer-based Systems. CiteSeer (1999)

mDixon-Based Synthetic CT Generation via Patch Learning

Xin Song[1,2], Jiamin Zheng[1,2], Chao Fan[1,2], and Hongbin Yu[1,2(✉)]

[1] School of Digital Media, Jiangnan University, Wuxi 214122, China
1785613841@qq.com
[2] Jiangsu Key Laboratory of Media Design and Software Technology, Jiangnan University, Wuxi 214122, China

Abstract. We proposed a new method for generating synthetic CT on abdomen from modified Dixon (mDixon) MR data of abdomens to address the challenges of PET/MR attenuation correction (AC). AC is necessary in process of PET/MR but MR data lack photon attenuation, thus multiple methods are proposed to generate synthetic CT. However, these existing methods requires advantaged MR sequences which needs fine acquisition and huge cost consumption. To address this problem, we proposed a new method for generating synthetic CT using Patch Learning (SCG-PL). Global model of SCG-PL is transfer learning and patch model is semi-supervised classification. The advantages of our method can be summarized into two points. (1) Patch learning is a gradual learning process with gradually updating global model on remodeling patch model, so our SCG-PL method is gradually capable of generating synthetic CT. (2) Semi-supervised classification adopted in the process of patch learning, only small amount of labeled data is needed in SCG-PL, which greatly reduced the workload of radiologists. The experimental results indicate that proposed SCG-PL method can effectively generate synthetic CT image from challenging abdomen images using mDixon MR sequence data only.

Keywords: Synthetic CT generation · mDixon-based MR · Abdomen · Attenuation correction (AC) · Patch learning (PL)

1 Introduction

Benefit from the superior soft tissue contrast and high anatomical resolution, Magnetic Resonance Imaging (MRI) scans have been widely applied in radiotherapy [1, 2]. With the evolution of modern medical imaging technologies, PET/MR, combining with advantages of two imaging modalities, can offer further information which cannot be presented in PET/CT and has potential applications. Meanwhile, compared with CT, the characteristic of using no ionizing radiation MRI has an absolute advantage. Yet there are rarely MRI or PET/MR used in radiotherapy procedure individually due to the lack of the geometric integrity and precision in mapping Relative Electron Density (RED) for dose calculations.

Solely using MRI or PET/MR for radiotherapy has become an interesting goal in radioactive medicine. Here, the primary problem to be solved is attenuation correction

© ICST Institute for Computer Sciences, Social Informatics and Telecommunications Engineering 2020
Published by Springer Nature Switzerland AG 2020. All Rights Reserved
Y.-D. Zhang et al. (Eds.): ICMTEL 2020, LNICST 327, pp. 411–421, 2020.
https://doi.org/10.1007/978-3-030-51103-6_37

in radiotherapy. To achieve this goal, multiple methods are proposed. However, the diversity and the real-time movement of human organs and tissues directly cause the failure of methods, such as atlas-based [3, 4] and template-based [5]. Standing out from the proposed methods, the synthetic CT generation method by distinguishing the tissue type of all voxel in MRI images.

At present, the limitations of the development of synthetic CT generation lie primarily in two areas, i.e. target body sections and methods of generation. From the site of body sections, most studies have focused on brain and breast, these body sections have common characteristic of simple organization and less deformation. Abdomen, as the challenging body section, with its complex anatomical structure and along with a large amount of organs deformation caused by massive human respiration is little studied. From the existing methods, some advantage sequences which are technically challenging like Ultrashort Echo Time (UTE) and Zero Echo Time (ZTE) are needed [6, 7].

Focusing on abdomen, on the basis of existing methods, we introduce the ensemble learning thought i.e. patch learning [8] to generate synthetic CT in the case of only using easily obtainable MR sequence. Global model of SCG-PL is transfer learning and patch model is semi-supervised classification. The advantage of our proposed synthetic CT generation on abdomen using patch learning method (SCG-PL for short) lies mainly in the following two points:

1) The combination of transfer learning and semi-supervised classification in patch learning enables the indistinguishable tissue type i.e. bone, to be well differentiated from voxels of Dixon sequence.
2) With the semi-supervised classification adopted in the process of patch learning, only small amount of labeled data is needed in SCG-PL, which greatly reduced the workload of radiologists.

The rest of this manuscript shows you our proposed SCG-PL method in detail. Related work is introduced in Sect. 2. Specifically, the method is explained in Sect. 3. Section 4 presents the experimental studies. Conclusion is given in Sect. 5.

2   Related Work

2.1   Patch Learning

Patch learning (PL) [8] is a strategy of gradually improving the performance of a machine learning model by iteratively updating the global model by remodeling the patch model. PL consists of the following three steps:

1) Train an initial global model using all training data.
2) Identify L patches from the initial global model, which contribute the most to the learning error, and train a (local) patch model for each patch.
3) Update the global model using training data that do not fall into any patch.

PL zooms into those parts that need more work without discarding the initial entire model, and become an excellent model until appropriate amount of patches remodeled, which is shown as Fig. 1.

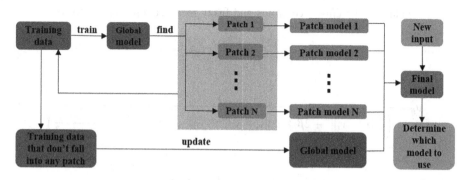

Fig. 1. Idea of PL.

2.2 KT-CM

Fuzzy c-means (FCM) [16] and maximum entropy clustering (MEC) [17] are classical soft partition clustering algorithms that aim to subdivide the data instances in a specific dataset into C pairwise disjointed clusters. Quadratic weights and Gini-Simpson diversity-based fuzzy clustering.

(QWGSD-FC) is an effective soft-partition clustering algorithm, which is intended to simultaneously maintain the advantages of some classical methods. As we all know, conventional soft-partition clustering approaches may not be powerful enough in intricate data circumstances. To address these challenges, based on QWGSD-FC, KT-CM was proposed by combining transfer learning with soft partition clustering. KT-CM offers us sufficient, extra knowledge of the data in the source domain as the prior information, except for the direct information in the target domain.

Let $X_T = \{x_{1,T}, \ldots, x_{N_T,T}\} \in R^{N_T \times d}$ denote the data set in the target domain, where the number of data instances is denoted as N_T and the dimension is denoted as d. C_T represents the number of clusters in the target domain. $U_T = [\mu_{ij,T}]_{C_T \times N_T}$ denotes the generated membership matrix, where the elements indicate the membership degree of the $x_j (j = 1, \ldots, N_T)$ data instance belonging to cluster $i(i = 1, \ldots, C_T)$.

The framework of KT-CM can be reformulated as

$$min \left(J_{KT-CM} = \sum_{i=1}^{C_T} \sum_{j=1}^{N_T} \mu_{ij,T}^2 + \beta \sum_{i=1}^{C_T} \sum_{j=1}^{N_T} \mu_{ij,T}^2 + \gamma \sum_{i=1}^{C_T} \sum_{j=1}^{N_T} \mu_{ij,T}^2 \|\hat{v}_{i,S} - v_{i,T}\|^2 \right)$$

$$s.t. \, 0 \le \mu_{ij} \le 1, \sum_{i=1}^{C_T} \mu_{ij} = 1$$

(1)

Where $\beta > 0$ denotes the Gini-Simpson diversity measure, $\gamma \ge 0$ represents the transfer optimization parameter, $\hat{v}_{i,S}$ represents the ith cluster centroid in the source domain, and $v_{j,T}$ represents the jth cluster centroid that is estimated in the target domain.

The update of the cluster centroids and fuzzy memberships are as follows

$$v_i = \frac{\sum_{j=1}^{N} \mu_{ij}^2 + \gamma \hat{v}_i \sum_{j=1}^{N} \mu_{ij}^2}{(1+\gamma) \sum_{j=1}^{N} \mu_{ij}^2} \tag{2}$$

$$\mu_{ij} = \frac{1}{\left(2\|x_j - v_i\|^2 + 2\beta + 2\gamma\|\hat{v}_i - v_i\|^2 \times \sum_{K=1}^{C} \frac{1}{2\|x_j - v_k\|^2 + 2\beta + 2\gamma\|\hat{v}_K - v_K\|^2}\right)} \tag{3}$$

2.3 Laplacian Support Vector Machine (LapSVM)

LapSVM, a semi-supervised classification algorithm based on manifold regularization, mainly research on how to use a small number of labeled samples and a large number of unlabeled samples for training and classification, which is an extension of the traditional SVM. The LapSVM learning model contains the intrinsic geometry information of the sample because of the introduction of sample manifold regularity item.

Assume the sample set $S = \{x_i, i = 1, \ldots, n\}$, x_i represents the ith sample and n represents the number of samples. Set $L = \{x_i, i = 1, \ldots, m\}$ represents the labeled sample, m represents the number of the labeled sample. Set $U = \{x_i, i = 1, \ldots, u\}$ represents the unlabeled sample, u represents the number of the unlabeled sample. Let y_i represent the category of the ith sample, $y_i \in \{-1, 1\}$. The LapSVM model is defined as follows

$$f^* = \min_{f \in H_k} \sum_{i=1}^{m} max(1 - y_i f(x_i), 0) + \gamma_A \|f\|_A^2 + \gamma_I \|f\|_I^2 \tag{4}$$

Where, $f = [f(x_i), x_i \in S]^T$ is an n-dimensional column vector on the training data set. $\|\cdot\|_A^2$ is the Ambient Norm defined in the regenerative nuclear Hilbert space (RKHS), H_k is the RKHS related to the kernel function. Parameter γ_A is weight, which controls the complexity of $\|f\|_A^2$ in RKHS. $\|f\|_I^2$ is the intrinsic regular term, which keeps the inner manifold structure of the sample distribution. Parameter γ_I is the weight of the function in the low-dimensional manifold, which controls the complexity of inner geometry function. By calculating the Lagrangian multiplier, the classifier is as follows

$$f^* = \sum_{i=1}^{n} \alpha_i^* K(x_i, x) \tag{5}$$

Where, α_i^* is the Lagrangian multiplier and K is the kernel matrix. Solving the Lagrangian multiplier

$$\alpha^* = (2\gamma_A I + 2\gamma_I KL)^{-1} J_L^T Y \beta^* \tag{6}$$

Where, I is the identity matrix and L is the Laplacian matrix. $Y \in R^{m \times n}$ is a diagonal matrix composed of identification samples $y_i (i = 1, \ldots, m)$. $J_L \in R^{m \times n}$ is a block matrix $[IO]$ formed by labeled samples and unlabeled samples. β is the Lagrange multiplier, which is

$$\beta^* = \begin{array}{c} max \\ \beta \in R^m \end{array} \sum \beta^i - \tfrac{1}{2} \beta^T Q \beta$$
$$Q = YJ_L K (2\gamma_A + 2\gamma_I KL)^{-1} J_L^T Y \qquad (7)$$
$$s.t. \sum_{i=1}^{m} \beta_i y_i = 0; 0 \le \beta_i \le 1, i = 1, \ldots, m$$

3   The Proposed SSG-PL Method

The proposed SCG-PL method includes data prepare stage, constructing Patch Learning (PL) model and generating synthetic CT image. In data prepare stage, we extract features of four types of MR data to form seven-dimensional MR feature data for each subject and obtain the referenced class prototypes of 4 key tissue types via conventional FCM from prior knowledge of referenced MR data. After data preparing, we construct PL model to obtain multiple 4-class classifiers. Global model is constructed as KT-CM and patches whose error impacts most become candidate patches, which need to be remodeled. Patch model is constructed as LapSVM, which is semi-supervised classification, and then updating the PL model. Finally, we organically combine multiple 4-class classifier to generate target synthetic CT image. Next, we detail each stages as follows.

3.1   Data Prepare

Feature extraction is of vital importance in machine learning and pattern recognition. The quality of features has a crucial impact on generalization performance. Considering the unavoidable noise in acquiring MR image, we adopt a principle of convolutional kernel to extract local texture features, which learns from the convolution layer in convolutional neural network (CNN) [10]. We extract local texture features from four types of abdominal MR data, i.e., fat, water, in-phase (IP) and opposed-phase (OP) [11], as the input to our SCG-PL method for each subject. Except for above features, position feature is adopted as well for better distinguish with the consideration of the similarity of signals of air and bone in MR data. The scan voxel size and pixel slices of 3D MR image are $0.98 \times 0.98 \times 5$ mm^3 and 512×512 pixel slices of Z-axis respectively. Considering isotropism, we design a grid partition strategy of 5×5 voxels assembled. Each voxel position feature is determined by spacing of size $4.9 \times 4.9 \times 5$ mm^3. Thus position feature (x, y, z) can be expressed as the indexes of grid, $1 \le x \le 103$, and $1 \le y \le 103$. Combining all the features we obtain seven dimension MR feature data as the input data to our method.

To generate final synthetic CT image from MR image as accurate as possible, we need referenced prior knowledge, i.e., referenced class prototypes of bone, air, fat tissue and soft tissue. Thus, paired CT and MR images are necessary and each pair needed to deformably registered before. The work and data flows of Phase 2 are shown in Fig. Suppose there are n pairs CT and MR images to be referenced, we take one pair as example for detail. For reference pair 1, bone class centroid can be first determined since we can recognize bone data in MR image with CT data whose Hounsfield Unit (HU) is 300, and takes the average value of bone data as bone class centroid. Then we apply FCM to the leftover data, i.e., data without bone, for clustering and obtain three clusters centroids later. At this point, we obtain four class centroids of pair 1. And n referenced pairs produce n class centroids, class prototypes are the average value of class centroids for every class. Thus we obtain the reference class prototypes of four key tissue types.

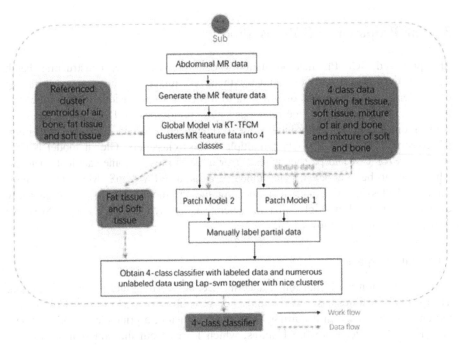

Fig. 2. Illustration of the work and data flows for generating 4-class classifier

3.2   Constructing PL Model to Obtain 4-Class Classifier

4-class classifier via PL model is necessary for generating synthetic CT, as shown in Fig. 2. Global model for constructing a 4-class classifier would naturally be KT-CM considering the existing historical referenced cluster centroids. The preliminary result of clustering would roughly be 4 classes, i.e., bone, air, soft tissue and fat tissue, after applying KT-CM to MR feature data. Owing to the fact that mDixon sequences are expert in reflecting fat tissue, the gained cluster of fat tissue after transfer clustering is relatively accurate. The leftover three clusters include soft tissue, bone and air. Soft tissue can be subdivided into one tissue approximate to soft bone and other soft tissue, which means soft bone and some soft tissue are mixed and hard to distinguish. In transfer clustering, the cluster of soft tissue not close to soft bone is believable as well. Actually, the signals of air and hard bone are both very low in MR data thus one of the clusters from transfer clustering is the mixture of them.

In order to gain the final 4-class classifier to generate precise synthetic CT image, patch models are reconstructed in these mixture parts. Benefiting from the prior knowledge of subject's CT image, we can acquire labeled examples. However, it is practically infeasible for radiologist to label the whole data manually because of huge time consumption. Thus, semi-supervised classification can be used in patch model and LapSVM is ideal. With numerous unlabeled data and a limit amount of mark data, which is practically feasible for radiologist, it's capable of obtaining final 4-class classifier.

3.3   Generating Synthetic CT Image Through Multiple 4-Class Classifiers

We assemble multiple results of 4-class classifiers via the strategy of voting to decide voxel type. Considering numerous data in MR feature data, it's infeasible to directly take the entire data as input to our method because of huge time consumption. Thus, we propose sampling-KNN mechanism, which involves randomly sampling the MR feature data and using K nearest neighbor (KNN) to restitute the whole prediction results, in our method to accelerate the whole process. Sampling-size denotes as ss.

Specific CT value are set to 380, −700, 98 and 32 corresponding with bone, air, fat tissue and soft tissue with the referring to [12] to reconstruct a synthetic CT image.

4   Experiment Results

4.1   Setup

In this section, we assess the effectiveness of the proposed PL-TL-SC method for generating synthetic CTs. Thus, ten subjects were recruited using a protocol approved by the University Hospitals Cleveland Medical Center Institutional Review Board.

Moreover, three existing methods are in comparison with our method, i.e., the all-water method (AW) [13], transfer fuzzy C-means clustering (TFCM) and the support machine (SVM). Three metrics, i.e., the mean absolute prediction deviation (MAPD), the root mean square error (RMSE), and R [14, 15], are used to evaluate the effectiveness of our method.

We adopt Leave-one-out strategy to generate ultimate result. That is, we treat one subject as test dataset and the other leftover subjects as train dataset. Each result for generating synthetic CT of one subject are assembled from results of the remainder classifiers, which excludes the classifier of the test one.

Our experimental studies were carried out on a computer with an Intel i5-4590 3.3 GHz CPU, 8 GB of RAM, Microsoft Windows 10 (64 bit), and MATLAB 2017a.

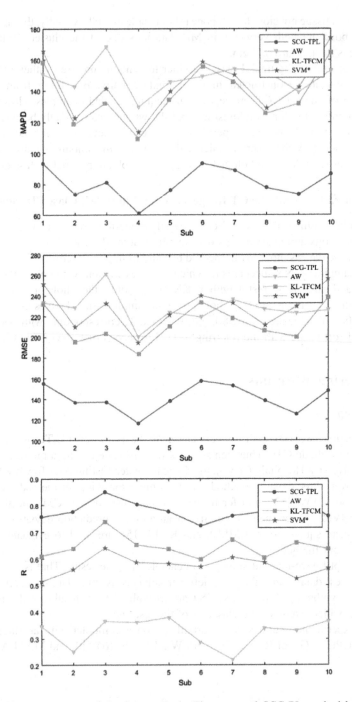

Fig. 3. Performance curves of the four methods. The proposed SCG-PL method has a lower mean absolute prediction deviation (MAPD), a lower root mean square error (RMSE), and a higher correlation (R) than the other three methods, including AW, TFCM and the SVM.

4.2 Results

We display the experimental results in Table 1 for each subject. In order to visually and intuitively present our method, the performance curves regarding to MAPD, RMSE and R metrics are shown in Fig. 2. As shown in Table 1, and we display the synthetic CT images from Sub8 with four methods above (Figs. 3 and 4).

Table 1. Performance comparison of generating synthetic CTS among the proposed SCG-PL and other methods

Sub	MAPD				RMSE				R			
	SCG-PL	AW	TFCM	SVM	SCG-PL	AW	TFCM	SVM	SCG-PL	AW	TFCM	SVM
1	**93.50**	150.45	159.02	165.59	**155.50**	233.49	232.64	251.39	**0.76**	0.35	0.61	0.52
2	**73.46**	142.80	118.63	122.24	**136.97**	228.49	195.38	210.21	**0.78**	0.25	0.64	0.56
3	**81.07**	168.21	132.39	142.04	**137.37**	261.03	203.70	232.85	**0.85**	0.36	0.74	0.64
4	**61.11**	129.55	109.36	113.25	**116.72**	200.63	183.44	194.80	**0.80**	0.36	0.65	0.59
5	**76.08**	145.55	134.63	139.75	**137.97**	224.10	210.92	221.79	**0.78**	0.38	0.64	0.58
6	**93.35**	149.09	156.05	158.22	**157.70**	219.24	234.02	240.17	**0.72**	0.29	0.60	0.57
7	**89.01**	154.07	145.61	150.10	**153.15**	235.96	218.61	232.73	**0.76**	0.22	0.67	0.61
8	**77.61**	152.41	125.44	128.56	**138.71**	227.09	206.17	211.88	**0.77**	0.34	0.60	0.59
9	**73.38**	139.04	131.70	142.41	**125.31**	222.99	200.56	229.57	**0.82**	0.33	0.66	0.52
10	**86.34**	153.00	164.96	174.12	**148.21**	225.90	238.13	255.79	**0.76**	0.37	0.64	0.56
Average	**80.49**	148.42	137.78	143.63	**140.76**	227.89	212.36	228.12	**0.78**	0.32	0.64	0.57

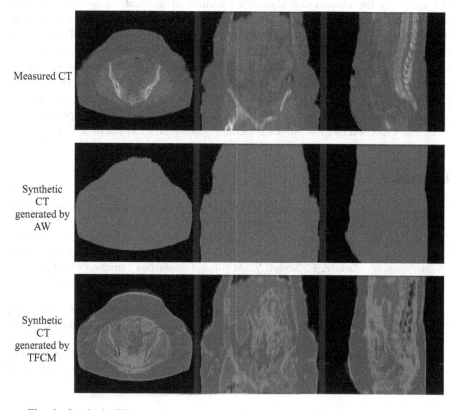

Fig. 4. Synthetic CT images on a representative subject (Sub 8) using four methods.

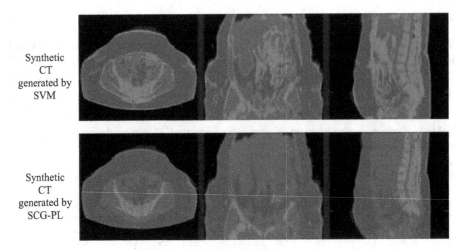

Synthetic CT generated by SVM

Synthetic CT generated by SCG-PL

Fig. 4. (*continued*)

5 Conclusion

We study on the challenging abdominal section in medical image for generating synthetic CT image with mDixon MR image only. Recognizing the similarity between the situation we face and the feature of Patch Learning, we propose the effective SCG-PL method with techniques such as transfer learning for global model and semi-supervised classification for patch model. Patch Learning is a process of gradual improvement with remodeling indistinguishable domains to gradually obtain good performance of classification. This exactly corresponds the complicated situation in part of abdominal section because of deformable tissues and varying air position over time. The proposed method proves preferable generation of synthetic CT image.

References

1. Zaidi, H., Alavi, A.: Current trends in PET and combined (PET/CT and PET/MR) systems design. PET Clin. **2**(2), 109–123 (2007)
2. Zaidi, H., Mawlawi, O., Orton, C.G.: Simultaneous PET/MR will replace PET/CT as the molecular multimodality imaging platform of choice. Med. Phys. **34**(5), 1525–1528 (2007)
3. Hofmann, M., Steinke, F., Scheel, V.: MRI-based attenuation correction for PET/MRI: a novel approach combining pattern recognition and atlas registration. J. Nucl. Med. **49**(11), 1875–1883 (2008)
4. Farjam, R., Tyagi, N., Deasy, J.O.: Dosimetric evaluation of an atlas-based synthetic CT generation approach for MR-only radiotherapy of pelvis anatomy. J. Appl. Clin. Med. Phys. **20**(1), 101–109 (2019)
5. Kops, E.R., Herzog, H.: Template based attenuation correction for PET in MR-PET scanners. In: IEEE Nuclear Science Symposium Conference Record, pp. 3786–3789 (2008)

6. Liu, L., Cao, Y., Fessler, J.A.: A female pelvic bone shape model for air/bone separation in support of synthetic CT generation for radiation therapy. Phys. Med. Biol. **61**(1), 169–182 (2015)
7. Berker, Y., Franke, J., Salomon, A.: MRI-based attenuation correction for hybrid PET/MRI systems: a 4-class tissue segmentation technique using a combined ultrashort-echo-time/Dixon MRI sequence. J. Nucl. Med. **53**(5), 796–804 (2012)
8. Wu, D., Mendel, J.M.: Patch learning. IEEE Trans. Fuzzy Syst. **1906**, 00158 (2019)
9. Qian, P., et al.: Knowledge leveraged transfer fuzzy c-means for texture image segmentation with self-adaptive cluster prototype matching. Knowl.-Based Syst. **130**, 33–50 (2017)
10. Chen, Y., Jiang, H., Li, C.: Deep feature extraction and classification of hyperspectral images based on convolutional neural networks. IEEE Trans. Geosci. Remote Sens. **54**(10), 6232–6251 (2016)
11. Eggers, H., Brendel, B., Duijndam, A.: Dual-echo dixon imaging with flexible choice of echo times. Magn. Reson. Med. **65**(1), 96–107 (2011)
12. Schneider, W., Bortfeld, T., Schlegel, W.: Correlation between CT numbers and tissue parameters needed for Monte Carlo simulations of clinical dose distributions. Phys. Med. Biol. **45**(2), 459–478 (2000)
13. Andreasen, D., Edmund, J.M., Zografos, V.: Computed tomography synthesis from magnetic resonance images in the pelvis using multiple random forests and auto-context features. In: Medical Imaging: Image Processing. International Society for Optics and Photonics, vol. 9784, p. 978417 (2016)
14. Sekine, T., ter Voert, E.E.G.W., Warnock, G.: Clinical evaluation of zero-echo-time attenuation correction for brain 18F-FDG PET/MRI: comparison with atlas attenuation correction. J. Nucl. Med. **57**(12), 1927–1932 (2016)
15. Delso, G., Wiesinger, F., Sacolick, L.I.: Clinical evaluation of zeroecho-time MR imaging for the segmentation of the skull. J. Nucl. Med. **56**(3), 417–422 (2015)
16. Wen, J., Tian, Y., Yehang, S.: Improved evidential fuzzy c-means method. J. Syst. Eng. Electron. **29**(1), 187–195 (2018)
17. Tao, X., Wang, R., Chang, R.: Density-sensitive fuzzy kernel maximum entropy clustering algorithm. Knowl.-Based Syst. **166**, 42–57 (2019)

Embedded 3D Printing Based on High Elastomeric Strain Wireless Sensor

Hongwei Wang$^{(\boxtimes)}$, Yue Wu, Xiaogang Ren, and Zhiying Cao

The Affiliated Changshu Hospital of Soochow University
(Changshu No.1 People's Hospital), Suzhou 215500, Jiangsu, China
1040558124@qq.com

Abstract. In view of the high degree of personalization of embedded 3D printing products, traditional 3D printing is not applicable. This paper presents an embedded three-dimensional printing technology based on high elastic strain wireless sensor. The whole method framework includes mechanical system, control module and visual module. Firstly, three non-collinear points on the high elastic strain wireless sensor are used to align the guide plate and the model. Then, according to the position and direction of the guide hole on the high elastic strain wireless sensor, the mechanical system is controlled to guide the model guide hole to move to the center of the visual module. The characteristic parameters such as roundness, length-width ratio, diameter and center distance of the guide hole are analyzed to determine whether the guide hole is qualified. The experimental results show that compared with the traditional three-dimensional printer, the three-dimensional printer designed in this paper shortens the production cycle and improves the print resolution.

Keywords: High elastomeric strain · Wireless sensor · Embedded · 3D printing products · Quality manufacturing

1 Introduction

3D printing is an additive manufacturing technology that can be used to quickly complete the industrial design or mold manufacturing. With the rise of this technology, some breakthrough methods can be implemented [1, 2]. Great breakthroughs have been achieved in the major subject of Beijing Municipal Science and Technology Commission "Development and manufacturing of the embedded 3D printing tumor medical guide plate forming equipment and manufacturing equipment engineering prototype" undertaken by Beijing University of Technology [3, 4]. In this treatment method, guide plate model is designed based on the conditions of the patients, and the therapeutic guide plate is generated through the embedded 3D printing technology. The doctor inserts the therapeutic probe along the guide hole on the guide plate into the patient's tumor location to implement the radiotherapy. Hence, the forming of the guide hole has a significant effect on the therapeutic effect [5–7]. However, the forming of the guide hole cannot guarantee the direct use [8, 9], which does not match the rapid development of the 3D printing technology. At present, the embedded 3D printer manufacturing is still in the blank stage and lack of systematic manufacturing means [10].

© ICST Institute for Computer Sciences, Social Informatics and Telecommunications Engineering 2020
Published by Springer Nature Switzerland AG 2020. All Rights Reserved
Y.-D. Zhang et al. (Eds.): ICMTEL 2020, LNICST 327, pp. 422–437, 2020.
https://doi.org/10.1007/978-3-030-51103-6_38

The embedded 3D printing products have very strong personalized features. Take the frontal tumor guide plate in this paper as an example. Each guide plate is designed for the individual patient. The location and orientation of each guide hole will be different from those of the other guide plates. Hence, the assembly line product manufacturing scheme is not applicable to the manufacturing of personalized product based on the embedded 3D printing [11, 12].

In recent years, computer vision technology has developed rapidly, and the vision-based 3D printer manufacturing technology has gradually been developed for the product application. For example, the machine vision system for the red pepper contaminated by aflatoxin, the gilthead freshness intelligent manufacturing system based on machine vision [7], the application of machine vision technology to make tiny cracks on the surface of biscuits. In this scheme, one plane of the object to be tested is fabricated, and the plane to be fabricated for each product is the same, so that the camera and the object to be tested can remain relatively fixed in the locational relationship, which cannot meet the multi-angle manufacturing requirements of embedded 3D printing products. Tile automatic visual quality control intelligent system based on machine vision [11, 12]. Button battery online manufacturing system based on machine vision. These systems are coupled with the conveyor belt to make product manufacturing more automated. It is possible to manufacture multiple products one by one, while still maintaining a relatively static relationship for each product [13]. Hence, it can be seen that the 3D printer manufacturing based on machine vision is currently a research hotspot, but the current manufacturing systems are mostly based on the integration of the computer vision and control modules. The products manufactured are relatively simple, the manufacturing path is fixed, and only one surface is fabricated in most of the products to be tested or the products are fixed so that the part to be tested and the location of the vision system remains unchanged. In this way, it is highly difficult to adapt to the products that are not in the design template. In case of the manufacturing based on the embedded manufacturing 3D by using the traditional quality manufacturing system, it will inevitably result in reduced manufacturing efficiency or even inability to manufacture the printing products due to the high degree of freedom of embedded 3D printing products [14].

In order to solve the above problems, a systematic embedded 3D printer manufacturing framework is put forward in this paper, and a prototype system is developed to verify the principle. The system is based on the model guidance, mechanical system coordination, supplemented by visual monitoring to achieve embedded 3D printer quality manufacturing. The main idea is to make use of the guide plate and corresponding three non-collinear points on the high elastomeric strain wireless sensor to align the coordinates of the guides and the model of the three, and then in accordance with the guide hole, the high elastomeric strain on the triangular surface of the wireless sensor. Subsequently, the geometric feature is used to extract the center coordinate and the normal vector of the guide hole and plans the manufacturing path. Finally, through the image of the guide hole, in accordance with the roundness, length-width ratio, diameter, and pitch of center hold in the guide hole, whether the guide hole is qualified is determined. In the traditional 3D printer, the design method and system put forward in this paper can shorten the manufacturing cycle and improve the printing resolution.

2 High Elastomer Strain Wireless Sensor

High elastomer strain wireless sensor is composed of elastomer and resistance strain gauge. Elastomer is the basis of high precision force sensor. Strain gauge is the core of sensor elastomer has many forms and complex structure. To design high precision force sensor, the geometry and material of elastomer must be reasonably selected and designed according to its range and accuracy. The guiding principle of designing high precision force transducer is to pursue good natural linearity, improve the output sensitivity of the transducer, and make the transducer have strong lateral resistance. At the same time, the transducer has simple structure, easy sealing, easy processing and so on.

Like other strain elastomers, the working principle of double-circular-hole beam elastomer is based on the deformation principle of metal elastomer and the strain principle of resistance strain gauge, that is, the physical phenomenon that resistance strain gauge changes resistance value with the deformation of elastomer under the action of external weight and force, resistance strain gauge is firmly bonded to the elastic surface of double-circular-hole beam with thermosetting resin binder. Four resistance strain gauges are connected to the Whiston bridge circuit. before the external force acts, the bridge is pre-balanced under the external force, the resistance strain gauge changes with the deformation of the elastomer with double circular holes, so that the strain bridge produces a voltage output signal proportional to the weight or external force.

According to Hooke's law, within the limit range of elasticity, the deformation of an elastic body is proportional to the stress, that is, to the force it is subjected to:

$$\varepsilon = \frac{\sigma}{E} \tag{1}$$

In the formula, ε represents the deformation of an elastic body. E represents the stress of an elastomer. σ represents Young's modulus of elastomeric materials

When the resistance strain gauge deforms with the deformation of the elastomer, the resistance strain gauge changes its geometric size and physical properties, thereby changing the resistance value:

$$\frac{\Delta R}{R} \bigg/ \frac{\Delta I}{I} = K = I + 2M + \frac{\Delta P}{p} \bigg/ \frac{\Delta I}{I}$$

$$\Delta R = kR\varepsilon = \frac{kR\sigma}{E} \tag{2}$$

In this formula, $\Delta R/R$ represents the relative resistance change of the resistance strain gauge. R represents Poisson's ratio of foil for resistance strain gauge. K represents the sensitivity of resistance strain gauge. P represents resistivity of Strain Resistance Materials.

The elastomer structure of the sensor is shown in Fig. 1. It is a double-circular-hole double-beam with one fixed end and one stressed at the other end. The double-beam is processed from the whole material, so the structure is very stable and reliable. When the external force or weight acts on the elastomer of the double-circular-hole beam, it will deform. In order to simplify the calculation of the sensor, the double-circular-hole double-beam can be simplified as one fixed end and the other end only along the force side. For statically indeterminate rectangular double beams which move towards the direction but cannot rotate, the combined calculation method should be used in the calculation.

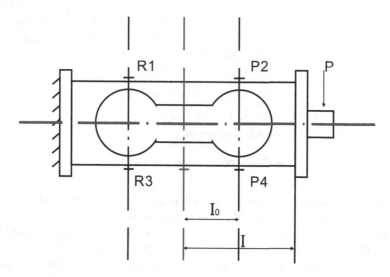

Fig. 1. Double circular holes and double beams

3   Overall Design Framework of the Embedded 3D Printing System

The embedded 3D printer quality manufacturing in this paper is used for the quality manufacturing of the guide hole shape. Its nature is the consistent manufacturing of the high elastomeric strain wireless sensor and the product. The machine vision method is a relatively efficient method, and the installation is simple, with relatively low cost. However, the traditional machine vision method cannot complete the manufacturing of the guide holes with different orientations in the direction of the guide plates. To address this issue, this paper put forward a method to guide the calculation of manufacturing paths through high elastomeric strain wireless sensors, in which the mechanical system coordinates with the machine vision to establish the shape and quality manufacturing framework of the embedded 3D printing product for the product location and quality manufacturing. The overall framework of the system is shown in Fig. 2 as the following.

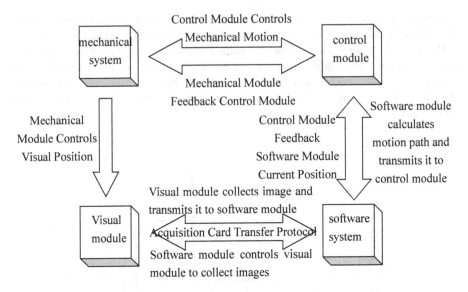

Fig. 2. Overall framework of the system

In the above figure, the software module is used to analyze the high elastomeric strain wireless sensor to calculate the motion path, so that the orientation of the guide hole can be parallel to the optical axis of the camera. In addition, it is located at the center of the camera, and then transmits the motion path to the control module, and the control module transmits a signal to the mechanical module motor, which controls the movement of the mechanical module. The location of the vision module relative to the guide is changed accordingly. At this point, the software module can be used to control the vision module to collect images and feed back to the software module. After the motion is in place, the mechanical module sends the feedback signal to the control module, and the control module gives feedback on the current mechanical module location of the software module in accordance with the signal.

4   Hardware Design of the Embedded 3D Printer System

The embedded 3D printer system hardware system is mainly responsible for the image acquisition and transmission, the guide plate location and camera location movement, the current location information reading and so on. In accordance with the above tasks, the hardware system structure designed in this paper is shown in Fig. 3 as the following. The hardware system mainly includes the mechanical structure and the electric control system.

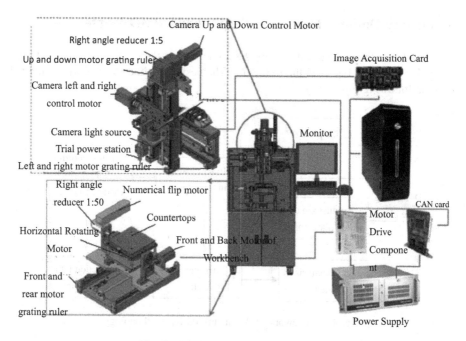

Fig. 3. Hardware system of the system

4.1 Mechanical Structure of the Embedded 3D Printing

The embedded 3D printing model can be simplified into a hemisphere. In this way, the guide hole is inserted on the hemisphere. In order to move the guide hole to the center of the camera and the orientation being parallel to the optical axis of the camera, the design based on the principle of the five-axis digital control machine tool includes two rotating shafts, θ_1 and θ_2, and a mechanical system with three translational axes, xyz, in which a single industrial camera is mounted on the y axis to form a CCD manufacturing mechanism with the x axis; θ_1 and θ_2 are mounted on the z axis, where θ_1 can drive the rotation of θ_2, and the above three constitute a three-dimensional work desk.

4.2 Electric Control System of the Embedded 3D Printing

The embedded 3D printing based on model guidance and visual monitoring the front surface of the tumor guide intelligent quality manufacturing system electrical system is mainly composed of the upper computer, motor control components, image acquisition equipment. The upper computer uses a general-purpose PC, which runs the self-programmed embedded 3D printing frontal surface tumor guide intelligent quality manufacturing software in the win7 system. The motor control component is based on the CAN protocol communication industrial computer. The CAN card adopts the CAN-PCI-02 by copley. The image acquisition equipment adopts the Siliconsoftware_mico-renableiv_vd4-cl image acquisition card, and the camera used is the SP2000 HD camera.

5   Quality Design Method for the Embedded 3D Printer

Combined with the mechanical system, the framework of the guide plate quality design method is shown in Fig. 4 as the following. The design method can mainly be divided into the following three parts:

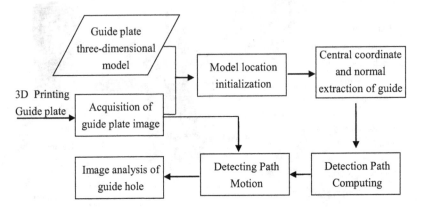

Fig. 4. Framework of the quality design methodology

(1) Initialization of the model location. In order to ensure that the model is consistent with the product, we must first ensure that the model coordinates and the guide coordinates are consistent. Hence, it is required to initialize the model location to align the high elastomeric strain wireless sensor coordinates of the guide with the world coordinates of the guide plate.

(2) Computation of the guide hole path. Each embedded 3D printing product has a corresponding model, which is the advantage of the embedded 3D printing products. It is possible to pre-estimate the location of the target to be tested through the model to determine the manufacturing path. This is efficient and convenient to manufacture by computer vision technology alone, which is also the significance of the model guidance put forward in this paper.

(3) 3D printer manufacturing. After the guide hole is moved to the location to be tested, the quality manufacturing is carried out by means of the images.

For an embedded 3D printed tumor guide to be manufactured, the guide points and the three approximate points on the model are first selected. In accordance with the high elastomeric strain wireless sensor, the world coordinates of the three points on the guide plate under the mechanical system are identified, and then the transformation matrix in the process of converting the model coordinates to the world coordinates of the guide plate is obtained. Subsequently, the other points of the guide are aligned with the process transformation to complete the model coordinates and the world coordinates of the guide plate. Then, the center coordinate and the normal vector of the guide hole cross section in the guide model are extracted and converted into the mechanical path recognizable motion path. The Dijkstra design method is used for optimization to control machinery system movement in the manufacturing path.

In this way, a guide hole can move to the location to be tested and stop, where the images are collected and analyzed to correct the location of the guide hole so that the center of the guide hole is located at the center of the image. Finally, the contour roundness, length-width ratio, diameter, distance from the previous hole and other information of the guide hole is determined to determine whether the pilot hole is qualified.

5.1 Location Initialization of the Embedded 3D Printer Model Based on the High Elastomeric Strain Wireless Sensor

In normal cases, the model of embedded 3D printing is obtained by high elastomeric strain wireless sensor editing software design or scanned by 3D scanner. Hence, each model can be considered as unconstrained from the coordinates of each vertex except for special requirements. In the embedded 3D printer of this paper, the locations and directions of different models in the same coordinate system are very different. In this way, a single matrix transformation formula can be used to align the high elastomeric strain wireless sensor with the physical location is impossible. If three non-collinear points on one object are known, the attitude of the object can be basically determined. In this paper, the world coordinates of the three non-collinear points of the guide are determined through the high elastomeric strain wireless sensor [9] and the corresponding points on the high elastomeric strain wireless sensor are picked up by the mouse, which are converted to the world coordinates in accordance with the calculation formula of binocular stereoscopic image. In combination with the mechanical structure, the world coordinate calculation formula of the marker point is deducted as the following.

$$\begin{bmatrix} x_w \\ y_w \\ z_w \\ 1 \end{bmatrix} = \begin{bmatrix} 1 & 0 & 0 & -c_x \\ 0 & 0 & -1 & c_y - D \\ 0 & 1 & 0 & -c_z \\ 0 & 0 & 0 & 1 \end{bmatrix} \begin{bmatrix} \frac{ft_x - X_r t_z}{f(X_r - X_1)} X_1 \\ \frac{ft_x - X_r t_z}{f(X_r - X_1)} Y_1 \\ \frac{ft_x - X_r t_z}{X_r - X_1} \\ 1 \end{bmatrix} \tag{3}$$

In the above equation: x_w, y_w and z_w stand for the world coordinate of the mechanical system lower guide mark; c_x, c_y and c_z stand for the mechanical system horizontal rotation center, which are the coordinates of the camera coordinate system where the image center is located; D stands for the distance from the horizontal rotation center to the front and back flip center; f stands for the camera focal length. The points on the model are shifted to the location of the actual world coordinate of the guide. Firstly, the 3 points on the model are rotated in accordance with the rotation in any space in the space, and then the triangle formed by the 3 points on the model is rotated to the corresponding edge of the triangle formed by the three points on the guide plate. Finally, the three points on the model are translated to the approximate locations of the three points on the guide plate. In accordance with this transformation process, the other points of the model are transformed to align the guide plate model with the actual location of the guide plate.

5.2   3D Printer Manufacturing and Path

5.2.1   Design Method for the Three-Dimensional Coordinate and Normal Vector Extraction of the Guide Hole

In this paper, quality manufacturing of the guide holes based on the embedded 3D printer is mainly carried out. Each guide plate corresponds to two high elastomeric strain wireless sensors. One is the embedded 3D3D printer model, and the other is the matching guide pin model. Based on the characteristics of the high elastomeric strain wireless sensor STL, the model is formed by a triangular face, and the three vertex coordinates and the normal vector of each triangular face are known. By extracting the triangular faces of the end faces of the guide holes manually, the geometric features of the triangular face and other triangular faces are identified accordingly. Hence, the three-dimensional coordinate and normal vector extraction design method of the guide hole based on geometric feature filtering and clustering is adopted as the following.

Firstly, based on the length and area of the triangular face, the triangular faces of the guide pin model that do not conform to the cross-section features of the guide holes are filtered out, and then the K-means cluster design method is adopted to cluster in accordance with the normal vector to obtain the triangular face sets in the same direction. The K-means clustering of the inner centroid of the triangle is performed inside the set, and each set of triangular faces which may be the cross section of the guide hole is segmented, and the central coordinates and normal vectors of each set are obtained. The guide plate model is estimated by a set of central coordinates and normal vectors. The location of the guide hole in the cross section, only the triangular face at the estimated location is used for the next calculation. Subsequently, the discrete curvature of the remaining triangular faces is calculated by the triangular mesh vertices discrete curvature calculation method based on the Voronoi diagram as the following.

$$K = \frac{2\pi - \sum \theta_i}{S} \tag{4}$$

In the above equation: K stands for the discrete curvature of V_0; θ_i stands for the angle between V_0 and V_{i+1}; S stands for the area of the triangle containing V_0.

In the experiment, if the discrete curvatures of the three vertices of the triangle are two positive and one negative, it is considered to be the required triangle, and vice versa. The calculation model is shown in Fig. 5 as the following.

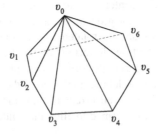

Fig. 5. Discrete curvature calculation model

The triangular surface after filtering is projected along a normal vector to a plane, and is made a circle by Hough transform. If it is a circle, it is considered to be a guide hole cross section. In accordance with the coordinates of the triangular surface in the three-dimensional coordinate system, the polygon center can be obtained. The method is used to obtain the center of the circle, and the average of the triangular surface normal vector is used as the normal vector of the cross section of the guide hole; otherwise, it is not considered to be the guide hole cross section.

5.2.2   3D Printer Manufacturing Path Calculation and Planning

As the mechanical motion cannot directly move in accordance with the coordinates and the normal vector, it is required to plan a path corresponding to the path of the mechanical device in accordance with the extracted guide hole coordinates and the normal vector. The purpose of the motion is to make the guide hole face the guide hole normal vector vertically to the camera plane. In addition, the center of the guide hole is at the center of the image, as shown in Fig. 6 as the following.

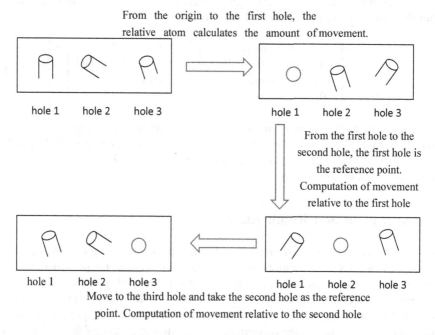

Fig. 6. Schematic diagrams of the guide hole in place

Firstly, the center coordinates of the similar normal vectors are grouped into one group. Subsequently, starting from the three-dimensional rotation matrix, the normal vector is decomposed to identify how the normal vector can be parallel to the y axis by the rotation of the y axis and the x axis, as shown in Fig. 7 as the following.

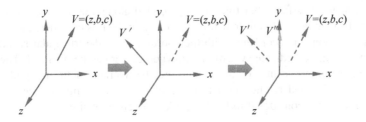

Fig. 7. Schematic diagram of the normal vector rotation

Two rotation angles α and β are as the following

$$\begin{cases} \beta = arccos \dfrac{c}{\sqrt{a^2+c^2}} \\ \alpha = arccos \dfrac{b}{\sqrt{a^2+b^2+c^2}} \end{cases} \qquad (5)$$

In accordance with the rotation angle, the coordinates of each point in the group after the rotation can be obtained as the following.

$$\begin{bmatrix} x' \\ y' \\ z' \\ 1 \end{bmatrix} = \begin{bmatrix} \cos \beta & 0 & -\sin \beta & 0 \\ 0 & 1 & 0 & 0 \\ \sin \beta & 0 & \cos \beta & 0 \\ 0 & 0 & 0 & 1 \end{bmatrix} \cdot \begin{bmatrix} 1 & 0 & 0 & 0 \\ 0 & \cos \alpha & \sin \alpha & 0 \\ 0 & -\sin \alpha & \cos \alpha & 0 \\ 0 & 0 & 0 & 1 \end{bmatrix} \begin{bmatrix} x \\ y \\ z \\ 1 \end{bmatrix} \qquad (6)$$

Finally, the path is planned in accordance with the design method to obtain the translation path. The aforementioned process is repeated to traverse each hole, and the angle of rotation and the distance of translation constitute the motion path of the mechanical system accordingly.

6 Experiment and Result Analysis

At present, there is no unified international standard for evaluating the quality of 3D printing. In addition to the control system designed in this paper, the print quality of the STL model is also affected by many factors such as the quality of the printing consumables, the mechanical transmission characteristics of the printer, and the print data generated by the slicing. In this paper, based on the actual needs, the printing accuracy, evaluation is carried out on the performance and control algorithm of the desktop level 3D printer designed in this paper. Comparison of the control algorithm put forward in this paper with the traditional algorithm shows that the proposed algorithm can reduce the time for establishing the topological relationship between the patches, which is simple to be implemented, stable and reliable.

6.1   Performance Test of the Embedded 3D Printer Prototype

(1) Model printing capacity, that is, whether a model can be printed out perfectly. The model printing capability is the basic requirement for the stability test of 3D printers. On the control pass, it mainly detects the anti-interference capacity of the printer and the sudden abnormal processing capability. For example, reading of print data, whether the mutual communication between modules in the printing process is stable, and effective processing of abnormal conditions. The test shows that the 3D printing control system designed in this paper can work continuously without failure under the condition of sufficient supplies. Figure 8 shows the diagram of the print molding effect. The comparison results suggest that the 3D printer model printing ability designed in this paper can complete the printing task and has relatively good stability.

Fig. 8. Model printing capability experiment

(2) Print dimensional accuracy. The so-called dimensional accuracy is the degree to which the actual size of the model fits the ideal size. The models for 3D printing are different. In this paper, the transverse dimensions of the model to be printed are experimentally compared, and the size of the circular axis and circular hole of the molded object is examined. Figure 9 shows a picture of the experimental model; Table 1 shows dimensional accuracy experimental data gives a comparison of the actual size of the experimental model to the ideal dimension. Based on the analysis of the data in Table 1, the absolute error of the inner diameter and the outer diameter of the model is within 0.2 mm in general. In addition, the relative error of the model with an outer diameter of 5 mm or more and an inner diameter of 10 mm or more can be controlled to be less than 1%.

Fig. 9. Diagram of the experimental model

Table 1. Dimensional accuracy test data

	Outer diameter (mm)						Inner diameter (mm)					
Ideal value	50	40	30	20	10	5	40	30	20	10	5	3
First measurement	49.73	39.34	29.87	20.2	10.04	3.17	49.73	39.94	29.87	20.2	10.04	3.17
Second measurement	49.76	39.85	29.82	20.4	10.07	5.32	39.86	29.72	19.96	9.72	4.84	2.75
Third measurement	49.93	39.83	29.77	20	9.88	5.22	39.86	29.83	19.47	9.87	4.83	2.85
Fourth measurement	49.9	39.96	29.85	19.96	9.94	5.3	39.81	29.66	19.81	9.86	4.78	2.82
Fifth measurement	49.81	39.82	29.89	20.1	9.96	5.22	39.82	29.97	19.86	9.84	4.7	2.73
Mean value	49.82	39.88	29.84	20.13	9.97	5.25	39.81	29.83	19.9	9.85	4.80	2.81
Absolute value	0.175	0.12	0.159	0.132	0.022	0.246	0.186	0.166	0.1	0.152	0.198	0.192
Relative error	0.35%	0.30%	0.53%	0.66%	0.22%	4.92%	0.47%	0.55%	0.50%	1.52%	3.96%	6.40%

6.2 Embedded 3D Printer Control Algorithm Test

The embedded 3D printing manufacturing method based on the high elastomeric strain wireless sensor put forward in this paper is compared with two commonly used methods, that is, the classification method of the geometric features based on the STL model and the stratification method of the topology information based on the STL model. In the experiment of this paper, the two methods are referred to as Method 1 and Method 2, respectively.

In the experiment, the methods are evaluated from three aspects: the model stratification time, the memory space and the volume error. The three methods are applied to six models with different degrees of complexity. The STL model is shown in Fig. 10 as the following. The six models are numbered as model 1–6 in turn, and the number of triangle patches included in each model is shown in Table 1 as the following. The experimental environment is OpenGL and C++ is used as the language of the method. Table 2 lists the model stratification time, the memory space and the volume error obtained by the three methods for the six experimental models when the model stratification thickness is 0.01 mm. Due to the different complexity of the models, the differences in the experimental values at various angles are also relatively large, as shown in the following.

Fig. 10. Six STL models

Table 2. Results of the 6 models under 3 methods

	Model		Stratification time/s			Volume error/mm^3		
No.	Description	Number of batches	Method 1	Method 2	Method put forward in this paper	Method 1	Method 2	Method put forward in this paper
1	Mobile phone	961	1	0.89	0.32	65.83	65.79	65.43
2	Teacup	1244	1.25	1.13	0.56	0.25	0.24	0.23
3	Character	1942	2.1	1.93	1.12	39.46	39.35	39.30
4	Mouse	9960	10.01	8.89	3.60	22.24	22.35	22.18
5	Lion	19358	12.4	11.7	5.51	0.72	0.65	0.66
6	Cassiopeia	23826	21.43	20.9	10.32	18.35	18.29	18.36

The method put forward in this paper has obvious advantages in the aspect of the stratification operation efficiency. In addition, the time consumption relative to Method 1 and 2 is also greatly reduced, which is only about 18% of Method 2. For some STL models with the simple structure, the method put forward in this paper takes only tens or even hundreds of milliseconds. For the models with the number of patches up to 2×10^4, it takes only a dozen seconds. Hence, the applicability is relatively strong. In comparison, Method 1 is more applicable to the models with the simple structure. However, for the STL models with the number of triangle patches above 2×10^3, the time required for stratification increases and the method cannot be applied. Compared with Method 1, the stratification time of Method 2 is slightly reduced, but the process of establishing the topology data is highly time consuming and takes up a relatively large memory. For the models with the number of patches below 2×10^4, the stratification can be quickly completed within 20 s. However, for the models with highly complexity, it takes a long time to complete the stratification.

Figure 11 shows the comparison of the method put forward in this paper, Method 1, and Method 2 in reducing the volume deviation. Compared with Method 1, the method put forward in this paper can reduce the volume deviation by about 3.6% on average, which can even achieve about 7% compared with Models 2 and 5. However, there is slight fluctuation in the processing of Model 10. Compared with Method 2, the method put forward in this paper can reduce the volume deviation by about 1.16% on average, except for the individual model (Model 6). The magnitude of the decline is relatively small, which is only about 0.69%. The experimental results are within the expected range. Although the volume deviation becomes larger, the stratification efficiency has been significantly improved, as shown in the following.

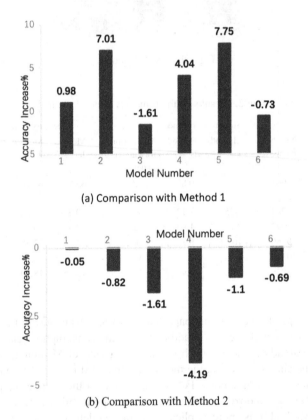

(a) Comparison with Method 1

(b) Comparison with Method 2

Fig. 11. Comparison of the accuracy performance of different methods

7   Conclusions

With the continuous deepening of the 3D printing technology and applications, embedded 3D printer manufacturing has become a hot issue. In this paper, preliminary study is carried out on this issue. In view of the problem in the embedded 3D printer manufacturing, a kind of embedded 3D printer manufacturing framework based on the high elastomeric strain wireless sensor is put forward. The design method can be used to implement the embedded 3D printer manufacturing with very little manual intervention. In addition, the framework can be extended to the shape manufacturing of other embedded 3D printing products.

References

1. Muth, J.T., Vogt, D.M., Truby, R.L.: Embedded 3D printing of strain sensors within highly stretchable elastomers. Adv. Mater. **26**(36), 6307–6312 (2014)
2. Li, F., Smejkal, P., Macdonald, N.P., Guijt, R.M., Breadmore, M.C.: One-step fabrication of a microfluidic device with an integrated membrane and embedded reagents by multimaterial 3D printing. Anal. Chem. **89**(8), 4701–4707 (2017)
3. Sun, K., Wei, T.S., Ahn, B.Y., Seo, J.Y., Dillon, S.J., Lewis, J.A.: 3D printing of interdigitated Li-ion microbattery architectures. Adv. Mater. **25**(33), 4539–4543 (2013)
4. Ladd, C., So, J.H., Muth, J., Dickey, M.D.: 3D printing of free standing liquid metal microstructures. Adv. Mater. **25**(36), 5081–5085 (2013)
5. Gebler, M., Schoot Uiterkamp, A.J.M., Visser, C.: A global sustainability perspective on 3D printing technologies. Energy Policy **74**, 158–167 (2014)
6. Fantino, E., Chiappone, A., Roppolo, I., Manfredi, D., Bongiovanni, R., Pirri, C.F., et al.: 3D printing: 3D printing of conductive complex structures with in situ generation of silver nanoparticles. Adv. Mater. **28**(19), 3711–3712 (2016)
7. Abbadessa, A., Blokzijl, M.M., Mouser, V.H.M., Marica, P., Malda, J., Hennink, W.E., et al.: A thermo-responsive and photo-polymerizable chondroitin sulfate-based hydrogel for 3D printing applications. Carbohydr. Polym. **149**, 163–174 (2016)
8. Spackman, C.C., Frank, C.R., Picha, K.C., Samuel, J.: 3D printing of fiber-reinforced soft composites: process study and material characterization. J. Manuf. Process. **23**, 296–305 (2016)
9. Parekh, D.P., Ladd, C., Panich, L., Moussa, K., Dickey, M.D.: 3D printing of liquid metals as fugitive inks for fabrication of 3D microfluidic channels. Lab Chip **16**(10), 1812–1820 (2016)
10. Castro, N.J., Patel, R., Zhang, L.G.: Design of a novel 3D printed bioactive nanocomposite scaffold for improved osteochondral regeneration. Cell. Mol. Bioeng. **8**(3), 416–432 (2015)
11. Javan, R., Zeman, M.N.: A prototype educational model for hepatobiliary interventions: unveiling the role of graphic designers in medical 3D printing. J. Digit. Imaging **31**(9), 1–11 (2017)
12. Pragnya, K., Keerti, K.: 3D printing high density ceramics using binder jetting with nanoparticle densifiers. Mater. Des. **155**, 443–450 (2018)
13. Gross, B., Lockwood, S.Y., Spence, D.M.: Recent advances in analytical chemistry by 3D printing. Anal. Chem. **89**(1), 57–70 (2017)
14. Zhang, B., Pei, X., Zhou, C., Fan, Y., Jiang, Q., Ronca, A., et al.: The biomimetic design and 3D printing of customized mechanical properties porous Ti6Al4V scaffold for load-bearing bone reconstruction. Mater. Des. **152**, 30–39 (2018)

Fruit Image Recognition Based on Census Transform and Deep Belief Network

Qi Xin[1], Shaohai Hu[1(✉)], Shuaiqi Liu[2,3(✉)], Hui Lv[4],
Shuai Cong[5(✉)], and Qiancheng Wang[2]

[1] College of Computer and Information,
Beijing Jiaotong University, Beijing 100044, China
shhu@bjtu.edu.cn
[2] College of Electronic and Information Engineering, Hebei University,
Baoding 071000, China
shdkj-1918@163.com
[3] Machine Vision Technology Creation Center of Hebei Province,
Baoding 071000, China
[4] Beagledata Technology (Beijing) Co., Ltd., Beijing 100089, China
[5] Industrial and Commercial College, Hebei University,
Baoding 071000, Hebei, China
congshuai@hbu.edu.cn

Abstract. Fruit image recognition plays an important role in the fields of smart agriculture and digital medical treatment. In order to overcome the disadvantage of the deep belief networks (DBN) that ignores the local structure of the image and is difficult to learn the local features of the image, and considering that the fruit image is affected by the change of illumination, we propose a new fruit image recognition algorithm based on Census transform and DBN. Firstly, the texture features of fruit images are extracted by Census transform. Secondly, DBN is trained by Census features of fruit images. Finally, DBN is used for fruit image recognition. The experimental results show that the proposed algorithm has a strong feature learning ability, and the recognition performance is better than the traditional recognition algorithm.

Keywords: Fruit image recognition · Deep belief network · Census transform

1 Introduction

Nowadays, the sharp increase in the amount of image data makes the number of images in general image recognition tasks become larger and larger, which also makes it difficult for traditional methods to meet people's needs. As a new subject in machine learning, deep learning has a lot of achievements in various fields. Compared with the artificial feature extraction method, the data features acquired through deep learning model are more representative of the rich inner information of big data and good features can be learned automatically without manual feature extraction. Therefore, deep learning is the future and will receive more attention in big data analytics [1–3].

© ICST Institute for Computer Sciences, Social Informatics and Telecommunications Engineering 2020
Published by Springer Nature Switzerland AG 2020. All Rights Reserved
Y.-D. Zhang et al. (Eds.): ICMTEL 2020, LNICST 327, pp. 438–446, 2020.
https://doi.org/10.1007/978-3-030-51103-6_39

Fruit image recognition plays an important role in the fields of smart agriculture and digital medical [4, 5]. With the rapid development of smart agriculture and digital medical in recent years, fruit image recognition has attracted more and more researchers' attention. In order to meet the needs of large-scale and efficient fruit recognition and classification, researchers began to identify fruit image with different algorithms. For example, in [6], the authors proposed a fruit classification based on six layer convolutional neural network and the classification accuracy is higher than that of traditional single feature. In [7], the authors present an automatic fruit recognition system for classifying and identifying fruit types and which is capable of automatically recognize the fruit name with a high degree of accuracy. And in [8], the authors proposed Kiwifruit recognition method at night based on fruit calyx image, whose recognition rate reached 94.3%. And in [9], the authors proposed a fast and accurate object recognition method especially for fruit recognition to be used for mobile environment. They combined color, shape, texture and intensity into their associated code fields to generate an object code that could be used as a search key for the feature database. And in [10], the authors given a fruit recognition method via image conversion optimized through evolution strategy principal component analysis and achieved a better recognition effect through the pretreatment, training and recognition of fruit images, with an average recognition rate of over 92%.

In general, the process of fruit image recognition system mainly focuses on pretreatment and feature extraction. In this kind of recognition system, fruit image acquisition is mostly conducted by placing the collected fruit in a strictly defined background in order to ensure that the recognition system is less interfered by the outside world, so as to improve the recognition accuracy of the system. However, the image in the actual environment is easily affected by the factors such as illumination change, fruit reflection and shielding, which in vary degrees impact the recognition accuracy of fruit image. In fruit recognition system, fruit features mainly include odor, color, shape and texture. While in the process of growth, different environment will lead to difference in shapes, sizes and colors. In addition, natural light intensity and shadow will also be different when fruit images are collected, which will affect the accuracy of image recognition. What's more, the complexity of the color and texture features of fruit images also makes the recognition more difficult. Therefore, better recognition algorithms are needed to solve this problem.

As a representative method in the deep learning, DBN is quite different from the previous algorithms in terms of training method structure. By adopting the idea of layered training, the training speed of DBN is greatly improved [11, 12]. In addition, the idea of layering also increases the system's ability to express complex functions. DBN usually takes pixel-level images as input and extracts the abstract features of the input images from bottom to top, from simple to complex, which is a process of automatic mining useful information in the data. However, general pixel-level images are easily affected by illumination and other factors, which affects the extraction of essential features of input samples in DBN. In order to improve the fruit image recognition performance of DBN under different illumination, we propose a new method combining Census transform with DBN to extract the texture features of images through Census transform to eliminate the influence of ununiform lighting on the recognition results.

2 Fundamentals of Census Transform

Census transform [13] is a locality-based non-parametric transformation algorithm, which is mainly used to characterize the local structure features of images and can well detect the edge and corner features of images. It uses the relationship between the gray value of the neighborhood pixel and center pixel as similarity base to eliminate the influence of ununiform lighting. And, the implementation of non-parametric transformation is simple and real-time, which has been widely popularized and applied in engineering implementation. The algorithm flow is shown in Fig. 1.

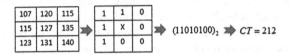

Fig. 1. The flow of Census transform algorithm

In Fig. 1, a pixel matrix is usually taken as the mask, whose central gray value is 127 and the gray value of each neighborhood pixel is shown in Fig. 1. We can get an 8-bit binary bit string through sequentially combine the eight digits obtained by the gray value comparison (from top to bottom, from left to right). Convert the binary bit string into a decimal number, and the gray value (whose range is [0, 255]) of the center pixel after Census transform is calculated.

Census transform replaces the original gray value of a pixel with its Census transform value, making these Census transform values mutually constrain and correlate [14]. These correlations make the Census transform value imply global feature information by passing local potential edge information to each other. In addition, the transformed values store information in a certain order, preserving the texture structure information between the local neighborhoods of the image, so that the global and local features of the image are not damaged, and the transformed features are easy to distinguish [15]. Therefore, we extract fruit features with the above algorithm.

3 Deep Belief Networks

Deep belief network [16] is a deep learning model with efficient learning algorithm proposed by Hinton. It combines unsupervised and supervised machine learning models and has become the main framework of deep learning algorithms since then. Deep belief network is superimposed by several restricted Boltzmann machines (RBM). Restricted Boltzmann machine [17] (RBM) is developed from a generated random neural network Boltzmann machine based on the principle of statistical mechanics. RBM consists of two layers that are visible layer and hidden layer. There is no connection within the layer and full connection between layers in the connection of neurons.

As shown in Fig. 2, the entire network of RBM is a bipartite graph, where $v = (v_1, v_2, \ldots, v_{n_v})^T$ and $h = (h_1, h_2, \ldots, h_{n_h})^T$ represent the state vector of the visible layer and the hidden layer, v_i represents the state of the i-th neuron in the visible layer, h_j represents the state of the j-th neuron, n_v and n_h represent the number of neurons contained in the visible layer and hidden layer, respectively. Let $a = (a_1, a_2, \ldots, a_{n_v})^T \in R^{n_v}$ and $b = (b_1, b_2, \ldots, b_{n_h})^T \in R^{n_h}$ respectively represent the bias vectors of the visible layer and the hidden layer, a_i represents the bias vector of the i-th neuron in the visible layer and b_j represents the bias vector of the j-th neuron in the hidden layer. And $W = (w_{i,j}) \in R^{n_h \times n_v}$ represents the weight matrix between the hidden layer and the visible layer and $w_{i,j}$ represents the connection weight between the i-th neuron in the hidden layer and the j-th neuron in the visible layer.

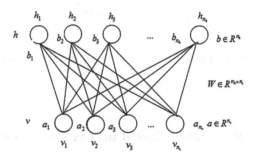

Fig. 2. RBM model

Even if the parameters of the model w_{ij}, a_i and b_j can be obtained through training, we still cannot effectively calculate the distribution determined by these parameters. However, the special structure based on the RBM model, that is, the visible layer and the hidden layer, are conditionally independent. When the state of all neurons on the visible layer is given, the probability that a certain neural unit on the hidden layer is activated (that is, the value is set to 1) is

$$P(h_j = 1|v, \theta) = \sigma(b_j + \sum_i v_i w_{ij}) \tag{1}$$

where $\sigma(\cdot)$ is the Sigmoid activation function and it is defined as

$$\sigma(x) = \frac{1}{(1 + \exp(-x))} \tag{2}$$

Similarly, we can get

$$P(v_i = 1|h, \theta) = \sigma(a_i + \sum_j w_{ij} h_j) \tag{3}$$

The relationship between the values of all nodes in a layer and the values of a single node is:

$$P(h|v) = \prod_{j=1}^{n_h} P(h_j|v); P(v|h) = \prod_{i=1}^{n_v} P(v_i|h) \tag{4}$$

The main task of RBM learning is to get the value of the optimal parameter $\theta = \{a_i, b_j, w_{ij}\}$ to fit the distribution of the given data sample. Assuming the training data sample to be $S = \{v^{(1)}, v^{(2)}, \ldots, v^t\}$, we usually obtain the RBM parameters $\theta = \{a_i, b_j, w_{ij}\}$ by maximizing the RBM logarithm likelihood $L(\theta)$ on the training data sample. The specific formula is as follows:

$$\theta^* = \arg\max_{\theta} \sum_{i=1}^{t} \log P(v^{(i)}|\theta) \tag{5}$$

In order to obtain the optimal parameters θ, we use the stochastic gradient ascent method to find the maximum of $\sum_{i=1}^{t} \log P(v^{(i)}|\theta)$.

4 Fruit Image Recognition Based on Census Transform and DBN

The parameter setting rules adopted in this paper are as follows: the size of the fruit image in the training and test data is 32×32, and the pixels of image is directly taken as the input of DBN, and the input layer of DBN is defined as 1024 units. In order to improve the computational efficiency and reduce the sampling error of the gradient estimation, it is necessary to use batch learning and divide the data set into small batches containing dozens or hundreds of samples in advance. Due to the limited data samples, each batch of data in this chapter is set to 10 samples. The fine-tuning learning rate of RBM is set as 0.01 and the number of iterations is set as 100. The activation function adopted by the neural network is Sigmoid function. As DBN can only be used for feature learning instead of classification function, we connect a softmax regression classifier at the last layer to classify the abstract feature attributes (network weight optimality of the hidden layer) that DBN eventually learns. The experiment shows that when the number of DBN layers is 2 and the number of hidden layers is 500, the identification accuracy of the whole model is the highest.

Based on the advantages of Census transform and deep belief network, we propose a fruit image recognition algorithm based on Census transform and deep belief network. The flow chart of the algorithm is shown in Fig. 3.

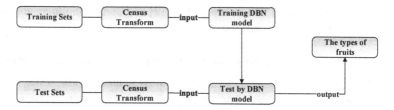

Fig. 3. The flow chart of the proposed method

The detail steps of the proposed algorithm are as following:

Firstly, the texture features of fruit images are extracted through Census transform to eliminate the influence of light changes on feature extraction. Census transform can effectively retain the local features of the image, which makes deep belief network effectively learn the local features of the image and reduce the learning of adverse feature description to deep belief network;

Secondly, the obtained Census features of the fruit images were used to train the depth belief network and obtain the relevant parameters of the network;

Finally, the deep belief network is used for fruit image recognition.

5 Experimental Results

Since the fruit images collected in our paper were collected under three different lighting environments and the fruit reflected light under different lighting conditions, the recognition results of DBN network were affected by ununiform lighting. In order to eliminate the influence of ununiform lighting. Figure 4 is an example of the illumination invariance of the fruit image obtained by our fruit data. Although there are illumination variations in Fig. 4(a) and Fig. 4(b), the renderings obtained by Census transform are almost the same in Fig. 4(c) and Fig. 4(d).

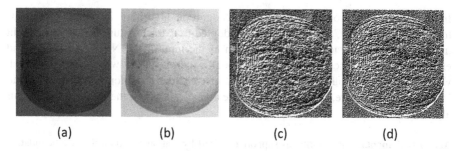

(a)   (b)   (c)   (d)

Fig. 4. The performance of Census transform on illumination invariance

In order to better illustrate the effectiveness of the proposed algorithm which named Census+DBN, different recognition algorithms are used for fruit image recognition, such as CNN proposed in [6], DBN proposed in [17] and Census+Softmax proposed in [20]. The experimental results are shown in Table 1.

Table 1. The performance comparison of different fruit image recognition methods

Method	CNN	DBN	Census+Softmax	Census+DBN
Recognition accuracy rate	0.94	0.95	0.90	0.98
False recognition rate (%)	0.84	6.88	13.4	0.17

It can be seen from Table 1 that CNN and Census+DBN have achieved good recognition results in fruit image recognition, with recognition accuracy rates 0.94 and 0.98, while the false recognition rates are 0.84% and 0.17% respectively, which indicates that Census+DBN is as robust to light as CNN. Compared with the recognition results of Census+Softmax method, it shows that the deep learning method has a strong feature learning ability and the recognition performance is better than the traditional recognition algorithm. In the two deep learning methods adopted in this paper, although CNN achieves a recognition accuracy rate of 0.94 and false recognition rate of 0.84%, the training time of CNN is too long. One iteration of CNN spends 160 s without GPU acceleration in our experiment. Although convergence of CNN can be achieved with a few iterations, it cannot meet the requirements of practical application. Although DBN takes 2.57 s for one iterate, illumination variation will affect DBN's feature extraction of images and the recognition performance of the algorithm. In this paper, we effectively eliminate the influence of uniform lightning of the image by using Census transform. At the same time, the recognition performance is better than CNN and the iteration time is 1.75 s since the Census transform is a high-speed transform without parameters. So, the proposed fruit image recognition by combining Census transform and DBN can satisfy the requirement of real time application.

6 Conclusion

In this paper, we propose a new fruit recognition algorithm by combing Census transform and DBN. Census transform is adopted to extract texture features from fruit images and the extracted features are taken as the input of DBN. The feasibility and effectiveness of the method are proved through experiments. However, in our method, we do not consider other features of fruits images such as color, shape, texture and intensity. In the future, we can add these features to improve the performance of our algorithm.

Acknowledgments. This work was supported in part by National Natural Science Foundation of China under Grant 61572063 and 61401308, Natural Science Foundation of Hebei Province under Grant F2020201025, F2016201142, F2019201151 and F2018210148, Science Research Project of Hebei Province under Grant BJ2020030 and QN2017306, Foundation of President of Hebei University under Grant XZJJ201909. This work was also supported by the High-Performance Computing Center of Hebei University.

References

1. Liu, S., Wang, J., Lu, Y., et al.: Multi-focus image fusion based on residual network in nonsubsampled shearlet domain. IEEE Access **7**(4), 152043–152063 (2019)
2. Liu, S., Liu, T., Gao, L., et al.: Convolutional neural network and guided filtering for SAR image denoising. Remote Sens. **11**(6), 702–720 (2019)
3. Chen, Y., Hu, X., Fan, W., et al.: Fast density peak clustering for large scale data based on kNN. Knowledge-Based System, 104824 (2019)
4. Altaheri, H., Alsulaiman, M., Muhammad, G., et al.: Date fruit dataset for intelligent harvesting. Data Brief **26**, 104514 (2019)
5. Soltan, M., Elsamadony, M., Mostafa, A., et al.: Nutrients balance for hydrogen potential upgrading from fruit and vegetable peels via fermentation process. J. Environ. Manage. **242**, 384–393 (2019)
6. Lu, S., Lu, Z., Aok, S., et al.: Fruit classification based on six layer convolutional neural network. In: 2018 ICDSP, China, Shanghai, pp. 1–5. IEEE (2018)
7. Zawbaa, H.M., Hazman, M., Abbass, M., et al.: Automatic fruit classification using random forest algorithm. In: 2014 14th International Conference on Hybrid Intelligent Systems, Arab, Kuwait, pp. 164–168. IEEE (2014)
8. Fu, L., Sun, S., Vázquez-Arellano, M., et al.: Kiwifruit recognition method at night based on fruit calyx image. Trans. Chin. Soc. Agricult. Eng. **33**(2), 199–204 (2017)
9. Kim, J., Vogl, M., Kim, S.D.: A code based fruit recognition method via image convertion using multiple features. In: 2014 International Conference on IT Convergence and Security (ICITCS), China, Beijing, pp. 1–4. IEEE (2014)
10. Vogl, M., Kim, J.Y., Kim, S.D.: A fruit recognition method via image conversion optimized through evolution strategy. In: 2014 IEEE 17th International Conference on Computational Science and Engineering, China, Chengdu, pp. 1497–1502. IEEE (2014)
11. Hinton, G.E., Salakhutdinov, R.R.: Reducing the dimensionality of data with neural networks. Science **313**(5786), 504–507 (2006)
12. Dunjko, V., Briegel, H.J.: Machine learning & artificial intelligence in the quantum domain: a review of recent progress. Rep. Prog. Phys. **81**(7), 074001 (2018)
13. Zabih, R., Woodfill, J.: Non-parametric local transforms for computing visual correspondence. In: Eklundh, J.-O. (ed.) ECCV 1994. LNCS, vol. 801, pp. 151–158. Springer, Heidelberg (1994). https://doi.org/10.1007/BFb0028345
14. Lai, X., Xu, X., Lv, L., Huang, Z., Zhang, J., Huang, P.: A novel non-parametric transform stereo matching method based on mutual relationship. Computing **101**(6), 621–635 (2019). https://doi.org/10.1007/s00607-018-00691-3
15. Lai, X., Xu, X., Zhang, J., et al.: An efficient implementation of a census-based stereo matching and its applications in medical imaging. J. Med. Imaging Health Inform. **9**(6), 1152–1159 (2019)
16. Mureşan, H., Oltean, M.: Fruit recognition from images using deep learning. Acta Universitatis Sapientiae, Informatica **10**(1), 26–42 (2018)
17. Ahmad, M., Ai, D., Xie, G., et al.: Deep belief network modeling for automatic liver segmentation. IEEE Access **7**, 20585–20595 (2019)
18. Le Roux, N., Bengio, Y.: Representational power of restricted Boltzmann machines and deep belief networks. Neural Comput. **20**(6), 1631–1649 (2008)

19. Larochelle, H., Bengio, Y., Louradour, J., et al.: Exploring strategies for training deep neural networks. J. Mach. Learn. Res. **10**, 1–40 (2009)
20. Cornejo, J.Y.R., Pedrini, H.: Automatic fruit and vegetable recognition based on CENTRIST and color representation. In: Beltrán-Castañón, C., Nyström, I., Famili, F. (eds.) CIARP 2016. LNCS, vol. 10125, pp. 76–83. Springer, Cham (2017). https://doi.org/10.1007/978-3-319-52277-7_10

Weather Radar and Antenna Design

Wind Turbine Clutter Suppression for Weather Radar Using Improved Ridge Regression Approach

Yv Ji, Xu Yao, Xiaodong Wang, and Mingwei Shen[✉]

College of Computer and Information Engineering, Hohai University,
Nanjing 211100, China
smw_hhu1981@163.com

Abstract. The problem of clutter suppression is gaining importance because of many disadvantages. However, conventional clutter suppression methods cannot eliminate the great disturbances to radar system caused by wind turbines. An improved ridge regression algorithm is investigated to accurately estimate the spectral moment of the weather signal contaminated by wind turbine clutter (WTC) in this paper. Firstly, a weighted regression model is introduced to solve the problem that the strong collinearity of the data in the regression model leads to unstable parameter estimation. Then the optimal regression parameter in the model is obtained by generalized cross validation (GCV) to improve the estimation accuracy of weather signal. Theoretical analysis and simulation results show that the spectral moment recovered by the proposed algorithm has better accuracy and stability in lower SNR.

Keywords: Weather radar · Ridge regression · Clutter suppression

1 Introduction

In recent years, the clutter suppression technology of wind turbine by weather radar has be widely concerned by more and more countries. By using simulation and measured data, scientists from various countries have analyzed the distribution characteristics of clutter and weather echo of wind turbine under different working modes of weather radar both in the time-domain and frequency-domain in detail. Accordingly, different suppression algorithms are proposed as followings: the adaptive spectrum processing algorithm proposed by Kong [1]; the range-doppler regression (RDR) algorithm proposed by Nail [2], the turbine clutter suppression method based on the adaptive filter (such as wiener filter, etc.) proposed by Yan [3], and the signal separation method proposed by Frank [4]. Regression algorithm is widely used in clutter suppression for the simple model and high operational efficiency. However, due to the limitations of wind farm scale, fan speed, weather radar working mode and other practical conditions, the above algorithm cannot take into account clutter suppression of wind turbines and lossless recovery of weather information.

In view of the above problems, in this paper, based on the RDR algorithm, an improved ridge regression algorithm is studied. By introducing the weighted regression

Y.-D. Zhang et al. (Eds.): ICMTEL 2020, LNICST 327, pp. 449–455, 2020.
https://doi.org/10.1007/978-3-030-51103-6_40

model, and selecting generalized cross validation (GCV) in solving the optimal regression parameter achieve the effect of optimizing the solving process. Moreover, the calculated ridge regression coefficient for the collinearity and stronger robustness is ensured, and the high precision of the recover of contaminated weather echo signal information is realized.

2 Weather Radar Signal Model

Assuming that the wind turbine clutter(WTC) signal is received by the ith range bin of the weather radar, and the received signal under the kth pulse is [5]:

$$C_i(k) = s_i(k) + w_i(k) + n_i(k), \, k = 1, \ldots, K \tag{1}$$

where K is the pulse number, $s_i(k)$ is the weather signal, $w_i(k)$ is the wind turbine clutter, $n_i(k)$ denotes the noise.

Adjacent range bins do not contain WTC signals, and the received signal under the kth pulse is:

$$C_j(k) = s_j(k) + n_j(k), \, k = 1, \ldots, K \tag{2}$$

The dynamic clutter caused by the high-speed rotation of the blades of wind turbines has a wide doppler spectrum, so WTC and weather targets will produce serious overlap in time and frequency domains, that is, the weather signal will be submerged in the clutter of wind turbines, and the spectral moment information of the weather signal of the polluted range bin will be seriously disturbed.

3 Improved Ridge Regression Approach

3.1 The Principle of Ridge Regression

Ridge regression is an improved method of least square estimation. Regularization is carried out on the original regression model and the solution process was optimized by adding penalty term. The general regression model is defined as follows:

$$Y = X\beta \tag{3}$$

where Y is the real values of spectral moment parameters, β is the regression coefficient to be found, and X is a fitting matrix defined as:

$$X = \begin{bmatrix} 1 & r_{i-10} & r_{i-10}^2 \\ \cdots & \cdots & \cdots \\ 1 & r_{i-1} & r_{i-1}^2 \\ 1 & r_{i+1} & r_{i+1}^2 \\ \cdots & \cdots & \cdots \\ 1 & r_{i+10} & r_{i+10}^2 \end{bmatrix}_{20 * 3} \tag{4}$$

where r_i is the slant distance size of scattered points of the ith range bin, the dimension of $X_{L \times M}$ is $L \times M$. L is the amount of selected units which is defined as L = 20, that is to take 10 range bins on both sides of the ith range bin with WTC contamination, and M is the degree of polynomial which is defined as M = 3.

A fitting estimation matrix including distance units contaminated by WTC is denoted as X^*:

$$X^* = \begin{bmatrix} 1 & r_{i-10} & r_{i-10}^2 \\ \cdots & \cdots & \cdots \\ 1 & r_{i-1} & r_{i-1}^2 \\ 1 & r_i & r_i^2 \\ 1 & r_{i+1} & r_{i+1}^2 \\ \cdots & \cdots & \cdots \\ 1 & r_{i+10} & r_{i+10}^2 \end{bmatrix}_{21*3} \tag{5}$$

Ridge regression cost function is given by:

$$\min \left\{ \|X\beta - Y\|^2 + k_{ridge} \|\beta\|^2 \right\} \tag{6}$$

where k_{ridge} is ridge regression parameter. Ridge regression coefficient is solved as follows:

$$\beta = (X^T X + k_{ridge} I)^{-1} X^T Y \tag{7}$$

The estimated value of the corresponding parameter Y is given by:

$$\tilde{Y} = X^* \cdot \beta = X (X^T X + k_{ridge} I)^{-1} X^T Y \tag{8}$$

The algorithm uses the weather information of unpolluted range bins, introduces an improved ridge regression algorithm, and then estimates the average doppler velocity, velocity spectrum width, power and other effective information of weather signals in the polluted range bin through efficient fitting.

3.2 Weather Information Extraction

The average doppler velocity of the echo signal is estimated as follows [6]:

$$V = \left(\frac{\lambda}{4\pi/\text{PRF}} \right) \angle R_{01} \tag{9}$$

The average spectral width is estimated as:

$$\sigma = \frac{\lambda}{2\sqrt{2}\pi/PRF} \left| \ln \frac{R_{01}}{|R_{11}|} \right|^{1/2} \tag{10}$$

The average power is estimated as:

$$p = R_{0l} \tag{11}$$

where R_{0l} Represents zero delay auto-correlation parameter of echo signal sequence of the lth unit to be processed, and R_{1l} represents the first order auto-correlation parameter of the echo signal sequence of the lth range bin. Ultimately the estimated spectral moment parameters of the weather signal in the unit polluted by WTC can be obtained as V_{est}, σ_{est} and p_{est}.

3.3   Improved Ridge Regression

An improved ridge regression uses the weighted regression model, and solves the optimal regression parameter by introducing GCV. Suppose the singular value decomposition of the X in the regression model is:

$$X = UDV \tag{12}$$

where U is a positive definite matrix of order L, and V is that of M. The dimension of D is L × M, the diagonal element of which is the square root of the eigenvalue λ_n ($n = 1, 2, ...M$) of $X^T X$. Define $M(k_{ridge}) = X(X^T W X + k_{ridge} I)^{-1} X^T W Y$, where W is the fitting weight corresponding to each range bin. The fitting weight of the jth range bin nearby is:

$$W_j = \begin{cases} \frac{1}{2}[1 + \cos((\pi/L_{prox})|j - i|)], & 0 \le |j - i| \le L_{prox} \\ 0, & \text{otherwise} \end{cases} \tag{13}$$

where L_{prox} is the distance threshold, and $L_{prox} = 10$.

The validation function of generalized cross validation is as follows:

$$\text{GCV}(k_{ridge}) = \frac{L^{-1}\left\|Y - \tilde{Y}\right\|^2}{[1 - L^{-1}\text{tr}(M)]^2} \tag{14}$$

which can be simplified as:

$$\text{GCV}(k_{ridge}) = \frac{L^{-1}\sum_{i=1}^{L}(k + \gamma_i)^{-2}\bar{y}_i^2}{[L^{-1}\sum_{i=1}^{M}(k + \gamma_i)^{-1}]^2} \tag{15}$$

where $U^T Y = (\bar{y}_1, \bar{y}_2, ..., \bar{y}_L)$, and γ_n is defined by:

$$\gamma_n = \begin{matrix} \lambda_i, & i = 1, 2, ...M \\ 0, & \text{otherwise} \end{matrix} \tag{16}$$

The obtained optimal ridge regression parameter k^*_{ridge} is substituted into the improved ridge regression model, and the estimated values of spectral moment parameters of the weather signal in the range bin polluted by WTC are input to obtain the fitting values of spectral moment parameters of meteorological signals in the range bin polluted by WTC, which contains v_{ridge}, σ_{ridge} and p_{ridge}.

4 Simulation Results and Performance Analyses

The validity of the proposed algorithm is verified by computer simulation. The main simulation parameters of the radar system are listed in Table 1.

Table 1. Simulation parameters.

Parameters	Values
Carrier frequency f_0	5.5 GHz
Pulse repetition frequency	1000 Hz
Radar height	1000 m
Wind turbine height	880 m
Wind turbine rotate speed	15 r/min
Wind turbine blade length	26 m

RDR algorithm [2] proposed by Nail is selected for performance comparison in this paper. Firstly, the solution of optimal ridge regression parameters is verified. GCV is introduced into ridge regression algorithm to solve the optimal ridge regression parameter. Due to the fitting matrix of every scatter points is in contact with the weather signal model of oblique distance only, the optimal value of the ridge regression parameter given by GCV is a constant.

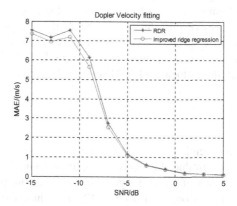

Fig. 1. Comparison of average doppler velocity fitting errors.

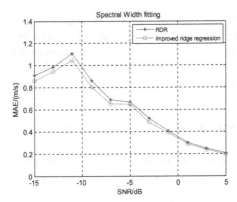

Fig. 2. Comparison of average spectral width fitting errors.

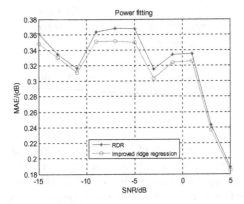

Fig. 3. Comparison of average power fitting errors.

The polynomial fitting method is adopted to estimate spectral moment of weather signals in the polluted range bin. As shown in Fig. 1, 2 and 3, when the SNR is lower than 5 dB, the fitting error of this algorithm is better than that of RDR. That is, this algorithm can still estimate spectral moment with higher accuracy under the condition of low SNR, and the recovered weather signal returns have higher accuracy, which effectively suppressing the influence of WTC on weather information.

5   Conclusion

This paper proposes the weighted ridge regression model for suppression of WTC for weather radars. GCV is introduced to solve the optimal regression parameter, and the spectral moment parameters of weather signals in the range bin polluted by WTC are fitted with high accuracy. Compared with RDR algorithm, the fitting error of the algorithm in this paper is smaller under the condition of low SNR, and the recovered spectral moment information is more accurate.

Acknowledgement. This work was supported in part by National Natural Science Foundation of China (No. 41830110, No. 61771182).

References

1. Kong, F., Zhang, Y., Palmer, R.: Characterization of micro-Doppler radar signature of commercial wind turbines. In: Proceedings of SPIE - The International Society for Optical Engineering Spie Defense + Security, pp. 1–7. Baltimore, MD (2014)
2. Nai, F., Palmer, R.: On the mitigation of wind turbine clutter for weather radars using range-Doppler spectral processing. IET Radar, Sonar Navigation 7(2), 178–190 (2013)
3. Kong, F., Zhang, Y., Palmer, R.: 9.3 Wind turbine clutter mittigation for weather radar: a feasibility study by means of scaled measurement. In: American Meteorological Society, Seattle, WA (2011)
4. Uysal, F., Selesnick, I., Isom, B.M.: Mitigation of wind turbine clutter for weather radar by signal separation. IEEE Trans. Geosci. Remote Sens. 54(5), 1–10 (2016)
5. Perfetti, B., Zheng, J., Kaveh, M.: Signal processing for wind turbine interference mitigation in Doppler weather radars: Data synthesis, clutter detector performance, and spectral interpolation in range-azimuth-Doppler, In: Ist International Conference on Radar Systems. IET, Glasgow (2013)
6. Hood, K., Torres, S., Palmer, R.: Automatic detection of wind turbine clutter for weather radars. J. Atmosph. Oceanic Technol. 27(11), 1868–1880 (2010)

Wind Turbine Clutter Mitigation for Weather Radar by Extreme Learning Machine (ELM) Method

Mingwei Shen[1($\boxtimes$)], Xu Yao[1], Di Wu[2], and Daiyin Zhu[2]

[1] College of Computer and Information Engineering, Hohai University,
Nanjing 211100, China
smw_hhu1981@163.com
[2] Key Laboratory of Radar Imagine and Microwave Photonics,
Nanjing University of Aeronautics and Astronautics, Nanjing 210016, China

Abstract. Because of its overall performance, the Extreme Learning Machine (ELM) has been very concerned. This paper introduces the ELM algorithm into the clutter mitigation for weather radar, and proposes a wind turbine clutter mitigation method. Firstly, building training samples. Secondly, the model parameters for ELM are examined and optimized aim to improve its overall performance. Finally, the optimized ELM algorithm is used to recover the weather signal of the contaminated range bin. Simulation results show that the proposed algorithm can realize the precise recovery of the weather signal.

Keywords: Weather radar · Extreme Learning Machine · Clutter suppression

1 Introduction

Because of the importance of renewable energy, the use of wind farms is increasing. This continued growth seriously threatens the performance of most radar systems, especially the weather radar. The echo signal from the wind farm is called the wind turbine clutter (WTC) [1]. The focus of this paper is the mitigation of WTC on weather radar [2].

In recent years, many researchers have been dedicated to suppressing WTC and proposing some mitigation methods. The common method is spatial interpolation. Unfortunately, the weather data is the typical spatial-temporal data. The spatial interpolation only utilizes the spatial continuity in the range domain of the weather signal, but ignores the correlation in the Doppler domain.

In this paper, we creatively introduce the ELM into the mitigation of WTC. Compared with existing algorithms, ELM has two advantages: First, the recovery of the weather signal can be achieved with a small error; Second, ELM has low complexity and fast learning ability.

© ICST Institute for Computer Sciences, Social Informatics and Telecommunications Engineering 2020
Published by Springer Nature Switzerland AG 2020. All Rights Reserved
Y.-D. Zhang et al. (Eds.): ICMTEL 2020, LNICST 327, pp. 456–461, 2020.
https://doi.org/10.1007/978-3-030-51103-6_41

2 Weather Radar Signal Model

Assume the lth range bin contains both WTC and the weather signal. The weather radar echo sampling in the n th pulse is expressed as

$$x_l(n) = w_l(n) + c_l(n) + s_l(n) + z_l(n), \ n = 1, 2, \ldots K \tag{1}$$

where K is the pulse number, $w_l(n)$ represents the wind turbine clutter, $c_l(n)$ is the ground clutter, $s_l(n)$ is the weather signal, $z_l(n)$ denotes the noise.

The weather signal is the distributed target, so the weather signal of a range bin is formed by the echo of multiple scattering particles [3]. Sum up the scattering particle echo vector of the lth range bin, then the weather signal return in the nth pulse can be expressed as

$$s_l(n) = \sum_{u=1}^{U} A_u e^{j(n-1)\omega_t} \tag{2}$$

where U is the number of scattering points in the lth range bin, A_u is the amplitude of the weather target particle u.

3 Wind Turbine Clutter Mitigation

3.1 Extreme Learning Machine for Weather Radar

For N different samples (t_i, y_i), the mathematical model of ELM can be expressed as

$$\sum_{j=1}^{L} \beta_j f(\omega_j \cdot t_i + b_j) = o_i, i = 1, \ldots, N \tag{3}$$

Where L is the number of hidden layer nodes, $f(x)$ is the activation function, β_j is the coefficient between the output layer node and the jth hidden layer node, ω_j is the coefficient between the input layer node and the jth hidden layer node, and b_j is the deviation of the jth hidden node. o_i is the network output of the ith sample. And there are

$$\sum_{j=1}^{L} ||o_j - y_j|| = 0 \tag{4}$$

There exist ω_j, b_j, β_j

$$\sum_{j=1}^{L} \beta_j f(\omega_j \cdot t_i + b_j) = y_i \tag{5}$$

Write the above formula in matrix form

$$H\beta = Y \tag{6}$$

Where

$$H = \begin{bmatrix} f(\omega_1 t_1 + b_1) & \cdots & f(\omega_L t_1 + b_L) \\ \cdots & \cdots & \cdots \\ f(\omega_1 t_N + b_1) & \cdots & f(\omega_L t_N + b_L) \end{bmatrix} \tag{7}$$

$$\beta = [\beta_1^T, \ldots, \beta_L^T]^T \tag{8}$$

$$Y = [y_1^T, \ldots, y_N^T] \tag{9}$$

Where H is called the hidden layer output matrix. The ω_j and b_j can be randomly selected, and the β can be obtained

$$\beta = H^+ Y \tag{10}$$

Where H^+ is the Moore-Penrose generalized matrix inverse of the H.

3.2 Training Sample Design

The input of the network is as follows

$$\begin{bmatrix} 1 & R_1 & R_1^2 \\ 1 & R_2 & R_2^2 \\ \cdots & \cdots & \cdots \\ 1 & R_i & R_i^2 \end{bmatrix} \quad i = 1, 2, \ldots, N \tag{11}$$

Where R_i is the range of the ith sample.

In this paper, the radial velocity and spectral width estimation of the weather signal are respectively selected for prediction, and their outputs are shown as follows

$$[v_1, v_2, \ldots, v_i] \tag{12}$$

$$[o_1, o_2, \ldots, o_i] \tag{13}$$

Where v_i, o_i are respectively the radial velocity and spectral width estimation of the weather signal of the ith training sample.

4 Simulation Results and Performance Analyses

The validity of the proposed method is verified by computer simulation. Radar simulation parameters are shown in Table 1. We select the 25th range bin as the range bin contaminated by WTC and select the 15th–35th range bins as samples.

Table 1. Simulation parameters.

Parameters	Values
Carrier frequency f_0	5.5 GHz
Pulse repetition frequency	1000 Hz
Radar height	1000 m
Wind turbine height	880 m
Wind turbine rotate speed	15 r/min
Wind turbine blade length	26 m

We compare the radial velocity estimation and spectrum width estimation before and after applying the proposed method. We choose two-dimensional joint interpolation for comparison. In order to make the experimental results more accurate, we conduct 100 independent Monte Carlo experiments.

Figure 1 and Fig. 2 respectively show the radial velocity and spectral width estimation before and after the application of ELM algorithm. As shown in Fig. 1 and Fig. 2, ELM algorithm can greatly reduce the error of radial velocity and spectral width estimation caused by WTC contamination. It further shows that ELM algorithm is suitable for clutter suppression of wind turbine for weather radar.

Figure 3 and Fig. 4 respectively show MAE in radial velocity and spectral width estimation in different SNR. From Fig. 3, we note that the MAE of ELM algorithm is less than two - dimensional joint interpolation, When the SNR is 20 dB, the MAE of ELM algorithm is 0.01 m/s, and the two-dimensional joint interpolation is 0.03 m/s, which reduced by 0.02 m/s. And in the Fig. 4, when the SNR is 10 dB, the MAE of ELM algorithm is 0.2 m/s, and the two-dimensional joint interpolation is 1.9 m/s, which reduced by 1.7 m/s.

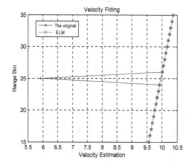

Fig. 1. The radial velocity estimation before and after the application of ELM

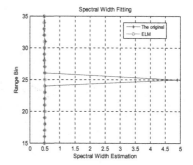

Fig. 2. The spectral width estimation before and after the application of ELM

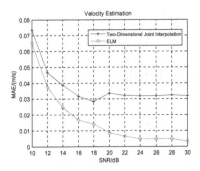

Fig. 3. MAE in radial velocity estimation in different SNR

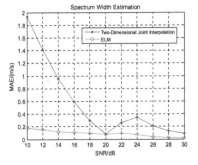

Fig. 4. MAE in spectral width estimation in different SNR

Figure 5 and Fig. 6 respectively show the mean values of radial velocity and spectral width estimation in different SNR. From Fig. 5, we note that the mean value of radial velocity estimation is sensitive to noise. And we can see that the radial velocity estimation of the ELM algorithm has a small deviation from the truth value. From Fig. 6 that the MAE in spectral width estimation of the ELM is small and finally converges to the truth value.

We analyze the complexity of two-dimensional joint interpolation and ELM. The computational complexity of the two-dimensional joint interpolation in range-Doppler domain is $O(N(N_f + K)n)$, while the ELM is $O(Nn)$. Under the simulation parameters, the number of selected range bins is $N = 21$, the selected Doppler frequency bins is $N_f = 22$, the number of experiments is $n = 100$. Calculation results show that computational complexity of the ELM is greatly reduced compared with the two-dimensional joint interpolation, which confirms the proposed algorithm is suitable for engineering application.

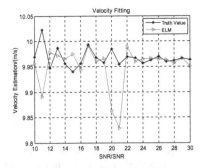

Fig. 5. Mean values of radial velocity estimation in different SNR

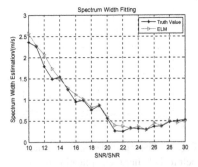

Fig. 6. Mean values of spectral width estimation in different SNR

5 Conclusion

This paper proposes a WTC suppression method for weather radar based on the ELM. Through the network training, the weather signal prediction model is constructed. Simulation results illustrate the proposed method can effectively suppress the WTC and realize the precise recovery of the weather signal. Increasing the efficiency and precision of ELM will be further explored in future works by improving the optimization algorithm.

Acknowledgement. This work was partly financed by National Natural Science Foundation of China (No. 41830110, No. 61771182).

References

1. Gallardo-Hernando, B., Munoz-Ferreras, J.: Wind turbine clutter observations and theoretical validation for meteorological radar applications. IET Radar Sonar Nav **5**(2), 111–117 (2011)
2. Hubbert, J.C., Dixon, M., Ellis, S.M.: Weather radar ground clutter. Part I: identification, modeling, and simulation. J. Atmosph. Ocean. Technol. **26**(7), 1165–1180 (2009)
3. Naqvi, A., Yang, S.T., Ling, H.: Investigation of doppler features from wind turbine scattering. IEEE Antennas Wirel. Propagat. Lett. **9**(7), 485–488 (2010)

Author Index

Printed in the United States
by Booksurge

Printed in the United States
By Bookmasters